1976

This book may be kept

CHEMICAL FALLOUT

Current Research on Persistent Pesticides

A Proceedings Publication of the
Rochester Conferences on Toxicity

Organized by the
Department of Radiation Biology and Biophysics
School of Medicine and Dentistry
University of Rochester
Rochester, New York

Conference Committee
George G. Berg
Morton W. Miller
Victor G. Laties

Second Printing

CHEMICAL FALLOUT

Current Research on Persistent Pesticides

Edited by

MORTON W. MILLER

and

GEORGE G. BERG

Department of Radiation Biology and Biophysics
School of Medicine and Dentistry
University of Rochester
Rochester, New York

With a Foreword by

Aser Rothstein

Department of Radiation Biology and Biophysics
School of Medicine and Dentistry
The University of Rochester
Rochester, New York

CHARLES C THOMAS · PUBLISHER
Springfield · *Illinois* · *U.S.A.*

Published and Distributed Throughout the World by

CHARLES C THOMAS • PUBLISHER

Bannerstone House

301-327 East Lawrence Avenue, Springfield, Illinois, U.S.A.

Natchez Plantation House

735 North Atlantic Boulevard, Fort Lauderdale, Florida, U.S.A.

© *1969, by* **CHARLES C THOMAS • PUBLISHER**

ISBN 0-398-01313-6

Library of Congress Catalog Card Number: 69-19178

First Printing, 1969
Second Printing, 1972

With **THOMAS BOOKS** *careful attention is given to all details of manufacturing and design. It is the Publisher's desire to present books that are satisfactory as to their physical qualities and artistic possibilities and appropriate for their particular use.* **THOMAS BOOKS** *will be true to those laws of quality that assure a good name and good will.*

This publication is the proceedings of the
1968 Rochester Conference on Toxicity.

Printed in the United States of America
00-2

CONTRIBUTORS

GEORGE G. BERG
Department of Radiation Biology and Biophysics
University of Rochester
School of Medicine and Dentistry
Rochester, New York

FREDRIK BERGLUND
National Institute of Public Health
Stockholm, Sweden

MATHS BERLIN
Institute of Hygiene
University of Lund
Stockholm, Sweden

PHILIP A. BUTLER
Bureau of Commercial Fisheries Biological Laboratory
U. S. Department of Interior
Sabine Island
Gulf Breeze, Florida

THOMAS W. CLARKSON
Department of Radiation Biology and Biophysics
University of Rochester
School of Medicine and Dentistry
Rochester, New York

ANTHONY DE FREITAS
National Research Council
Division of Biosciences
Ottawa, Canada

WILLIAM B. DEICHMANN
Department of Pharmacology
University of Miami School of Medicine
Coral Gables, Florida

v

WILLIAM F. DURHAM
Chief, Pesticides Research Laboratory
Communicable Disease Center
Perrine, Florida

DAVID W. FASSETT
Laboratory of Industrial Medicine
Eastman Kodak Company
Kodak Park Works
Rochester, New York

HAROLD C. HODGE
Department of Pharmacology
University of Rochester
School of Medicine and Dentistry
Rochester, New York

ELDRIDGE G. HUNT
Department of Fish and Game
Pesticides Investigations
Sacramento, California

A. R. ISENSEE
Crops Research Division
Agricultural Research Service
U. S. Department of Agriculture
Beltsville, Maryland

ARNE JERNELÖV
Institutet för Vatten och Luftvårdforskning
Stockholm, Sweden

ALF G. JOHNELS
Naturhistoriska Riksmuseet
Stockholm, Sweden

VICTOR G. LATIES
Department of Radiation Biology and Biophysics
University of Rochester
School of Medicine and Dentistry
Rochester, New York

MORTON W. MILLER
Department of Radiation Biology and Biophysics
School of Medicine and Dentistry
University of Rochester
Rochester, New York

TOR NORSETH

Department of Radiation Biology and Biophysics
University of Rochester
School of Medicine and Dentistry
Rochester, New York

DAVID PEAKALL

Department of Pharmacology
Upstate Medical Center
Syracuse, New York

JACK RADOMSKI

Professor of Pharmacology
University of Miami
Research and Teaching Center of Toxicology
Coral Gables, Florida

ROBERT W. RISEBROUGH

Institute of Marine Resources
University of California
College of Agriculture
Berkeley, California

JOHN ROBINSON

Tunstall Laboratory
Broad Oak Road
Sittingbourne, Kent, England

ROBERT L. RUDD

Department of Zoology
University of California
Davis, California

DONALD A. SPENCER

National Agricultural Chemical Association
Washington, D. C.

WILLIAM STICKEL

U. S. Department of Interior
Bureau of Sport Fisheries and Wildlife
Patuxent Wildlife Research Center
Laurel, Maryland

TSUGUYOSKI SUZUKI

Department of Human Ecology
Faculty of Medicine
University of Tokyo
Tokyo, Japan

ROBERT VAN DEN BOSCH

Division of Biological Control
College of Agriculture Sciences
University of California at Berkeley
Berkeley, California

JAROSLAV VOSTAL

Department of Pharmacology
University of Rochester
School of Medicine
Rochester, New York

RICHARD WELCH

Department of Biochemical Pharmacology
Burroughs Wellcome and Company
Tuckahoe, New York

IRMA CALVERT WEST

California Department of Public Health
Bureau of Occupational Health
Berkeley, California

GUNNEL WESTÖÖ

National Institute of Public Health
Stockholm, Sweden

J. HENRY WILLS

Pesticide Control Board
Department of Health
Albany, New York

CHARLES F. WURSTER, JR.

Department of Biological Sciences
State University of New York
Stony Brook, New York

ROGER G. YOUNG

Toxicology Training Program
New York State College of Agriculture
Cornell University
Ithaca, New York

FOREWORD

THE DEPARTMENT OF Radiation Biology and Biophysics grew out of a wartime project that was part of the Manhattan District. Part of our job, very secret at the time, was to determine as rapidly as possible the toxicity of uranium compounds and of other heavy elements such as polonium and plutonium. We had to become "instant experts" in toxicology. After the war was over and we became an academic unit of the Medical School, supported by the United States Atomic Energy Commission, most of our educational programs were concerned with radiobiology. We also inherited a small program in biophysics which has since grown considerably. But we still had in our department a number of staff whose primary discipline was toxicology rather than radiation biology or biophysics. One of the leaders of the department, Dr. Harold Hodge, retained an abiding interest in toxicology even after he left us to organize a separate pharmacology department. Out of his and our shared interests grew a doctoral program in toxicology sponsored jointly by the Pharmacology Department and the Department of Radiation Biology and Biophysics. Although the program is only in its second year, we have a training grant to support students in toxicology, and primarily through the efforts of Dr. Hodge and the Department of Pharmacology, we also have a grant for establishing a Toxicological Center at the Medical School. This grant was awarded by the National Institutes of Health.

In summary, we have at Rochester a large and expanding interest in research and education in the field of toxicology. This conference is one means of expressing this interest and of focusing on one very important area in the field of toxicology. Over one hundred scientists throughout the world were contacted for their suggestions about the organizational details of the meeting. A majority of the replies suggested that a small, closed meeting with ample time for discussion would be the most beneficial.

The conference was organized into five sessions with the topics arranged in a sequence that follows pesticides from their entry into the ecological systems to their ultimate effects, if any, on the human population. We plan to convene such a conference on a yearly basis, with the two subsequent topics to be concerned with "Heavy Metals and Their Effects on Cells and Subcellular Elements" and "Inhalation Toxicity."

I wish to express the department's appreciation to Drs. Morton W. Miller, George G. Berg, and Victor G. Laties for serving as the Conference Committee; to Mrs. Barbara J. Linkhorst and Mrs. Dorris B. Nash for acting as the committee's secretaries; to members of the department's secretarial staff for the typing of manuscripts; to Mrs. Florence E. Marsden for handling the participants' travel arrangements; to our department's photography and illustration section for efficiently producing illustrations at short notice; and to Dr. William G. Aldridge for assisting with the conferences electronic gear.

Grateful acknowledgment is extended to the Division of Nuclear Education and Training of the United States Atomic Energy Commission, which granted funds in partial support of the conference.

ASER ROTHSTEIN

INTRODUCTION:
A VIEW FROM THE PODIUM

THIS CONFERENCE should have been entitled "Birds of Omen." To begin with, birds — pheasants, hawks, and eagles — alerted Swedish naturalists to the appearance of a threat to life in the form of new sources of mercury poisoning. Chemists and toxicologists joined in uncovering the existence of a serious public health problem, similar to a pattern of contamination with mercury which had caused human deaths in Japan. The Swedes acted in a decisive and highly principled way to restore a clean environment by restricting or banning many industrial and agricultural uses of mercurials. Still, some Swedish lakes will not be fit for fishing in this century, and the consumption of fish from other lakes will have to be restricted.

The conference was organized as a double confrontation. We wished to put together two problems, that of organomercurials and that of organochlorines, and to bring together two groups of experts — the ecologists on the one hand, and the toxicologists and public health physicians on the other. The distribution of organomercurial poisons is a local matter. The distribution of organochlorine poisons is worldwide. Both are persistent pollutants and control must come in advance of trouble — once you have too much, it is too late. We hoped that the story of organomercurials would serve as a model for examining the use of organochlorines, and we wanted to promote an exchange of views between experts in generally antagonistic specialties.

This program entailed a calculated risk. There was a possibility that the mercury experts and the DDT experts would not be interested in each others' data. Worse, there was a chance for the kind of hostility that blocks communication in a field where a scientific controversy cannot be divorced from a conflict of policy and where lives are ultimately staked on the policy decisions.

xi

The outcome of the conference was more favorable than we dared to expect. First, we were fortunate to secure a very representative attendance from a circumscribed professional group. Thirty-seven scientists attended the closed sessions as chairmen, participants, and invited guests. Second, the papers showed useful similarities between the problems posed by organochlorine compounds and those posed by the mercurials. Third, there was a creative interchange between specialties, and where the discussion was heated, the heat generated light — see, for example, the discussion in Chapter 6 by Van den Bosch, or Stickel's brilliant summation of the problems of reproductive damage in Chapter 26.

The bread-and-butter work reported by the participants was concerned with the measurement of relations between dose and effect. If new chemicals are to be produced and used, someone must undertake the task of advance testing, and certify that some manner of use is proper and harmless. A great deal of information is presented concerning no-effect levels and lowest known damaging doses. This kind of work is to chemical industry and agriculture what the testing of models and prototypes is to the aircraft industry, and suffers from the same limitation — you cannot guarantee that you found the flaw. The test pilot tortures the prototype plane without mishap, but when the commercial models are in the air a wing falls off. Then comes the crash investigator who asks whether the operator failed in following instructions, or whether the test pilot did miss a structural weakness. Of the twenty-four speakers, I would describe fifteen as test pilots and nine as crash investigators. Some of the test-piloting was extremely ingenious in discovering hidden, nonlinear failures of the system — see Chapter 18 by de Freitas and Chapter 20 by Welch *et al.* Some crash investigators called for junking the most popular agrochemical system now in use. Others said it couldn't be done, but no one said it shouldn't be done.

The first session dealt with the ecological fate of pesticides. The highlight of the session was the demonstration by Risebrough of a global pathway of distribution which converts organochlorines into a classical low-altitude fallout. No matter where these compounds are spread, they rise as vapors and are promptly spread worldwide, to fall out as dust. As in the case of mercurials so here, the omen was brought by birds and traced along the food chain of bird populations. In both cases a major share of pollution comes from industrial wastes — in one case from mercury used in the wet process of making paper pulp, and in the other from polychlorinated biphenyls used in the manufacture of plastics. The sea is the ultimate target and we are contaminating it at an increasing rate. What are the prospects of recovery if the contamination is stopped? The study of a lake showed that recovery is slow. Pesticides are trapped in the biotic phase of the ecosystem more effectively than would be expected from passive physico-chemical distribution. The peak of concentration is at the apex of the food web, a position which is shared at Clear Lake by a population of migratory birds and a population of resident fishermen. The study of soil showed a dramatic contrast between the ecology of degradeable and nondegradeable pesticides at the point of application. There was virtually no recovery from overloads of arsenicals even in a quarter of a century. Removal of organochlorines and organomercurials was slow, and both classes of compounds were taken out of the soil primarily by volatilization. Cross-referenced with Risebrough's data, this meant that there was mainly a translocation of poison from soil to sea. DDT accumulates largely as its somewhat less toxic degradation product, DDE, but mercurials discharged in harmless concentrations are both accumulated in nature and transformed into a more potent poison. This pollution was cleaned up by Sweden, although this meant revamping the whole process of newsprint manufacture in both Sweden and England. The discussion brought out the fact that the cleaner industrial process was also more economical. The old process is still in use in the United States.

The second session narrowed the topic from ecosystems to single populations. The handling of the crop population hinges on the interplay of three social subsystems: the economic, which maximizes dollar yields from investment; the political—administrative, which balances economic yields against toxicity to man; and the agro-ecological, which is the only one presently concerned with the stability of the crop ecosystem. Examples were given of the way in which the interplay of economics and politics overrides the university-based ecologist. The outcome favors spoiler agriculture and leads inevitably to a collapse of the crop; to change this, we must abandon toxicity to man as our administrative yardstick for the misuse of pesticides. This was a strong conclusion to be proposed to a predominantly toxicological conference, but it was well accepted by the participants. Who would take over the policing tasks now vacant or held by public health officials? Van den Bosch discussed the need for a profession of agro-ecology practiced under the same constraints and with the same privileges as medicine and committed to the health of the ecosystem which supports the crop.

Butler spoke of estuaries, thus rounding off the ecological survey of Session I, and presented evidence of damage by organochlorine pesticides to populations of oysters, shrimp, and fish. Agricultural run-off caused the differences in pollution observed from estuary to estuary and from season to season.

Bird populations are incidental victims of pesticides. A hot controversy over the impact of organochlorines on birds took up half of this session and continued throughout the conference. The exposure of birds to mercurials was traced through the combined resources of museum biology and of neutron activation analysis. Bird feathers provided a precise record of exposure of bird populations to mercury over a whole century. The method resolved one season from another within a year, and was geographically as precise as our knowledge of migratory patterns of birds. Other experiments traced mercury from industrial wastes through fish to water-feeding birds, and showed how the pollutant persists in the aquatic ecosystem after it is shut off at the source. This pattern of retention should be cross-referenced with the pattern of retention of organochlorines presented in Session I.

Sessions III and IV got down to the core issue of mechanism of damage in individuals. The brain is the target organ for mercurials as it is for organochlorines. Much careful work on mercurials was devoted to the knotty problem of extrapolating from behavioral signs in animals to brain damage in man. A tentative threshold dose was suggested for methylmercury poisoning, although there was no assurance that the lowest brain concentration that produced detected injuries was in fact a threshold dose. The highly toxic organomercurials may be merely vehicles for the efficient concentration of mercury at the target organ, where the ultimate damage is probably due to a biotransformation which releases mercuric ions locally. Discussion brought out the existence of a protective mechanism, which may account for a threshold to mercury damage, at least in the kidney. Both protection and damage involve binding to sulfhydryl groups on proteins. The investigations of mercurials were capped with a study of their intracellular distribution.

Much less is known about physiological and biochemical actions of organochlorine pesticides. There were correlations between body burden of organochlorines in people and some diseases, without any indication that the pesticide made the illness any worse. Household use of DDT was found to be a major source of human exposure in Florida, leading to fourfold increases in body burden. A variety of organisms have enzymes that degrade DDT to DDE, and this conversion serves as the mechanism of resistance in house flies. There is evidence, however, that monkeys and man lack this pathway. In pheasants, DDT kills by brain damage, but there is a remarkably poor correlation between signs of brain damage on one hand, and the concentrations found in the brain on the other. Stress contributes to susceptibility to DDT. In rats, one kind of cold stress could change a chronically tolerated dose of DDT into a toxic dose. Wurster discussed a biochemical mechanism of DDT death, based on interference with the calcium binding sites on excitable plasma membranes, where calcium acts as a stabilizer. He then presented an exhaustive review of evidence that the principal damage done by organochlorine compounds is to reproduction, rather than to survival. Much lower doses would be involved here, and the endocrine balance would be the target rather than the brain function. The last paper by Welch, Levin, and Coney (Chapter 20) brought about a dramatic confrontation of toxicology with ecology which became, to many of us, the high point of the conference. Welch demonstrated physiological actions of chlorinated hydrocarbon insecticides at

extremely low doses — two orders of magnitude below previously
established effective concentrations. There were two mutually rein-
forcing actions: a dampening of production and an acceleration of
removal of estrogens. Welch was conservative in interpreting his
findings, and explained the need for further experiments during
discussion (he had only preliminary data on birds). It became
immediately clear, however, that he furnished the key to the ecological mystery of the thinned eggshells. Eggshells collected in
natural history museums around the world were shown to be
much thinner beginning at a date which coincided closely with
the first, wartime mass production and broadcasting of DDT. This
could be a meaningless coincidence. On the other hand, it could
be a true cause-effect relationship, and an indication of global
ecological damage by organochlorine pesticides. In the latter case,
it would be high time to consider banning organochlorines in favor of short-lasting, and perhaps more selectively toxic compounds.

The argument linking cause and effect began with the demonstration of a mechanism that gives a global distribution to organochlorines from local sources (Session I), included experiments on
inducing eggshell-thinning in the laboratory (Session II) and led
to a theory of endocrine disturbance by organochlorines which
linked eggshell-thinning with reproductive failure (Session III).
This argument remained unconvincing as long as the proponents
had to admit that they were postulating a large physiological effect
from implausibly small doses. Hence the profound impact of
Welch's paper which brought the toxicology of organochlorines
into the parts per billion range, and focused attention on an endocrine mechanism which controls both eggshell thickness and reproductive behavior.

Following Session III there was an extended discussion period
on the fundamental problem of defining "damage." Spencer explained the United States Government machinery for the control
of pesticides (Appendix).

The closing Session V dealt with protecting the human population from damage. To protect people one must first do some
tests on people, and this kind of research was presented in the
closing paper and discussed at length. Berglund and Berlin pre-

sented the work and the philosophy used in setting an official Allowable Daily Intake in Sweden for mercurials in food. Durham presented the counterpart of this work in the form of studies used in the United States to set limits of human exposure to organochlorines. West moved from the extrapolations of theory to the frustrating fight to keep people from being killed in practice. As health officer in a state which uses a fifth of the United States' production of pesticides, she talked of the human end of the runaway situation, whose agricultural origins were described in Session II. Ibsen's drama comes to life again in the public health officers' struggle with inadequate regulations, professional apathy, and commercial pressures.

The closing General Discussion allowed more time for the question, Are organochlorine pesticides threatening the survival of bird species? The lessons of the conference were distilled into a moving appeal by Rudd for a change in our ethical commitment.

Birds were prominent in the research presented at the conference for two reasons. They serve as a sensitive model of damage to man, since some of them are more subtly dependent than we are on a set of sex hormones which they share with us. They serve as a sensitive predictor of a crash of the ecological system, because they are large, highly visible flyers and hence a convenient model for the fate of the small and often hidden flyers, the insects. In the latter case, it is significant that persistent pesticides are damaging meat-eating birds while sparing grain-eaters, because of differences both in diets and in reproductive patterns. This is a model for the way in which the use of persistent pesticides ultimately destroys pest-eating organisms, both birds and insects, and promotes outbreaks of insect pests.

It is a striking fact that insects and fungi were hardly spoken of at all. Had we called a conference on the subject of selective pest killers, every paper would have dealt with the target organisms. We examined, instead, our practices of using persistent, nonspecific poisons in an attempt to kill pests. *Qui tacent clamant:* this slience indicts us.

GEORGE G. BERG

CONTENTS

CHEMICAL FALLOUT

Current Research on Persistent Pesticides

SESSION I
DISTRIBUTION, ACCUMULATION AND REMOVAL IN NATURE

Chapter 1

CHLORINATED HYDROCARBONS IN MARINE ECOSYSTEMS

R. W. RISEBROUGH

ABSTRACT

Polychlorinated biphenyls, chlorinated hydrocarbons which are extensively used in industry and agriculture, are now widely distributed in marine ecosystems of the Pacific Ocean. Birds contain higher concentrations of these chemicals than fish; tissues of the peregrine falcon have contained the highest amounts which have so far been recorded. The polychlorinated biphenyls have not yet been detected in samples of airborne particulates but their observed distribution in the sea indicates that they are dispersed by wind currents and that their fallout pattern is similar to that of the DDT compounds. Their effects upon natural populations, including man, are as yet unknown. Among the chlorinated hydrocarbon "pesticides," DDE is accumulating in significant amounts in marine food chains and is also present in highest concentrations in marine birds, especially the procellariiform species. Its distribution indicates that coastal areas are not the primary sources of contamination. A quantitative approach to the problem of aerial transport of chlorinated hydrocarbons to the sea has been made by analyzing the airborne particulates which contribute to marine sedimentary deposits. The results obtained to date indicate that wind transport can account for the observed distribution of DDT compounds in California waters and that the amount of chlorinated hydrocarbons entering the tropical Atlantic as fallout from the Northeast Trades is comparable to that entering the sea from a major river system.

REPORT

THE ACCUMULATING EVIDENCE that no part of the world is now free of "pesticide" residues, the products of atomic explosions, or of a variety of industrial pollutants has induced subtle but profound changes in our concept of our position in the global ecosystem and in our individual conceptions of remoteness and isolation. There has never been any question that local ecosystems could be irreversibly changed by the introduction of synthetic

5

chemicals, whether these introductions be purposeful or incidental. Since the sea, however, is the dominant feature of the
worldwide ecosystem, interpretations of environmental contamination in terms of local ecosystems, whether terrestrial or aquatic,
may therefore be misleading. The accumulation of significant
amounts of several pollutants in marine organisms, pollutants
which are nonpolar and therefore water-insoluble but lipid-soluble, not only elicits an uncertainty about the long-term utilization
of the sea as a source of human food, but has suddenly raised the
question of the ultimate survival of a number of species of sea
birds. These species comprise a very large fraction of the world's
wildlife and doubts about their future would have been considered preposterous and untenable only three or four years ago.

Perhaps because fish kills have provided the most dramatic
evidence of insecticide contamination, distribution studies in the
United States and elsewhere have frequently consisted of monitoring the residue levels in major river systems (2, 7). In spite of
the low solubility of chlorinated hydrocarbons in water (4), their
tendency to pass into the vapor phase and to leave the region of
application (1, 5, 6, 17, 22), their persistence in soils despite drainage (3, 14, 24), and their presence in very low concentrations in
some streams draining areas of intensive application (23), large
amounts of these stable persistent pesticides are transported to the
sea by way of rivers with high silt contents. Thus, the Sacramento
and San Joaquin rivers, which drain the Central Valley of California, one of the most heavily pesticided areas of the world, annually
bring about 1,900 kilograms of chlorinated hydrocarbons into San
Francisco Bay (2, 27), and the Mississippi contributes about
10,000 kg to the Gulf of Mexico (7, 27).

When the Institute of Marine Resources began a study of the
distribution of chlorinated hydrocarbons in marine fish, we expected to find, therefore, much higher levels in the fish from San
Francisco Bay than in the fish from the Pacific Ocean (30). The
data of Table 1-I, however, show that collections of the Northern
anchovy and of the English sole from the coastal waters contained
as much as or in two instances significantly more total DDT residue than did the collections from San Francisco Bay. Total DDT

TABLE 1-I
DDT RESIDUES* IN COLLECTIONS OF NORTHERN ANCHOVY
(*ENGRAULIS MORDAX*) AND ENGLISH SOLE (*PAROPHRYS VETULUS*)
FROM SAN FRANCISCO BAY AND CALIFORNIA COSTAL WATERS

Species Locality, Date	Number	Mean Wt (g)	p,p'-DDE†	Total DDT- ppm†
Northern Anchovy				
San Francisco Bay	17	12.4	0.21	0.59
July 29, 1965		(6-22)	±0.05	±0.11
San Francisco Bay	29	4.0	0.11	0.33
November 4, 1965		(2-5)	±0.05	±0.04
Monterey	30	26.5	0.68	0.90
November 30, 1965		(21-35)	±0.19	±0.22
Morro Bay	29	25.9	0.45	0.74
June 16, 1965		(19-33)	±0.12	±0.22
Port Hueneme	15	11.6	2.44	3.04
February 24, 1966		(2.8-21.5)	±0.77	±1.00
Terminal Island	44	11.7	10.2	14.0
Los Angeles, June 25, 1965		(6.5-20)	±2.1	±1.9
English Sole				
San Francisco Bay	18	14.3	0.14	0.55
July 29, 1965		(5-35)	±0.02	±0.07
San Francisco Bay	33	17.3	0.13	0.55
November 4, 1965		(7.5-53)	±0.03	±0.12
San Francisco lightship	15	253	0.12	0.19
December 1, 1965		(175-306)	±0.03	±0.04
Monterey	15	195	0.53	0.76
February 15, 1966		(89-262)	±0.13	±0.16

*From R. W. Risebrough, D. B. Menzel, D. J. Martin, Jr., and H. S. Olcott, in preparation. DDT residues include the two isomers of DDT, p,p'-DDT and o,p'-DDT and their metabolic derivatives: p.p'-DDE, o,p'-DDE, p,p'-DDD and p,p'-DDMU.
†Wet-weight parts per million, means, standard errors, 95% confidence limits.

concentrations were highest in the collection from waters off Los Angeles, but anchovies from the Channel Islands area off Port Hueneme also contained significantly more residue than the anchovies of San Francisco Bay. Since no major river system enters the Pacific Ocean from southern California the source of the DDT could not be agricultural drainage waters. Similarly, the data of Table 1-II, which sets forth the distribution of total DDT residues and of polychlorinated biphenyls in several collections of marine fish, including shiner perch from San Francisco Bay, hake, jack mackerel and English sole from the coastal waters of California, bluefin and yellowfin tunas from waters off Baja California and Central America, and skipjack tuna from South America and the

TABLE 1-II
DDT* AND POLYCHLORINATED BIPHENYL (PCB) RESIDUES IN
MARINE FISH.†

Species, Locality, Date	Number	Mean Wt. (g)	Total DDT ppm	p,p'- DDE %	PCB	DDT/ PCB
Northern Anchovy						
Terminal Island	44	11.7	14.0	83	1.0	14
June 25, 1965		(6.5-20)	±1.9			
Shiner Perch						
San Francisco Bay	14	5.5	1.0	28	1.2	0.8
October 20, 1965		(4-8)	±0.1			
San Francisco Bay	10	26.7	1.4	35	0.4	3.5
October 20, 1965		(10-48)	±0.3			
San Francisco Bay	15	15.3	1.1	33	1.2	0.9
November 4, 1965		(8-49)	±0.1			
English Sole						
San Francisco Bay	18	14.3	0.55	25	0.11	5
July 29, 1965		(5-35)	±0.07			
San Francisco Bay	33	17.3	0.55	24	0.11	5
November 4, 1965		(7.5-53)	±0.12			
San Francisco lightship	15	253	0.19	63	0.05	4
December 1, 1965		(175-306)	±0.04			
Monterey	15	195	0.76	70	0.04	19
February 15, 1966		(89-262)	±0.16			
Jack Mackerel						
Channel Islands	31	81.8	0.56	57	0.02	28
November 22, 1965		(45-141)	±0.10			
Hake						
Puget Sound	22	281	0.18	23	0.16	1.1
January 29, 1966		(185-350)	±0.05			
Hake						
Channel Islands	6	384	1.8	68	0.12	15
February 24, 1966		(61-872)	±1.1			
Bluefin Tuna†						
Body muscle	7	—	0.56	45	0.04	14
			±0.24			
Liver	9	—	0.22	45	0.04	6
			±0.13			
Yellowfin Tuna						
Liver§	13	—	0.07	13	nd¶	>7
			±0.02			
Liver¶	13	—	0.62	30	0.04	15
			±0.19			
Skipjack Tuna						
Liver††	3	—	0.057	23	0.1	0.6
Body Muscle††	13	—	0.051	18	nd	>30
			±0.014			
Liver‡‡	25	—	0.056	11	nd	>30
			±0.023			
Liver§§	12	—	0.029	21	nd	>20
			±0.008			

*From Risebrough *et al.* (30). Concentrations in wet weight, ppm. Means, standard errors, 95% confidence limits.

content of the eggs, however, was low, and the ratio of total DDT to the maximum detectable amount of PCB was within the range of the ratios recorded in other species (Tables 1-III and 1-IV).

TABLE 1-III
DDT* AND PCB RESIDUES IN MARINE BIRDS† AND IN THE PEREGRINE FALCON***

Species, Locality, Date	Total DDT‡ ppm	% DDE	PCB	DDT/PCB
Cassin's Auklet§	5.8	98	0.16	36
Ancient Murrelet¶	0.75	90	0.15	5
Fulmar**	0.41	76	0.08	5
Fulmar**	3.4	89	0.34	10
Red Phalarope††	0.78	79	0.10	8
Rhinoceros Auklet‡‡	2.7	97	0.36	8
Slender-billed Shearwater§§	32.0	92	2.1	15
Sooty Shearwater¶¶	12.3	94	1.2	10
Sooty Shearwater¶¶	10.3	86	0.9	12
Peregrine Falcon***				
Breast muscle, second year female, migrant from Arctic	104	99	22	4.5
Breast muscle, immature California	13	99	10.5	1.2
Breast muscle, adult female, California	112	98	109	1.0

*From Risebrough *et al.* (29) and Risebrough, Kirven and Herman (28).
†Entire bird analyzed, except peregrine falcons.
‡Includes p,p'-DDT, p,p'-DDD, p,p'-DDE, p,p'-DDMU, p,p'-DDE; ppm, wet weight.
§*Ptychoramphus aleuticus*, adult female, Farallon Islands, April, 1966.
¶*Synthliboramphus antiquus*, Monterey Bay, Nov. 1, 1966.
**Fulmarus glacialis*, Monterey Bay, Nov. 1, 1966.
††*Phalaropus fulicarius*, Monterey Bay, Nov. 1, 1966.
‡‡*Cerorhinca monocerata*, Monterey Bay, Nov. 1, 1966.
§§*Puffinus tenuirostris*, Monterey Bay, Dec. 12, 1966.
¶¶*Puffinus griseus*, Monterey Bay, Nov. 1, 1966.
***From Risebrough, Kirven and Herman (28).

The ratios of total DDT to PCB appear to cast some light on the distribution and fallout patterns of the chlorinated hydrocarbons. In San Francisco Bay species, including two wintering peregrine falcons, Caspian terns, Western gulls and shiner perch, this ratio tends to be close to unity. Analyses of Western gull eggs in addition to those reported in Table 1-IV show that eggs from San Francisco Bay contained more PCB than eggs from the Farallon Islands, 27 miles west of the Golden Gate Bridge, which in turn contained more PCB than eggs from Baja California (28).

TABLE 1-IV
DDT AND PCB CONTENT IN EGGS OF SEVERAL BIRD SPECIES

Species, Locality	Number	Total DDT* micrograms	p,p'-DDE %	PCB* micrograms	DDT/PCB
Brandt's Cormorant† (*Phalacrocorax penicillatus*)					
Farallon Islands	17	326	91	113	2.9
Pelagic Cormorant (*Phalacrocorax pelagicus*)					
San Mateo Co. Calif.	2	128 (125-130)	90	62 (48-75)	2.1
Murre *Uria aalge*					
Farallon Islands	6	1945 (932-3621)	96	558 (364-1010)	3.5
Pigeon Guillemot *Cepphus grylle*					
Farallon Islands	1	110	95	20	5.5
San Mateo Co.	1	103	91	62	1.7
Cassin's Auklet *Ptychoramphus aleuticus*					
Farallon Islands	2	147 (127-167)	97	15 (12-18)	10
Western Gull *Larus occidentalis*					
Farallon Islands	1	423	95	118	3.6
San Mateo Co.	1	235	94	112	2.1
San Francisco Bay	1	458	87	480	0.95
Black-crowned Night Heron *Nycticorax nycticorax*					
San Francisco Bay	1	541	89	330	1.6
	1	869	99	24	36
Caspian Tern *Hydroprogne caspia*					
San Francisco Bay	2	1269 (1216-1322)	89	805 (660-950)	1.7 (1.3-2.0)
San Diego Bay	5	1430 (991-2430)	88	1010 (550-1600)	1.4
Forsters Tern *Sterna forsteri*					
San Diego Bay	2	665 (598-732)	89	114 (91-137)	5.8
Least Petrel *Halocyptena microsoma*					
Baja California	2	30 (23-37)	84	3.1 (1.2-5.0)	10
Peregrine Falcon‡ *Falco peregrinus*					
Baja California	1	4830	97	471	10

*Total micrograms.
†Samples pooled for PCB analysis, p,p'-DDE content ranged from 63 to 1240 micrograms.
‡From Risebrough, Kirven, Herman, Reiche, and Olcott (28).

Of the two night heron eggs so far analyzed from San Francisco Bay, one had an "ocean" profile: high DDE, low p,p'-DDT and DDD, high DDT/PCB ratio; the other had a "bay" profile, suggesting that the adult females had spent the previous months in different localities. The Caspian and Forster's terns nest side by side on the dikes in San Diego Bay, yet the DDT/PCB ratio is much lower in the Caspian terns, which feed primarily in San Diego Bay and along the coast, than in the Forster's terns, which feed along the brackish and freshwater canals of the Otai River drainage. In the pelagic bird species, in most of the fish from the ocean, and in other specimens from areas remote from sites of application such as Baja California, the ratios of DDT to PCB are of the same order of magnitude and most values are between 5 and 15. If both PCB and DDT were dispersed around the world by the same transport system, their relative concentrations in "remote" areas would very likely be similar. Thus, most of the values obtained for this ratio in birds from a remote area of the Gulf of California are approximately 10 (28). Eggs of the elegant tern, however, a bird which migrates to other regions after the nesting season, had a significantly different ratio. It will therefore be of considerable interest to compare these ratios, and the ratios of either to other chlorinated hydrocarbons, among individuals and species breeding in remote areas of the world. Thus resident species could be compared with migrant species which could in turn be compared with locally raised juveniles before these have left the area.

Global fallout of the radioactive products of atomic experiments such as strontium 90, iodine 131 and cesium 137 is highly dependent upon local precipitation patterns. This is also true of the naturally occurring radioactive nuclides such as lead 210, which is a member of the uranium 238 series and which has a tropospheric residence time of only twenty days (8). It could be expected, therefore, that the fallout of chlorinated hydrocarbons associated with airborne particulate material would follow the same patterns, whether these particles are the original carrier such as the mineral talc, or whether they have absorbed chlorinated hydrocarbons originally present in the air as vapors. What is not clear is whether nonpolar compounds such as PCB must necessarily become associated with airborne particulates before entering the aqueous part of the marine ecosystem.

A quantitative approach to the problem of aerial transport of chlorinated hydrocarbons over the sea has therefore been made by measuring their concentrations in airborne particulates. These particulates constitute a large fraction of the material entering the sedimentary layers of the sea floor (37) and calculations of their chlorinated hydrocarbon content would therefore permit an estimation of the minimal amount of chlorinated hydrocarbon fallout in the sea. Dust deposits in glaciers, which can be accurately dated with lead 210 (16) have been used in calculating particulate fallout over the sea (37) and will undoubtedly eventually be used also to estimate chlorinated hydrocarbon fallout. To date, nylon mesh screens coated with glycerin have been used to collect marine airborne particulate material. Dust collected on such screens has been shown to collect the mineral talc in concentrations much higher than those expected on the basis of its natural distribution (38). Talc has been extensively used as a diluent for insecticides, and the rate at which it has been deposited on glaciers shows a significant increase after 1940 (38). Analysis of the airborne dust collected from the onshore winds at the end of the Scripps Institution of Oceanography pier in La Jolla showed that insecticides, predominately p,p'-DDT, were present in concentrations ranging from 1 to 81 ppm, with an average value of 18.2 ppm corresponding to 7×10^{-11} grams (g) per cubic meter (M^3) of air (27). By a striking coincidence, analysis of thirty-three samples of particulate material in water of marshes, irrigation canals, streams, rivers and lakes of California yielded an average value of 14.7 ppm with a range from 1.80 to 78.00 (21). The glycerin screen method fractionates against those materials carried on particles less than several microns or which are present as vapors. This value is therefore a lower limit of the amount of chlorinated hydrocarbons in the air. No PCB peaks were observed in the chromatograms of the dust extracts, which were therefore pooled, concentrated and saponified in order to degrade DDT and DDD. PCB was not present in the saponified extracts, and a maximum concentration of 5 parts per billion (ppb) was calculated for the dust samples, ten thousand times lower than that of the total insecticides (27). Since the ratio of PCB to DDT observed in fish and birds is much higher, it appears that PCB remains in the vapor phase and does not adsorb to particulate matter. Additional work, however, is being carried out in order to clarify this point. We are also collecting dust-free samples from air which has been passed through cold dimethylformamide, which would retain any chlorinated hydrocarbons present in the vapor phase in the air.

The dust collected at the end of the Scripps pier also contains lead produced by internal combustion engines in concentrations averaging approximately 0.5 micrograms (μg) /m³ of air. This value is about ten times lower than the mean value found in American cities but is one hundred times higher than that in marine air from the open sea (9, 10). It will be of considerable interest to compare the fallout patterns of the chlorinated hydrocarbons with that of lead produced by internal combustion machines, which can be traced and identified by virtue of its isotope composition.

The winds at the Scripps pier in La Jolla are predominantly landward with an unknown fraction of air from nearby agricultural areas. Although a present shortage of knowledge about air circulation patterns and fallout rates in this region makes it difficult to calculate the amounts of chlorinated hydrocarbons brought to the coastal waters by winds, it seems apparent that such a transport system could account for the unexpected geographical distribution of the DDT compounds in the marine and freshwater fish of California.

The Science Research Council of the United Kingdom had earlier mounted a similar screen on the eastern tip of the island of Barbados in an attempt to collect extraterrestrial dust of meteorite origin. The Northeast Trades at Barbados have blown across 5000 kilometers (km) of the tropical Atlantic, and it was therefore considered highly unlikely that any dust accumulating on the screen could be of continental origin. This assumption was, however, in considerable error since significant quantities of airborne particulate material were collected on the screen which upon mineralogical and biological examination proved to be most likely from Africa and Europe. Fallout rates of the dust over the tropical Atlantic and the extent to which it contributed to the bottom sediments were calculated (11). Gram samples were made available to us for analysis. The concentrations of chlorinated hydrocarbons in the dust ranged from less than 1 ppb to 164 ppb, with an average of 41. Knowledge of the fallout rate made possible an estimate of the quantity of insecticides entering the tropical Atlantic between the Equator and 30° N. This figure, a minimum value since the recovery of very small particulate materials was low, was 600 kg/yr, which compares with the value of 1,900 kg entering the San Francisco Bay via the Sacramento and San Joaquin Rivers (27). It cannot be concluded that these insecticides had originated only in Africa; the DDT residues accumulating in the rare Bermuda petrel (39) might have come from any part of the world.

Most of the chlorinated hydrocarbon content of the particulate material in marine air consists of p,p′-DDT. In marine fish and birds, p,p′-DDE is the major component and it would therefore appear that this conversion of DDT to DDE occurs early in the marine food chains. In San Francisco Bay and Puget Sound, DDE constitutes a much lower proportion of the total DDT compounds and DDT and DDD are comparatively more abundant. Similarly, DDE comprises only 22 per cent of the total DDT entering San Francisco Bay in the San Joaquin River (2). In bottom mud where anaerobic conditions might prevail, DDT can be expected to be converted to DDD rather than DDE (20, 35). From mammalian and insect toxicity studies it had earlier been concluded that DDE is relatively harmless. Recent research at the Patuxent Wildlife Research Center, however, has indicated that the toxicity of DDE to birds is much higher than expected and may be about one half that of p,p′-DDT (34). Like p,p′-DDT and other chlorinated hydrocarbons, p,p′-DDE is capable of inducing liver epoxidase enzymes (15). Such enzymes degrade sex hormones and other steroids by hydroxylating them (25) and thereby may affect calcium metabolism. This mechanism would explain how chlorinated hydrocarbons could be responsible for the increasing number of instances of abnormal calcium metabolism in birds. A significant decrease in eggshell weight of several birds of prey in Britain after the Second World War is evidence for a fundamental change in the environment (26). No change in the physical environment which could elicit a physiological effect of this magnitude among several species of birds over a wide area has been recorded. A change in the chemical environment would be manifest only through a change in the chemical composition of the diet. Of the new chemicals introduced into the environment after the Second World War which are ingested and retained by birds, only the organochlorine compounds have as yet been linked with calcium metabolism. No abnormal calcium metabolism has as yet been observed in sea birds, nor has it been looked for.

Application of the persistent biocides and the release of nondegradable waste products into the environment have frequently been justified by arguments that point out that local populations of organisms remain relatively uncontaminated and that, especially in the case of the chlorinated hydrocarbon insecticides, even the persistent compounds disappear with time. It is abundantly clear that these arguments have become misleading, irrelevant, and wrong. Pollutants do go everywhere. Those which are nonpolar, water-insoluble and which have finite vapor pressures will eventually appear in marine food chains. The DDT compounds and the polychlorinated biphenyls have already done so to an alarming degree.

DISCUSSION

RADOMSKI: How were the polychlorinated biphenyls detected in your assay? The commercial compound is a forest of peaks when analyzed. Which of these peaks were present in the analyzed biological material?

RISEBROUGH: There are several commercial preparations which differ in their average chlorine content. Each, however, is a mixture of compounds so that the chromatograms consist of clusters of peaks. The retention times of the peaks identified as PCB in extracts of fish and birds are given in the text. The profile of these peaks most closely matches that of the commercial compound containing 54% chlorine.

WURSTER: Theoretical considerations would predict a net transfer of chlorinated hydrocarbons from the land areas of the world into the oceanic basins where one would expect them to accumulate, especially considering that their half-life is probably about a decade in the environment. The data that you have from the Pacific Basin indicates that this is occurring. They check well with our data on sea birds in the North Atlantic [Wurster, C. F., and Wingate, D. B.: *Science, 159*:979, 1968].

The interesting thing is that they show the level in those two oceans approaching the level found in some bodies of fresh water and in San Francisco Bay, which drain rather heavily treated areas. Presumably the ocean basins will continue to become still more contaminated. The heaviest contamination of which I am aware in a sizeable lake is Lake Michigan, probably because it flushes only once per century. It serves as an example of things to come in the oceans.

In Lake George, trout fry were not viable after hatching from contaminated eggs containing several parts per million of DDT [Burdick, G. E., *et al.*: *Trans Amer Fish Soc, 93*:127, 1964]. The same thing was observed recently in Jasper National Park, where mortality of fry occurred at a few tenths of a part per million [Cuerrier, J-P., *et al.*: *Naturaliste Can, 94*:315, 1967]. This past spring, fry hatched from contaminated salmon eggs on Lake Michigan but about 20 per cent of them died from DDT poisoning at a time when there should be very low mortality.

It follows that if the contamination level in the world's major fisheries is approaching the levels associated with the collapse of fisheries in some freshwater areas, then we can soon expect a repeat performance in the oceans. What will happen to important fish such as swordfish or tuna? There have apparently been substantial declines in the take of some marine fish; I wonder if some fish are already producing nonviable fry caused by their burden of chlorinated hydrocarbon residues?

RISEBROUGH: My answer can consist only of two additional questions. Who has the responsibility of finding out what might happen to swordfish and tuna, and secondly, who is in fact going to find out what will happen? The answer to the first is clearly the manufacturers of DDT, dieldrin, PCB or any other persistent compound which ultimately accumulates in the global ecosystems. Unfortunately their major efforts to date in this direction have consisted of writing letters to *Science* which comment upon the relative abundance of robins. As for the second, the inherent limitations of government and international organizations prevent them from taking the initiative. It would appear that the answer must come from cooperative efforts of individual ecologists, naturalists and molecular biologists, perhaps those molecular biologists who are becoming somewhat alienated from the major trends in their field.

BUTLER: Geographic areas have been defined in which DDT levels in marine fish are frequently above 5 ppm in the gonad. In one area of the Gulf of Mexico where we know that reproduction of the speckled trout has declined, we assume that the decline is caused by DDT contamination. There are also indications that contamination with pesticides may be interfering with the reproduction of crabs in California.

HUNT: Did I understand you to say that contamination is greater in the marine ecosystem than it is in the inland water system?

RISEBROUGH: No, but they appear to be approximately equivalent. In the paper written by yourself and J. O. Keith (my reference #21), you reported on the chlorinated hydrocarbons in particulate material in freshwater systems in California. The average level was 14 ppm, and there were thirty-five samples with a range from 1 to 70 ppm. The average value of all samples of airborne particulate material which we collected off the Scripps Pier in La Jolla was 17 ppm with a comparable range from 1 to 80 ppm.

HUNT: Were the animals that live in these two systems also compared? Are you saying that the levels in marine fishes are higher than those in inland fishes?

RISEBROUGH: In the same paper you gave values for chlorinated hydrocarbon concentrations in California freshwater fish. I totaled the residues in whole fish or in flesh and found that almost all of these fell within the range of 0.2 to 2.0 ppm, the range within which most of the values recorded in marine fish collected off California shores also fall.

HUNT: We have a considerable amount of data on some species, striped bass, for example. The contamination can average as much as 35 or 40 ppm. I would say that certain inland species, particularly the carnivorous fish, would have levels considerably higher than ocean fish.

RISEBROUGH: Yes, of course, but these would reflect local sources of contamination rather than general fallout.

REFERENCES

1. ACREE, F., JR.; BEROZA, M., and BOWMAN, M.C.: *J Agr Food Chem, 11:* 278, 1959.
2. BAILEY, T.E., and HANNUM, J.R.: *J Sanit Eng Div Amer Soc Civil Engrs, 93:*27, 1967.
3. BARTHA, R.; LANZILOTTA, R.P., and PRAMER, D.: *Appl Microbiol, 15:*67, 1967.
4. BOWMAN, M.C.; ACREE, F., JR., and CORBETT, M.K.: *J Agri Food Chem, 8:* 406, 1960.
5. BOWMAN, M.C.; ACREE, F., JR.; LOFGREN, C.S., and BEROZA, M.: *Science, 146:*1480, 1964.
6. BOWMAN, M.C.; ACREE, R., JR.; SCHMIDT, C.H., and BEROZA, M.: *J Econ Entomol, 52:*1038, 1959.
7. BREIDENBACH, A.W.; GUNNERSON, C.G.; KAWAHARA, F.K.; LICHTENBERG, J. J., and GREEN, R.S.: *Public Health Rep, 82:*139, 1967.
8. BURTON, W.M., and STEWART, N.G.: *Nature, 186:*584, 1960.
9. CHOW, T.J., and JOHNSTONE, M.S.: *Science, 147:*502, 1965.
10. CHOW, T.J., Scripps Institution of Oceanography (personal communication).
11. DELANEY, A.C.; DELANEY, A.C.; PARKIN, D.W.; GRIFFIN, J.J.; GOLDBERG, E.D., and REIMANN, B.E.F.: *Geochim Cosmochim Acta, 31:*885, 1967.
12. Documentation of Threshold Limit Values. (Committee on Threshold Limit Values, American Conference of Governmental Industrial Hygienists) 1966, vol. 41.
13. ERRO, F.; BEVENUE, A., and BECKMAN, H.: *Bull. Envir Cont Toxic, 2:*372, 1967.
14. FAHEY, J.E.; BUTCHER, J.W., and TURNER, M.E.: *Pesticides Monit J, 1* (4) : 30, 1968.
15. GILLETT, J.W.; CHAN, T.M., and TERRIERE, L.C.: *Agr Food Chem, 14* (6) : 540, 1966.
16. GOLDBERG, E.D.: Symp. Radioactive Dating, I.A.E.A., Athens, Nov. 1962. I.A.E.A., Vienna, 1963, p. 121.

17. HARRIS, C.R., and LICHTENSTEIN, E.P.: *J Econ Entomol, 54*:1038, 1961.
18. HOLDEN, A.V., and MARSDEN, K.: *Nature, 216*:1274, 1967.
19. HOLMES, D.C.; SIMMONS, J.H., and TATTON, J.O'G.: *Nature, 216*:227, 1967.
20. JOHNSON, B.T.; GOODMAN, R.M., and GOLDBERG, H.S.: *Science, 157*:560, 1967.
21. KEITH, J.O., and HUNT, E.G.: Transactions of the Thirty-First North American Wildlife and Natural Resources Conference. Wildlife Management Institute, 1966.
22. LICHTENSTEIN, E.P.; ANDERSON, J.P.; FUHREMANN, T.W., and SCHULZ, K.R.: *Science, 159*:1110, 1968.
23. MOUBRY, R.J.; HELM, J.M., and MYRDAL, G.R.: *Pesticides Monit J, 1* (4) : 27, 1968.
24. NASH, R.G., and WOOLSON, E.A.: *Science, 157*:924, 1967.
25. PEAKALL, D.B.: *Nature, 216*:505, 1967.
26. RADCLIFFE, D.A.: *Nature, 215*:208, 1967.
27. RISEBROUGH, R.W.; HUGGET, R.J.; GRIFFIN, J.J., and GOLDBERG, E.D.: *Science, 159*:1233, 1968.
28. RISEBROUGH, R.W.; KIRVEN, M.N.; HERMAN, S.G.; REICHE, P., and OLCOTT, H.S. (In preparation).
29. RISEBROUGH, R.W.; MENZEL, E.B.; MARTIN, D.J., JR., and OLCOTT, H.S.: *Nature, 216*:589, 1967.
30. RISEBROUGH, R.W.; MENZEL, E.B.; MARTIN, D.J., JR., and OLCOTT, H.S. (In preparation).
31. SAX, N.I.: *Dangerous Properties of Industrial Chemicals*, New York, Reinhold, 1963.
32. TATTON, J.O'G., and RUZICKA, J.H.A.: *Nature, 215*:346, 1967.
33. THOMPSON, D.R.: Wisconsin Conservation Department Survey Report, February 14, 1966.
34. United States Department of the Interior, Fish and Wildlife Service; Wildlife Research Problems Programs Progress 1966 Resource Publication 43, Washington, D.C.
35. WEDEMEYER, G.: *Science, 152*:647, 1966.
36. WIDMARK, G.: *J Ass Off Anal Chem, 50*:1069, 1967.
37. WINDOM, H.L.: Ph.D. thesis. Univ. of Calif., San Diego, 1968.
38. WINDOM, H.; GRIFFIN, J.J., and GOLDBERG, E.D.: *Environ Sci Technol, 1*: 923, 1967.
39. WURSTER, C.F., JR., and WINGATE, D.B.: *Science, 159*:979, 1968.
40. The National Science Foundation, (grant GP 6362) provided financial support. I thank P. L. Ames, E. D. Goldberg, S. G. Herman, M. N. Kirven, H. S. Olcott, P. Reiche and R. L. Rudd for their assistance. The Monsanto Chemical Company generously provided samples of Aroclor 1254 and 1262. The fish extracts were originally prepared during a study sponsored by the U. S. Bureau of Commercial Fisheries.

Chapter 2

PESTICIDES AND THE WESTERN GREBE
A Study of Pesticide Survival and Trophic Concentration
at Clear Lake, Lake County, California

S. G. HERMAN, R. L. GARRETT and R. L. RUDD

ABSTRACT

One of the best-known examples of trophic concentration of persistent pesticide residues occurs in Clear Lake, California. The entire
lacustrine ecosystem contains chlorinated hydrocarbon residues, chiefly
of the DDT series. The effects of trophic concentration are most
obvious in the Western grebe *(Aechmophorus occidentalis)*, a fish-eating bird, in which both acute mortality and reproductive inhibition,
presumably attributable to high residue loads, have combined to cause
population declines. This report centers on probable pathways of
residue transfer and on the precise manner in which reproduction
might be affected. Various aspects of the population biology of the
Western grebe and its prey species are described. The breeding population of this colonially nesting species is approximately 150 pairs.
Regular aerial censusing shows variation in total numbers throughout
the year as well as differential distribution in the lake. Residue loads
in grebe tissues have remained relatively high over several years, averaging, as examples, in 1967 in DDD alone 544 ppm (wet weight) in
subcutaneous fat, 296 ppm (lipid weight) in eggs, and 546 ppm (lipid
weight) in the yolk sacs of hatchling grebes. All fishes contain DDD
and other residues. Residues in grebes vary seasonally and appear
strongly correlated with feeding rates and selection of prey types.
Comparison with another breeding population of grebes at Topaz
Lake, California reveals a different distribution and abundance of
tissue residues and a different seasonal variability. Maintenance of captive grebes has given precise information on rates of growth and on
feeding characteristics. Various components of physiological, biological,
and trophic concentration are described in detail. Studies will continue
for two additional reproductive seasons.

REPORT

Introduction

A REVIEW OF THE literature produced by investigations of pesti-
cide-wildlife relationships reveals a phylogeny of emphasis
and concern spanning the twenty-odd years elapsed since the com-

mercial introduction of chlorinated hydrocarbon insecticides. The picture is clearly an evolutionary pattern, combining trends of focus and divergence resulting today in widened, sophisticated, and intensified investigations.

Concern over the acute effects of these poisons dominated research for nearly a decade after their widespread introduction. Transferred and delayed effects were soon discovered, however, and some research shifted in these directions. The phenomenon of trophic concentration established that the problem was indeed an ecological one, requiring ecological approaches. To an increasing extent, ecology is a dominant theme characterizing pesticide studies, for it is in the biological community that the significance of pesticide use — both utility and hazard — lies. Pesticide investigations commonly involve two nominally separable components — distribution and significance. Although these two entities cannot be separated ecologically, they frequently are described independently.

Rapid advances in chromatographic technique and increased financial support have greatly improved our understanding of persistent pesticide distribution, especially in wild animals and in water. Even the most remote areas on this planet, including the polar regions and geographically obscure marine environments, harbor organisms with pesticide residues. These materials now contaminate not only living systems but permeate geological components including soil, water, and air.

The significance of residues in organisms is less well understood, although the literature documenting castastrophic and detrimental changes and reductions in wild animal populations, including vertebrates, is voluminous and rapidly growing (7). There is no longer any doubt that pesticides can and do adversely affect wild animals, and the prognosis is for more serious and widespread effects. Nor is there any indication that chlorinated hydrocarbon insecticides are being "phased out" or produced in smaller quantities. In the United States alone, 271 million pounds of the major chlorinated hydrocarbons, including DDT and the aldrin-toxaphene group, were manufactured in 1966. Although DDT production fell slightly after the publication of Rachel Carson's *Silent Spring* in 1962, the present trend is one of increase. The generally more toxic aldrin-toxaphene compounds are being produced in quantities which increase about 10 per cent annually (17). These latter compounds are also more persistent than DDT.

Studies of residue significance may be made in the laboratory or in the field. The most productive studies use both approaches. Laboratory studies including feeding trials have yielded much pertinent and important information, and continue to do so, but the very nature of the current widespread contamination requires that new emphasis be placed on the field aspects. "Pollution ecology" is a relatively new term, but one that clearly transcends its purely semantic utility.

The current work at Clear Lake has been in progress for nearly two years and is scheduled to continue until December 31, 1969. The majority of the data in this paper derive from the period between February and December of 1967. In this paper, we will describe the problem and its history and discuss what we believe to be the more relevant facets of our findings. We will also relate our conclusions to, and suggest lines of investigation for the future of, similar studies.

The Primary Study Areas and the History of the Problem

Clear Lake, Lake County, California, is an eutrophic, freshwater lake of 42,000 surface acres situated 1325 feet high in the Inner Coast Range approximately 100 miles north of San Francisco. It is shallow (average depth 22 ft.) and does not exhibit thermal stratification. The bottom is composed of soft and highly organic ooze. Lake level is regulated by a dam. Cool, wet winters and hot, dry summers characterize the climate of the Clear Lake basin. The waters of Clear Lake are quite turbid, averaging less than three feet Secchi disk visibility throughout the year (4). The fish fauna is rich and includes eight native and twelve exotic species (5). The lake is highly productive, supporting sizeable recreational and commercial fisheries.

Topaz Lake is a deep, freshwater lake of 4,000 acres at the eastern base of the Sierra Nevada sixty-five miles south of Reno, Nevada. The surface elevation is 5,000 feet. Lake level is also regulated by a dam. Summer and winter temperatures show great contrast; the surface of the lake sometimes freezes. The fish fauna includes several species of the family Salmonidae and at least two Cyprinids and one native Catastomid. Topaz Lake is also highly productive (11).

Since aboriginal times, inhabitants of the Clear Lake shore-line have been aware of the presence in the summer of a small but abundant midge, *Chaoborus astictopus* (Diptera: Chaoboridae), known locally as the Clear Lake gnat. This insect does not compete with man for food or fiber, does not bite, and does not adversely affect the public health. Objection to it results from the fact that the species is highly phototropic. Its creation as a major pest might well be dated to the introduction of the electric light to Lake County.

Clear Lake was treated with DDD (TDE) * three times over a period of nine years in an attempt to control the Clear Lake gnat. From September, 1949 until September, 1957, 120,726 pounds of DDD were applied directly into the lake from moving barges. Final concentrations of the insecticide were calculated to approximate 1 part DDD to 70 million parts of water (0.0143 ppm) after the first application and 1 part DDD to 50 million parts of water (0.020 ppm) after each of the subsequent two treatments. It is now known that coverage was not always uniform, and in fact concentrations must have varied widely in different parts of the lake after treatments.

The second lake treatment took place in September 1954. One hundred Western grebes, *Aechmophorus occidentalis,* were reported dead the following December. Pathological examination revealed no infectious disease. Additional reports of grebe mortality were received in March 1955 and December 1957. Post-mortem examinations were made of specimens submitted to the California Department of Fish and Game; again, no infectious disease was found. Early in 1958, a composite sample of fat from two "sick" Western grebes was found to contain 1600 ppm (wet weight) of DDD. Samples of several species of fish, more grebes,

---

*The nomenclature of pesticides and their metabolites mentioned in this paper is described in a later section.

and amphibians were analyzed. All contained DDD. It was concluded that DDD transferal had occurred through the food chain, poisoning the grebes, and that the nesting population of Western grebes at Clear Lake had declined as a result of the DDD treatments (10).

Alden H. Miller (10) estimated the pretreatment population of Western grebes on Clear Lake to have been in excess of one thousand pairs. Reproductive failure was evident shortly after the first treatment in 1949. No Western grebe nests were found in 1958 and 1959, in spite of intensive searches by Miller and other workers. Fewer than fifty individuals were seen on the entire lake during the breeding seasons of these years. During the 1960 season, no more than thirty grebes were seen on the lake, and neither nests nor young were seen. Department of Fish and Game personnel noted thirty-eight grebes in the vicinity of a nesting colony in 1961; sixteen nests were found on June 14 but could not be relocated. No young were observed. One juvenile was seen in 1962, and three were reported in 1963. Reproductive success remained low, and in 1966 the present study began with a reevaluation of the problem. Many nests and eggs were found, but only three young were seen, and chlorinated hydrocarbon residues in adults and eggs were high.

Objectives of the Current Project

This study has two primary objectives. The first is to evaluate the effects of the pesticide levels currently present in Western grebes and their eggs at Clear Lake. The second aim is to describe as precisely as possible the trophic and other pathways involved in the production of these levels in the grebes. These two phases are intimately related; our procedures integrate them.

Methods

Pesticide residue figures are notoriously difficult to evaluate and rationalize biologically. This is especially true of data obtained from material collected in the field. We strongly believe that any field investigation of pesticide distribution and effects must be firmly based on detailed ecological observations and measurements and that residue data must be kept in a position of secondary priority. To a large extent, residues found in an organism are the product of its life history. These aspects have been emphasized in this study.

Aerial Census

A census flight of standardized pattern covers the surface of Clear Lake once a month. This flight, in a light plane, takes approximately two hours and is made in the early morning. The Lake is divided into six sections, and each of these is flown in a circular pattern 150 to 200 feet above the surface. Two observers and the pilot are present. Grebe numbers and distribution are transcribed directly to a map by one of the observers.

Observation of Grebes

Detailed field notes are maintained in a standard manner. Of particular interest are grebe feeding habits, nuptial behavior, incubation behavior, colony formation, movements, and care of young. During the past two breeding seasons, the floating grebe nests have been marked with upright bamboo poles. Numbered tags are attached to these poles and records of clutch size, evidence of hatching, and behavior of adults are made on subsequent visits. Observations are made routinely at Clear Lake and Topaz Lake. In 1967, twenty-one days were spent observing the incubation behavior of six pairs of Western grebes at Eagle Lake in Lassen County, California.

Incubation of Grebe Eggs

To evaluate potential contamination effects on fertility of eggs and hatchability and viability of young, grebe eggs have been artificially incubated. In 1967, attempts were made to raise the young grebes hatched in the laboratory.

Maintenance of Captive Grebes

Since August of 1967, three captive Western grebes have been maintained in Davis. Two of these were captured when only a few days old, and the third was hatched in an incubator. Daily records of feeding rates have been kept. Other records include grebe weights, growth rates, and the ontogeny of behavioral patterns. Information on means of aging, plumage changes and other pertinent life history details has been obtained.

Sampling and Processing Procedures

GREBES. Grebes are sampled once a month at Clear Lake and are taken at less regular intervals from other localities, including Topaz Lake. Collecting is done with a shotgun and normally follows the aerial census by 6 to 24 hours. Ordinarily, ten grebes are taken every month during the fall and winter at Clear Lake; six per month are taken during the spring and summer. Each grebe is observed carefully before collection. Notes are made of its behavior, location on the lake, its relation to other grebes, and the time of collection. Collected grebes are refrigerated within 2 hours and transported to the laboratory in a portable ice chest. In the laboratory, they are refrigerated until processed. Processing is complete within four days of collection.

Each grebe is weighed to the nearest gram. Standard measurements, including culmen length, tarsometatarsus length, and wing length are taken. Notes are made of coloration. Nine separate tissues are taken from each bird. These are brain, breast muscle, leg muscle, liver, gonads, uropygial glands, subcutaneous fat, visceral fat, and thigh fat (the latter is taken from the surface of the carcass near the points of leg attachment). Testes in males are measured, weighed, and their color noted; in females the diameter of the three largest ova are recorded. Irregularities in organs are recorded when noted. The stomach is opened and the contents removed; recognizable prey items are measured and frozen separately. The intestines are saved. Finally, the left wing is saved. This is later dried in an expanded position to facilitate age determination. As it is excised, each tissue sample is placed in a sample bottle thrice prerinsed with hexane. The bottle is then covered with aluminum foil, capped, and placed in a freezer. Samples are stored at −20° C until processed by the chemist.

FISH. Fish samples are taken each month at Clear Lake when possible. A special effort is made to obtain species and size classes known to be taken by Western grebes. Samples are also taken at

Topaz Lake. Each fish is individually wrapped in aluminum foil as it is taken from the net. Fish are refrigerated until taken to the laboratory. Depending on the number of fish to be processed, they are then either frozen immediately or processed within 48 hours.

Identifications are made to species. Each fish is weighed, measured, and sexed; a few scales are removed for later aging. Fish are then frozen until handled by the chemist.

Eggs of Grebes. Grebe eggs were sampled at Clear Lake and Topaz Lake in 1966 and 1967. Whole eggs (exclusive of shells) were analyzed in 1967. Whole and partial clutches were taken at different times during the breeding cycle. Several eggs have been obtained from the oviducts of collected grebes. All eggs are individually analyzed and are examined for fertility before processing. Storage procedures are identical to those used for grebes and fish.

Water. Comprehensive water sampling is beyond the scope of this project as it is currently structured. Few water samples have been taken. Support is being sought to expand this important phase as well as sediment sampling, and it is hoped that these components of the ecosystem will be sampled extensively in the near future.

Other Samples. Samples of invertebrates and plankton are to be collected routinely in 1968 and 1969.

Analytical Technique

Residue analyses were made by Stoner Laboratories, Campbell, California. Extraction and cleanup procedures for gas chromatography were adapted from the pesticide residue analysis methodology established by the Pure Food and Drug Administration (16). Extracts were analyzed with a Dorhmann C-200A Micro coulometric detector. A 5 foot by $\frac{1}{4}$ inch outside diameter aluminum column packed with 5% Dow 11 on gas-chrom 60-80 mesh, at a block temperature of 250°C, a column temperature of 205°C, and 60 ml/min nitrogen flow was used. A standard was run adjacent to each sample. Quantitations were based on area under the peak.

TABLE 2-I

FREQUENCY OF OCCURRENCE OF FIVE PESTICIDES IN 230 GREBE TISSUE SAMPLES OF TEN KINDS FROM CLEAR LAKE

Tissue	N	DDT		DDD		DDE		DDMU		DDMS	
		No.	%	No.	%	No.	%	No.	%	No.	%
B	(33)	0	(0)	33	(100)	33	(100)	33	(100)	30	(90)
LM	(17)	0	(0)	17	(100)	15	(88)	17	(100)	17	(100)
L	(18)	0	(0)	18	(100)	18	(100)	18	(100)	18	(100)
BM	(62)	0	(0)	61	(98)	61	(98)	62	(100)	54	(87)
UG	(32)	2	(6)	32	(100)	32	(100)	32	(100)	32	(100)
SF	(18)	1	(5)	18	(100)	15	(83)	18	(100)	18	(100)
TF	(15)	0	(0)	15	(100)	14	(93)	15	(100)	15	(100)
VF	(18)	1	(5)	18	(100)	17	(94)	18	(100)	18	(100)
T	(10)	2	(20)	10	(100)	10	(100)	10	(100)	10	(100)
OV	(7)	0	(0)	7	(100)	7	(100)	7	(100)	7	(100)
Total		6		229		222		230		219	

B = brain; LM = leg muscle; L = liver; BM = breast muscle; UG = uropygial glands; SF = subcutaneous fat; TF = thigh fat; VF = visceral fat; T = testes; OV = ovaries. Figures indicate number and per cent of samples in which individual forms were found.

Results and Conclusions

More than a decade has passed since DDD was last applied directly into Clear Lake. Contamination levels remain relatively high, and this fact renders Clear Lake an ideal site for a modular examination of pesticide survival, kinetics, and effects.

In addition to the more than 120,000 pounds of DDD applied to the lake between 1949 and 1957, approximately 500,000 pounds of DDT were applied to the surrounding watershed during the period 1949-1964 (6). A fraction of this DDT has found its way into the lake. Water samples taken in 1967 verified the fact that DDT is still entering the lake by means of the two major inflowing streams. The historical importance of this continuing contribution must at this time be speculative, but there can be no doubt that DDT from watersheds around the lake continues to influence levels of contamination and their persistence in the lake.

DDD remains the most commonly recovered and abundantly found pesticide in the Clear Lake ecosystem (Tables 2-I and 2-II). Bridges *et al.* (3) have shown that DDT is altered to DDD under conditions similar to those in Clear Lake. Miskus *et al.* (12) have demonstrated that microorganisms present in Clear Lake water effect this conversion. Finley and Pillmore (9) first described the conversion of DDT to DDD in animal tissues, and Peterson and

TABLE 2-II
DOMINANCE OF FIVE PESTICIDES IN 230 GREBE TISSUE SAMPLES
OF TEN KINDS FROM CLEAR LAKE

Tissue	N	DDT		DDD		DDE		DDMU		DDMS	
		No.	%	No.	%	No.	%	No.	%	No.	%
B	(33)	0	(0)	14	(42)	6	(18)	12	(36)	1	(3)
LM	(17)	0	(0)	12	(70)	1	(5)	4	(23)	0	(0)
L	(18)	0	(0)	13	(72)	1	(5)	4	(22)	0	(0)
BM	(62)	0	(0)	49	(79)	5	(8)	7	(11)	1	(1)
UG	(32)	0	(0)	24	(75)	2	(7)	6	(18)	0	(0)
SF	(18)	0	(0)	16	(88)	1	(6)	1	(6)	0	(0)
TF	(15)	0	(0)	14	(93)	1	(6)	0	(0)	0	(0)
VF	(18)	0	(0)	15	(84)	1	(5)	2	(11)	0	(0)
T	(10)	0	(0)	6	(60)	0	(0)	4	(40)	0	(0)
OV	(7)	0	(0)	6	(85)	1	(15)	0	(0)	0	(0)
Total		0		169		19		40		2	

Figures show number and percentages of tissues in which individual forms were dominant.

(See Table 2-I for explanation of Letter-Symbols)

Robison (13) outlined the metabolic degradation pathway of DDT in the rat. DDT may go either to DDE, as it apparently does commonly in the marine environment (14), or to DDD. The striking prominence of DDD in Clear Lake may be explicable in terms of (a) the massive amounts of DDD put into the lake and (b) the fact that peculiarities of community metabolism in the lacustrine ecosystem convert the watershed contribution of DDT to DDD.

Bailey and Hannum (2) showed an inverse relationship between sediment particle size and pesticide concentration. The bottom of Clear Lake is soft, organic ooze. The persistence of pesitcides in the lake has been great. The large, confined, and dynamic biomass, which has acted to bind pesticides, and the nature of the bottom sediments, have enhanced the persistence of chlorinated hydrocarbons in the lake.

Life History of the Western Grebe

The Western grebe is a relatively large monotypic species that feeds almost exclusively on fish. It is confined to western North America. Breeding occurs on inland lakes. Wintering birds are found in the ocean and bays along the Pacific Coast from southern Alaska to Baja California. Spring and fall migrations are nocturnal. In California, most breeding colonies are found in the northern half of the state. Historically, Clear Lake has been an important reproductive site.

Nesting is colonial. Observations of captive grebes indicate that they are sexually mature and capable of breeding at the age of eight to ten months. The floating nests are built of plant material in emergent vegetation along the margins of lakes. Two to five eggs are normally laid; modal clutch size is three. Incubation apparently begins with the laying of the first or second egg, and lasts twenty-three to twenty-seven days. The young leave the nest within minutes of hatching, taking positions on the backs of the adults, under the folded wings. The young remain on the backs of the parents for ten days to two weeks, dismounting only to defecate and drink. During this period, the hatchlings are commonly held by one adult while the other forages for small fish, which make up the main diet of young grebes. After the age of two weeks, the juveniles are seen in the water with increasing frequency, and the subadults are indistinguishable from the adults from the age of eight to ten weeks on. Subadults are distinguishable from adults when dissected. We recognize four age classes: hatchlings, juveniles, subadults, and adults.

Western grebes eat fish and their own feathers. Other food items are taken only occasionally. When this study was begun, it was expected that grebe stomachs would yield much information on food habits. However, of 114 grebes taken between March and December, 1967, recognizable food remains were found in only eighteen grebes (Table 2-III). In Clear Lake, grebes select Centrarchids of rather small size, up to 13.5 cm in length and 12 g in weight. In Topaz Lake, where Centrarchids are not available, grebes take the native Cyprinids.

TABLE 2-III
RECOGNIZABLE FOOD REMAINS IN EIGHTEEN WESTERN GREBE
STOMACHS COLLECTED IN 1967

Date of Collection	Location	No. of Food Items	Identification	Length Range in Centimeters
April 22	CL	1	Centrarchidae	10.1
May 5	TL	1	Cyprinidae	20.5
June 11	TL	1	Cyprinidae	20.2
June 11	TL	1	Crayfish	—
July 21	TL	1	Crayfish	—
October 18	CL	1	Centrarchidae	6.0
October 18	CL	3	Centrarchidae	4.0-7.0
October 18	CL	1	Centrarchidae	6.5
October 18	CL	1	Centrarchidae	7.0
November 28	CL	6	Centrarchidae	3.5-7.0
November 28	CL	2	Centrarchidae	5.5-13.5
December 28	CL	5	Centrarchidae	6.0-7.0
December 28	CL	3	2 Cent., 1 Cyprin.	6.5; 10.0
December 28	CL	9	Centrarchidae	5.5-10.5
December 28	CL	4	Centrarchidae	6.5
December 28	CL	2	Centrarchidae	6.0

CL = Clear Lake; TL = Topaz Lake.

Feeding habits of grebes are difficult to evaluate quantitatively in the field. One of the most interesting aspects of the data now available is the absence of food from a large percentage of the stomachs. Records of feeding rates and weight fluctuations in

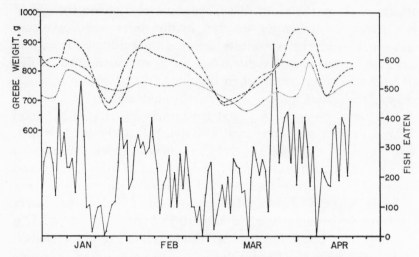

FIGURE 2-1. Combined feeding rates and individual weight fluctuations for three captive subadult Western grebes, January through April, 1968. Grebe weights in grams.

captive grebes have shed some light on this matter. Figure 2-1 shows daily feeding rates and weight fluctuations for three captive subadult Western grebes. Feeding rates are combined to include the three individuals. The live fish offered to grebes as food average 3 g each. The fluctuations occurred independent of food availability, as surplus fish were available at all times during the course of these records. We conclude that the patterns observed in captive grebes reflect adaptive characteristics also found in the wild population. Grebes are opportunistic feeders. Predators that must experience long periods of food shortage or absence have acquired the ability to endure periods of virtual famine. Grebes encountering large numbers of fish in a proper size and species range feed voraciously until satiated. Heavy reserves of depot fat which are built up during these times will allow the birds to pass later periods of low prey availability. Captive grebes, which have exhibited a high degree of endogenous behavioral control in all regards, have responded to a constant abundance of food in the manner demonstrated. The implications of this feeding pattern with reference to pesticide acquisition and storage are important.

Birds, finding and feeding upon heavily contaminated fish, may build up high concentrations of pesticides in a short time. If egg formation occurs at the peak of such a cycle, high residue concentrations may be expected in the eggs. If the wild population exhibits synchrony similar to that in the captive group, all eggs laid in a short period of time may exhibit similar residue values.

The Abundance and Distribution of Western Grebes on Clear Lake

A thorough understanding of the abundance and distribution of any species requires marking, releasing, and following a portion of the animals. Such a program is beyond the scope of this project. However, the monthly aerial censuses of grebes on Clear Lake, and observation and counts in other areas, have provided indications of the mechanics of these phenomena. Figure 2-2 and Figure 2-3 illustrate seasonal changes in adult grebe members and distribution on Clear Lake. The high peaks in the spring and fall months reflect the transient passage of grebes to and from other summer breeding sites. Transient grebes are frequently seen in large flocks in the northern and southern parts of the lake. These groups may be seen in the spring distribution maps.

In 1967, nest-building and egg-laying began in early May, at a time when transients were still using the lake. The nesting colo-

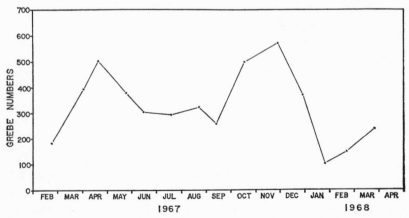

FIGURE 2-2. Seasonal changes in adult grebe numbers on Clear Lake, Lake County, California, February 1967 through March 1968.

ny is located near the shore of the northern portion of the lake
and is best seen as the area of major grebe concentration on the
June 15 map. Concentration in the breeding area occurs rapidly
and peaks at the height of breeding activity in June. Dispersal into
the open lake follows breeding. The breeding population in 1967
included about three hundred adults; a similar number was present in 1966. The slight rise in numbers in August is due to the
maturation of the young produced in 1967. The drop in September is due to postbreeding dispersal and is rapidly followed by the
appearance on the lake of grebes from other sites. The large increases in October and November correlate with the beginning of
freezing conditions in northern areas of high breeding density
and the early winter reduction in numbers coincides with the
arrival of large numbers of birds on San Francisco Bay and other
coastal sites. Many of the birds nesting on Clear Lake winter elsewhere, and it is unlikely that a permanently resident group occurs
on the lake.

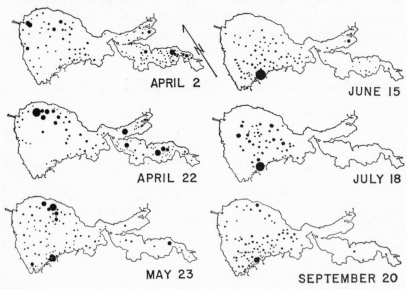

FIGURE 2-3. Seasonal changes in adult grebe distribution on Clear Lake, Lake
County, California. Diameters of dots increase with increases in group size:
smallest dots represent one grebe; the largest dot, on June 15, represents 200+
grebes.

Topaz Lake shows a simpler pattern of abundance and distribution. In 1967, only transient and summer resident birds were recognized. The lake is vacated during the winter, and the summer breeding population included approximately 125 adults.

Pesticide Residue Levels

All samples analyzed are examined for DDT, 1,1,1,-trichloro-2,2-bis (*p*-chlorophenyl) ethane; DDE, 1,1-dichloro-2,2-bis (*p*-chlorophenyl) ethylene; DDD, 1,1-dichloro-2,2-bis (*p*-chlorophenyl) ethane; DDMU, 1-chloro-2,2-bis (*p*-chlorophenyl) ethylene; and DDMS, 1-chloro-2,2-bis (*p*-chlorophenyl) ethane. A portion of the samples is analyzed for other chlorinated hydrocarbons.

DDD and its metabolites (DDMU and DDMS) and DDE are the most frequently recovered materials in Clear Lake. DDD usually appears in the greatest amounts and is most often the dominant metabolite in grebe tissues (Tables 2-I and 2-II). The same is true of grebe eggs and prey in Clear Lake. In Topaz Lake, DDE is the most commonly found and abundant material in grebes, grebe eggs, and prey.

We recognize three separate processes of chlorinated hydrocarbon concentration in natural systems. *Physiological* concentration is the distribution of pesticides in the individual host animal after transfer from the environment. *Biological* concentration is the transfer of pesticides from the environment to the organism by means of filtering and absorption. Several workers (1, 8) have demonstrated the importance of this process in fish, and it is in this class of vertebrates that it shows greatest relevance here. *Trophic* concentration derives from the ingestion of contaminated material by the host organism. Both biological and trophic routes are available to fish, but the latter means is undoubtedly of greatest importance to fish-eating birds and predators in general.

Physiological concentration has been examined in adult Western grebes from Clear Lake. The lipophylic nature of organic pesticides is well known. An intertissue comparison of lipid content and DDD concentration has been made of 187 tissue samples from sixty-two adult grebes taken on Clear Lake between March and December, 1967. Table 2-IV and Figure 2-4 demonstrate this

TABLE 2-IV
THE RELATIONSHIP OF DDD CONCENTRATIONS TO PER CENT OF
LIPIDS IN 187 GREBE TISSUES FROM CLEAR LAKE

Tissue	N	X % Lipids	SE	Range	X ppm DDD	SE	Range
LM	(17)	1.90	0.22	(1.04-4.89)	4.56	0.75	(0.28-13.2)
L	(18)	4.43	0.12	(3.46-5.60)	12.21	1.60	(2.25-30.1)
BM	(62)	5.40	0.27	(0.31-10.3)	14.58	1.31	(0.00-43.4)
OV	(7)	16.12	4.75	(1.64-35.3)	48.92	21.32	(8.70-128.0)
UG	(32)	28.22	2.52	(17.2-80.6)	108.92	11.88	(16.00-280.0)
SF	(18)	67.20	4.25	(26.0-86.0)	499.55	5.33	(52.00-968.0)
TF	(15)	86.53	1.18	(78.5-95.6)	561.08	24.81	(76.80-1070.0)
VF	(18)	73.25	2.56	(45.2-89.1)	570.25	87.30	(52.5-1490.0)

Means, standard errors, and ranges for per cent of lipids and ppm DDD in eight tissue categories. Wet-weight basis.

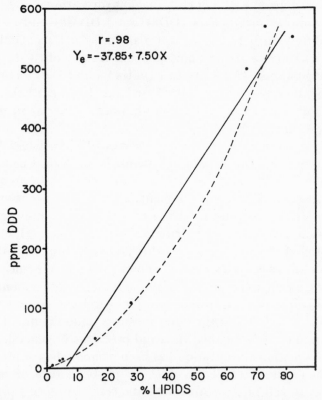

FIGURE 2-4. Physiological concentration of DDD. Mean values of wet weight concentrations of DDD in 187 samples of grebe tissues plotted against the mean per cent of lipid values in the same tissues. Data also shown in Table 2-IV.

relationship. Brain and testes figures are omitted from the graph. Brain samples are highly variable, probably because of the unique nature of brain lipids and also because of analytical problems encountered in extracting pesticides from brain. The testes analyzed were changing rapidly during the course of sampling and also proved difficult to handle with respect to extraction.

Wet-weight physiological concentration of DDD in grebe tissues is a function of lipid concentration. The correlation coefficient for the regression line in Figure 2-4 is 0.98, and it is significant at the 99.99 per cent level. The exponential curve predicts close to 0.0 ppm DDD at 0.0 per cent of lipids, although this datum point is impossible in nature. These data have obvious implications of value in choosing an index tissue for this and other studies and in making broad predictions of pesticide levels in various tissues. When tissues are compared on an intratissue basis the correlation fails to hold, however, due to differences in the total amounts of pesticides found in individual birds and consequent variations in degree of lipid saturation by pesticides.

Residue levels found in grebe eggs are also a product of physiological concentration. Several fully formed eggs have been found in the oviducts of collected grebes. When the pesticide levels in egg and breast muscle lipids are compared in the same bird, close qualitative and quantitative correlations exist. Figure 2-5 makes this comparison for four breast muscle-oviduct egg combinations. The means of comparison used is a "contaminant profile." This graph is a crude representation of a gas chromatograph. The vertical axis measures parts per million. The horizontal axis, in this case, shows DDT and several of its breakdown products arbitrarily and equally spaced. Lipid concentrations were chosen because the relative lipid content of the tissues compared varies; wet-weight concentrations would not, therefore, provide a valid comparison. The breast muscle residue figures for bird number 670628-3 are tentative. It is apparent that residue levels in egg lipids closely reflect levels found in breast muscle lipids. Factors regulating levels in the adult females during egg formation will determine pesticide levels found in the eggs.

Both biological and trophic concentration are involved in the

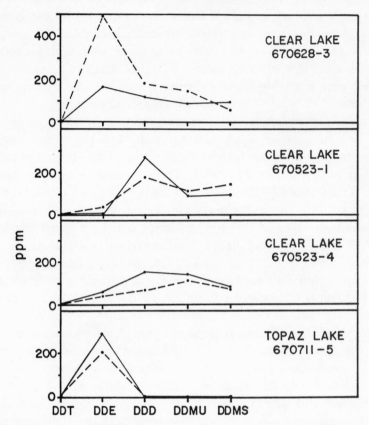

FIGURE 2-5. Contaminant profiles comparing pesticide levels in breast muscles (dashed line) and oviduct eggs (solid line) from single females. Lipid basis.

establishment of pesticide residue levels in fish. Hunt and Bischoff (11) demonstrated the general trophic relationships of residues in Clear Lake species. Carnivores show larger amounts than algal feeders; larger, older fish usually show higher amounts than smaller, younger fish of the same species. We wish ultimately to describe precisely the trophic pathways leading to the contamination of grebe tissues. To accomplish this aim, it will be necessary to evaluate the relative contributions of biological and trophic concentration at the fish level. A comprehensive analysis of fish

TABLE 2-V
DDD RESIDUES IN A SAMPLE OF BLUEGILL SUNFISH FROM CLEAR LAKE

Length in Centimeters	Weight in Grams	No. of Fish	X ppm DDD Wet Weight	Micrograms per Fish
3.2	0.5	21	1.25	0.63
3.8	0.9	31	1.96	1.76
4.8	1.5	20	1.77	2.66
5.8	3.5	10	1.76	6.16
11.9	29.0	1	0.77	22.33
13.4	46.0	1	0.65	29.90
15.2	74.0	2	1.28	95.46
18.6	106.0	1	0.66	69.96
19.0	128.0	1	0.64	81.92
20.0	155.0	1	1.82	282.10

has awaited a clear understanding of the feeding habits of grebes. Hence, fish sampling in 1967 was not extensive. Table 2-V shows the relationships of pesticide content and fish size in a limited sample of bluegill sunfish (*Lepomis macrochirus*: Centrarchidae) taken from Clear Lake on September 20, 1967. Several trends are indicated. The correlation between fish size and micrograms of DDD is close in the size classes taken by grebes, when the fish are feeding primarily on macroplankton. The correlation fails in the median size classes when the fish are large enough to prey on other fish. At the upper range where fish are mature as judged by size, residues are less influenced by size. If the major pesticide contribution to these fish were through biological means, a close size-residue relationship could not be expected because water filtered over the gills would be a function of time, not size. If most DDD reaches fish by means of trophic concentration, however, size would be a reflection of amount of food ingested, and a close relationship could be expected. It appears that trophic concentration is currently the major route of residue acquisition for fish in the size range taken by grebes.

Trophic concentration is the process by which the environmental transfer of pesticides from fish to grebes is effected. Residue levels in individual grebes at any point in time will be controlled by a complex of factors including physiological and ecological components.

Residue concentrations in adult grebes fluctuate temporally, Figure 2-6 shows these fluctuations in grebe breast muscle samples taken from Clear Lake and Topaz Lake in 1967. Breast muscle was chosen as an index tissue after a careful weighing of several criteria. Breast muscle has been used by other investigators; our use of it provides a basis for comparisons with other studies and other species. Because pesticide content is related to lipid content, a tissue with little variation in per cent of lipids was desirable.

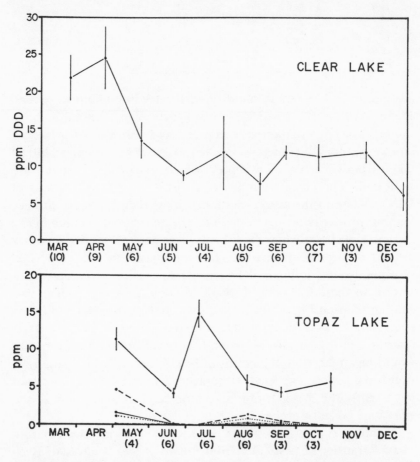

FIGURE 2-6. Temporal fluctuations in residue levels (wet weight) in adult Western grebe breast muscle. Sample size shown below months. *Top,* fluctuations in DDD in Clear Lake grebes. Vertical lines span two standard errors of the mean. *Bottom,* fluctuations in DDE (●——●, the highest values), DDD (● — — ●), DDT (●——●),DDMU (● - - - - - ●), and DDMS (●— - —- —●). Standard errors shown only for DDE.

Breast muscle satisfies this requirement (Table 2-IV). Breast muscle is available from every adult and subadult grebe collected and may be taken in a standard way from each specimen. Breast muscle also provides the chemist with a sufficient sample of tissue to allow reliable analysis. Finally, comparability with other tissues appears high.

Nine tissues from each of ten grebes collected in 1967 were analyzed. A computer program written to generate correlation coefficients compared residue concentrations in the lipids of each tissue with those in each other tissue for each of the five chlorinated hydrocarbons commonly recovered from Clear Lake grebes. Breast muscle consistently ranked highest in this comparison. In other words, pesticide levels in breast muscle lipids correlate well with levels found in other tissues.

Figure 2-6 considers the major material (DDD) found in Clear Lake birds and DDD, DDT and several of its metabolites in Topaz Lake grebes. Age of birds, environments visited, pesticide levels in diet, rates of elimination of pesticides by the grebes, and feeding rates are among the factors controlling levels in individual birds sampled. We conclude that the major ecological determinants of pesticide levels in grebes are prey selection and availability and feeding rates as they change in time and space.

Pesticide levels change on a temporal basis in both lakes. The birds sampled change in age only slightly during the sampling period; this change is not sufficient to account for residue changes, nor does it show a positive correlation with age. In Topaz Lake, DDE is the predominant pesticide found in prey, and fluctuations in grebes occur independent of fluctuations in levels of other metabolites. In Clear Lake, where levels in prey differ from those in predators quantitatively, but not qualitatively, levels fluctuate in a more coincident manner. Feeding rates observed in captive grebes show great daily fluctuations which, if present in the wild birds, account in part for variations in pesticide levels over the greater time period between samples. Temporal and spatial differences in prey selection and availability can also be expected to contribute to these fluctuations.

Further indications of the importance of prey selectivity and feeding rates as factors modulating trophic concentration come from an examination of longer-term fluctuations in grebe residue levels. Table 2-VI lists DDD residues found in the fat of Clear Lake grebes from 1958 until 1967. Figures for residues in fat are used here because fat was the common index tissue in the earlier studies. We are currently having analyzed fat samples from all grebes collected in 1967 and anticipate that this will provide us with a clearer picture of the changes in residue values on all scales.

TABLE 2-VI
DDD RESIDUES IN WESTERN GREBE FAT FROM BIRDS COLLECTED
1958 THROUGH 1967

Year	Range ppm DDD	Mean ppm DDD	No. of Analyses	No. of Grebes
1958	723-1600	1161	2	3
1959	1465	1465	1	3
1960	51-1150	537	5	13
1961	16-656	451	13	13
1962	1480-3130	1988	6	6
1963	220-1520	809	5	11
1964	—	—	—	—
1965	—	—	—	—
1966	112-499	267	12	12
1967	52-1490	544	51	18

Data from 1958 through 1963 courtesy of California Department of Fish and Game. Wet-weight basis.

It has recently become convenient, if not particularly accurate, to speak of "pesticide half-life." Borrowing terms from the field of radiation physics may prove to have its place, but the implication of simplicity inherent in this generality is no less dangerous than those present in other generalities applied to ecological problems. Many ecological data are lacking for the specimens represented in Table 2-VI, but it remains obvious that residue levels of DDD in grebe fat at Clear Lake have by no means followed an orderly route of reduction through time. It is also clear, from the 1967 data, that residue values in grebes can change on a short-term basis as well. Birds taken at different times in different years may, therefore, be expected to vary with regard to pesticide levels, and prey selection and feeding rates are likely the most important factors producing this variation.

Qualitative and quantitative differences in pesticide levels from Clear Lake and Topaz Lake are characteristically specific to the two locations. Figure 2-7 compares typical contaminant profiles of grebes and eggs collected in the two lakes. This geographic specificity again indicates the dynamic nature of residue level change in response to variables associated with trophic concentration.

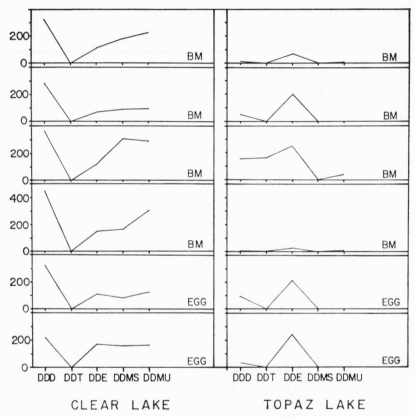

CLEAR LAKE   TOPAZ LAKE

FIGURE 2-7. Contaminant profiles comparing two separate geographic areas. Western grebe breast muscles and eggs from Clear Lake and Topaz Lake are shown. Lipid basis.

Reproductive Success in the Western Grebe

There exists strong inferential evidence that pesticides, mainly DDD, have been responsible for reproductive failure in Western

grebes at Clear Lake. An objective evaluation of these effects requires a thorough knowledge of other factors regulating reproductive success in areas not plagued with the stigma of pesticide
contamination. It is too early to make a precise evaluation, but
various aspects of this phase of the study have been clarified in
1967.

Fluctuating water levels, habitat availability, vegetation density
in the colony, predation on eggs, and human interference are
several variables that control breeding success. All of these act
and interact to influence nesting success. Pesticides may adversely
affect reproduction in wild birds in a variety of ways. They may
cause adult mortality, cause the death of embryos or newly hatched
young, or act in indirect ways such as removal of a food source
(15).

In 1967, 165 nests with eggs were marked and followed
through parts of the incubation cycle. Some of these represented
renesting attempts. More than 450 eggs were laid; of these, 89
were taken for incubation and residue analysis. We estimate that
only forty to sixty hatchlings reached subadulthood. DDD residues in eggs were high. Table 2-VII lists a representative sample
of residues in egg lipids. Lipid concentration figures are used here
because of great variations in the wet weights of the eggs collected

TABLE 2-VII
PESTICIDE RESIDUES IN SEVENTEEN WESTERN GREBE
EGGS COLLECTED AT CLEAR LAKE IN 1967 — PPM LIPID BASIS

Date of Collection	Clutch and Egg Number	DDT	DDE	DDD	DDMU	DDMS
May 23	1	0.0	41.0	177.0	113.0	146.0
	4	0.0	41.6	69.2	110.0	74.3
June 9	108A	0.0	154.0	588.0	206.0	150.0
	104B	0.0	170.0	219.0	161.0	158.0
	153A	0.0	141.0	112.0	0.0	0.0
	157B	0.0	37.6	187.0	74.8	138.0
	157C	0.0	27.6	181.0	86.3	129.0
	181B	0.0	112.0	322.0	124.0	81.5
	211A	7.3	42.8	175.0	34.4	0.0
	211B	0.0	94.8	162.0	50.4	0.0
	211C	12.2	66.5	190.0	65.2	0.0
	221A	0.0	170.0	202.0	0.0	0.0
	221B	0.0	170.0	235.0	49.6	37.2
June 28	3	0.0	494.0	254.0	152.0	58.8
July 31	6A	0.0	254.0	561.0	278.0	183.0
	6C	0.0	313.0	436.0	358.0	197.0
	4B	0.0	87.2	1007.0	223.0	255.0

in different conditions of dessication and decomposition. Several preliminary judgments are possible. There is a general tendency for DDD concentrations to increase as the nesting season advances. There is close qualitative and quantitative agreement in residue values from eggs taken from the same clutch. DDD is the most common material in fourteen of the seventeen eggs shown. The dominance of DDE in two of the eggs and the presence of DDT in two others indicate the presence of newly arrived birds in the breeding population.

The evidence that chlorinated hydrocarbons, particularly DDD, have adversely affected grebe numbers and reproduction at Clear Lake is particularly strong. A careful evaluation of all ecological factors potentially controlling reproductive success in 1967 indicates that none was sufficiently significant to explain the limited production of juvenal grebes. Pesticide residues are strongly implicated in this reproductive inhibition. We believe that mortality during the hatchling stage is the most important contribution to this failure. Of forty-eight Clear Lake eggs artificially incubated, twelve were dead when collected or died in the first week of incubation, eighteen died after one week of incubation, and eighteen hatched. Although problems were encountered in technique, especially during the critical hatching period, indications are that embryo viability was below that which we consider normal. All but three of the 1967 hatchlings died within 72 hours of hatching. Newly hatched young in the field are exposed to pesticides from their yolk sacs and in diet. These values are shown in Table 2-VIII. A comprehensive evaluation of all factors, including the effects of pesticides, must await a comparison of several reproductive seasons.

TABLE 2-VIII
DDD RESIDUES IN FIELD-COLLECTED HATCHLING WESTERN GREBES
AT CLEAR LAKE, JUNE 28, 1967 — PPM LIPID BASIS

Number	Yolk sac	% Lipids	Stomach contents	%Lipids
J4A	937.0	20.8	226.0	4.70
J4B	756.0	14.3	136.0	6.19
J7A	177.0	21.2	239.0	1.33
J7B	268.0	24.0	71.6	2.79
JB	594.0	15.7	37.7	12.00
J7	—	—	38.2	14.50

ACKNOWLEDGMENTS

Many individuals and agencies have assisted us in the conduct of this investigation. We are grateful to them. Specifically, we wish to thank Nancy E. Herman for preparation of figures and for maintenance of captive birds; Dr. S. F. Cook, Jr., for continuing professional counsel; Sterling Bunnell, Jr., M.D. for initial financial support; Elizabeth Stoner for helpful suggestions regarding analytical technique and unusual general interest in the project; E. G. Hunt and A. I. Bischoff of the California Department of Fish and Game for unpublished data and frequent counsel; Gordon Conway for assistance in computer programming, and J. O. Keith of the Bureau of Sports Fisheries and Wildlife for unpublished data and frequent assistance. We especially acknowledge the Division of Wildlife Research, Bureau of Sports Fisheries and Wildlife for financial support awarded to one of us (R. L. R.) to continue these investigations.

REFERENCES

1. ALLISON, D.T.; KALLMAN, B.J.; COPE, O.B., and VANVALIN, C.: Some chronic effects of DDT on cutthroat trout. USDI, Bur. Sport Fisher. Wildl., Res. Rept. 64, 1964, 30 p.

2. BAILEY, T.E., and HANNUM, J.R.: Distribution of pesticides in California. *Proc Sanit Eng Div Amer Soc Civil Engrs,* 1967, p. 27.

3. BRIDGES, W.R.; KALLMAN, B.J., and ANDREWS, A.K.: Persistence of DDT and its metabolites in a farm pond. *Trans Amer Fisheries Soc, 92:*421, 1963.

4. COOK, S.F., JR.; CONNERS, J.D., and MOORE, R.L.: The impact of the fishery upon the midge populations of Clear Lake, Lake County, California. *Ann Entom Soc Amer, 57:*701, 1954.

5. COOK, S.F., JR.; MOORE, R.L., and CONNERS, J.D.: The status of the native fishes of Clear Lake, Lake County, California. *Wasmann J Biol, 24* (1) : 141, 1966.

6. COOK, S.F., JR.: The Clear Lake gnat; its control, past, present, and future. *California Vector Views, 12* (9) :43, 1965

7. DUSTMAN, E.H., and STICKEL, L.F.: Pesticide residues in the ecosystem. In *Pesticides and Their Effects on Soils and Water.* American Society of Agronomy Special Publication No. 8, 1966, pp. 109-121.

8. FERGUSON, D.E.; LUDKE, J.L., and MURPHY, G.G.: Dynamics of endrin uptake and release by resistant and susceptible strains of mosquito fish. *Trans Amer Fisheries Soc, 95* (4) :335, 1966.

9. FINLEY, R.B., JR., and PILLMORE, R.E.: Conversion of DDT to DDD in animal tissue. *AIBS Bull, 13* (3) :44, 1963.
10. HUNT, E.G., and BISCHOFF, A.I.: Inimical effects on wildlife of periodic DDD applications to Clear Lake. *California Fish and Game, 46* (1) : 91, 1960.
11. LA RIVERS, I.: *Fishes and Fisheries of Nevada.* Nevada State Fish and Game Commission, 1962. 782 pp.
12. MISKUS, R.P.; BLAIR, D.P., and CASIDA, J.E.: Conversion of DDT to DDD by bovine rumen fluid, lake water, and reduced porphyrins. *J Agri Food Chem, 13* (5) :481, 1965.
13. PETERSON, J.E., and ROBISON, W.H.: Metabolic products of *p,p'*-DDT in the rat. *Toxic Appl Pharmacol, 6* (3) :321, 1964.
14. RISEBROUGH, R.W.; MENZEL, D.B.; MARTIN, D.J., JR., and OLCOTT, H.S.: DDT residues in Pacific sea birds: a persistent insecticide in marine food chains. *Nature, 216* (5115) :589, 1967.
15. RUDD, R.L.: *Pesticides and the Living Landscape.* Univ. of Wisconsin Press, 1964, 320 pp.
16. U. S. Dept. of Health, Education, and Welfare: *Guide to the Analysis of Pesticide Residues.* Vols. I and II.
17. United States Department of Agriculture: *The Pesticide Situation,* 1960-1967 (July, 1961-October, 1967) .

DISCUSSION

WURSTER: Your data on the persistence of DDD in Clear Lake suggest a half-life of about ten years for DDT in a natural environment. DDT in Maryland soil was also shown to have a half-life of ten or twelve years, although that did not include losses from the system that did not comprise a breakdown process but simply a transfer from one place to another [Nash, R. G., and Woolson, E. A.: *Science, 157*:924, 1967].

RUDD: By "extraordinary persistence," I mean levels that remain very high. Such terms as "half-life" have little reality in biological systems. I am coming more to the view that the residual levels in these species are a function of the biological phases in the ecosystem; it is a biological cycling, not just a physical transfer followed by a chemical alteration. We are seeing captive DDD in populations, if you want to put it that way.

WURSTER: The term "half-life" of course is jargon in this case, and I used it just because it was convenient. I meant the time it takes for half to no longer be present. A ten-year survival time should not be surprising when we consider degradation studies in water and soil.

RISEBROUGH: What strikes me after looking at the various chromatograms from Clear Lake are the similarities of the profiles. One can look at a gas chromatogram and identify it as material from Clear Lake. It is very different from the profiles from the ocean and San Francisco Bay. Presumably, DDD is first degraded to DDMU, then to DDMS, and so on. If considerable degradation occurs, why should the profiles from lake fish and migratory birds be the same?

HERMAN: There is no answer yet; we have only recently sent samples of grebes taken from San Francisco Bay for analysis. DDD becomes the predominant metabolite in grebe tissue shortly after the grebes arrive at Clear Lake and begin feeding. I think this is primarily a reflection of the relatively high levels of DDD in the Clear Lake fish, but it probably also indicates the rapidity with which profiles can change in organisms.

RISEBROUGH: Perhaps the grebes have a different profile at the very beginning of the nesting season.

HERMAN: Perhaps they do. We have always taken grebes which had been on the lake for a while.

RISEBROUGH: I gave in my paper an example of two black crowned night heron eggs from two different nests in San Francisco Bay; the profiles are very different. One is a typical Bay profile with a relatively high amount of PCB, a large amount of DDD, and a relatively low amount of DDE. The other one is a typical "ocean" profile: a high level of DDE and low levels of DDT, DDD, and PCB. I would conclude from this that one female bird had spent the winter in San Francisco Bay, the other along the coast.

HERMAN: That would depend, however, on how fast these things change.

In our paper, Figure 2-7 compares a contaminant profile from Clear Lake grebes and grebe eggs with several profiles from Topaz Lake grebes and their eggs. One can easily identify the origin of these birds according to their profile by simply aligning the metabolites in random order and plotting the parts per million.

Clear Lake has a characteristic profile as does Topaz Lake. We suspect that we will get a different profile from birds collected in marine situations. One Topaz Lake bird taken in May, when it should just have returned from the marine wintering area, had 62 ppm of PCB in lipids from a sample of breast muscle. Polychlorinated biphenyls are not commonly recovered from grebes taken after they have been present for some time on fresh water.

Chapter 3

PERSISTENCE OF PESTICIDE RESIDUES IN SOILS

P. C. KEARNEY, R. G. NASH and A. R. ISENSEE

ABSTRACT

Pesticide persistence in our environment is of current national interest. The effective life of pesticides in soils varies from a few weeks to several years. Their pesistence depends primarily on the structure and properties of the compound, and to a lesser degree on location and soil properties. Extremes in persistence are found with the insecticides. The highly toxic phosphates do not persist for extended periods of time in soils. In contrast, some of the chlorinated hydrocarbon insecticides may persist for four to five years under normal rates of application. Disappearance of most complex organic pesticides from soil follows a first-order reaction curve. Disappearance of simpler organic pesticides often exhibits a lag phase followed by rapid metabolism by soil microorganisms. Heavy metals and arsenic pesticides tend to accumulate at the application site.

REPORT

Introduction

THE PERSISTENCE of agricultural pesticide residues in the environment has caused considerable controversy in recent years. There is little doubt that Rachel Carson's *Silent Spring* did much to arouse concern over the frequency with which pesticide residues were encountered. Subsequent to the first alarms sounded on the possible dangers to the public caused by pesticides, a massive research effort was directed toward understanding the fate of these compounds in relationship to man and his surroundings.

The true significance of the impact of pesticide residues in the environment is becoming somewhat clearer. A voluminous amount of data has been collected on the occurrence, distribution and persistence of all classes of pesticide residues in plants, soils, water, and food. The results of these studies have been largely reassuring. Sophisticated analytical methodology has often failed to detect measurable residues in any links of the food chain. Where detected, residues exist at levels well below established tolerances.

A comprehensive examination of the research accomplishments on pesticides is far beyond the objective of this presentation. A summary of the persistence of many classes of pesticides is presented, followed by a consideration of the rate at which they disappear from one segment of the environment, *viz.*, agricultural soils. Soils represent the source of nutrients for crops and forage for grazing livestock. Progressive accumulation of pesticides in agricultural soils could reach some threshold level at which plant uptake might be significant. Therefore, the longevity and modes of dissipation of major organic insecticides and herbicides from soils are considered.

Persistence

Persistence, used in the context of pesticides, is a relative term. In the past, residual insecticides were considered beneficial, since they offered full season control against the ravages of destructive soil insects. In contrast, residual herbicides have damaged sensitive crops grown in rotation with resistant plants. Just how persistent are most pesticides used in modern agriculture?

The persistence (in months) of eleven major pesticide classes is shown in Figures 3-1 and 3-2. Data for these figures were developed

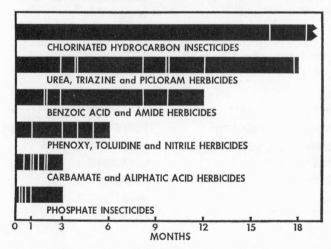

FIGURE 3-1. Persistence in soils of several classes of insecticides and herbicides.

from a review of approximately eighty sources concerned with pesticide persistence in soils. The persistence values represent the time required for the bioactivity to reach a level of 75 to 100 per cent of the control, or for a 75 to 100 per cent loss of the pesticide. In addition, the values shown are those resulting from normal rates of application and normal agricultural conditions. For general references, the reader is referred to the following review articles (1, 8, 21, 22).

A comparison of soil persistence of the major classes of insecticides and herbicides is shown in Figure 3-1. Each bar represents

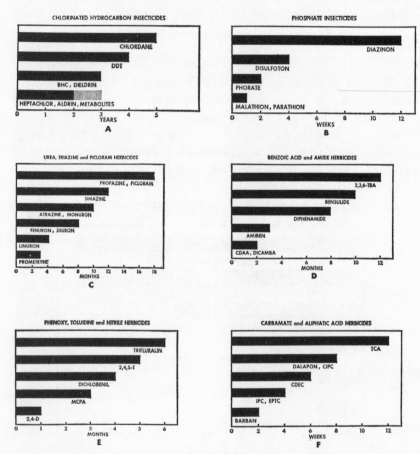

FIGURE 3-2. Persistence in soils of individual pesticides within a class.

one or more classes of pesticides. The open spaces represent the persistence of individual members within the classes. Thus, the phosphate insecticides are short-lived in soils and are dissipated within a few weeks. The organic herbicides exhibit a wide spectrum of persistence ranging from a few weeks for the carbamates and aliphatic acids to a year and a half for certain of the *s*-triazines. The most persistent pesticides are the chlorinated hydrocarbon insecticides. Most members of this class remain in the soil for more than one growing season.

A more detailed examination of individual pesticides is found in Figure 3-2. The time on the various graphs is presented in weeks, months or years, in order to illustrate the diversity in persistence. The classes of compounds have been subdivided into their individual members. Common and chemical names for the various insecticides and herbicides are found in Table 3-I. The length of each bar represents the average persistence of the compound (s) listed.

Extremes in persistence are found with the insecticides. Fortunately, the highly toxic phosphates do not persist for any extended period of time in soils (Fig. 3-2B). The most persistent phosphate insecticide, diazinon, remains for only three months. In contrast, some of the chlorinated hydrocarbon insecticides may persist for four to five years under normal rates of application (Fig. 3-2A). However, when large quantities of these compounds are applied, (e.g., for termite control around houses) toxic concentrations may remain two to three times longer than at normal rates (3). Heptachlor and aldrin are metabolized in soil to their epoxides, heptachlor epoxide and dieldrin, respectively, which are persistent. The herbicides also exhibit a wide range in persistence. The carbamate, barban, for example, persists for only two weeks (Fig. 3-2F) while the *s*-triazine herbicide, propazine, may persist for 18 months (Fig. 3-2C). The three-month persistence of another *s*-triazine, prometryne, indicates that large differences exist between herbicides within a particular class. A similar difference can be noted between two benzoic acid herbicides, dicamba and 2,3,6-TBA (Fig. 3-2D).

As has already been pointed out, the persistence values pre-

TABLE 3-I
COMMON AND CHEMICAL NAMES OF PESTICIDES

Aldrin	1,2,3,4,10,10-Hexachloro-1,4,4a,5,8,8a-hexahydro-1,4-*endo, exo*-5,8-dimethanonaphthalene
Amiben	3-Amino-2,5-dichlorobenzoic acid
Atrazine	2-Chloro-4-ethylamino-6-isopropylamino-*s*-triazine
Barban	4-Chloro-2-butynyl-*m*-chlorocarbanilate
Bensulide	N- (2-Mercaptoethyl) benzensulfonamide S- (0-0-diisopropyl phosphorodithioate
BHC	1,2,3,4,5,6-Hexachlorocyclohexane
Cacodylic acid	Dimethylarsinic acid
CDAA	α-Chloro-*N,N*-diallylacetamide
CDEC	2-Chloroallyl diethyldithiocarbamate
Chlordane	1,2,4,5,6,7,8,8-Octachloro-2,3,3a,4,7,7a-hexahydro-4,7-methanoindane
CIPC	Isopropyl *N*- (3-chlorophenyl) carbamate
2,4-D	2,4-Dichlorophenoxyacetic acid
Dalapon	2,2-Dichloropropionic acid
DDT	1,1,1-Trichloro-2,2-bis (*P*-chlorophenyl) ethane
Diazinon	*0,0*-Diethyl 0- (2-isopropyl-4-methyl-6-pyrimidinyl) phosphorothioate
Dicamba	2-Methoxy-3,6-dichlorobenzoic acid
Dichlobenil	2,6-Dichlorobenzonitrile
Dieldrin	1,2,3,4,10,10-Hexachloro-6,7-epoxy-1,4a,5,6,7,8,8a-octahydro-1,4-*endo,exo*-5,8-dimethanonaphthalene
Diphenamid	*N,N*-Dimethyl-2,2-diphenylacetamide
Disulfoton	*0,0*-Diethyl S-[2 (ethylthio) ethyl] phosphorodithioate
Diuron	3- (3,4-Dichlorophenyl) -1,1-dimethylurea
DSMA	Disodium methanearsonate
EPTC	Ethyl *N,N*-dipropylthiocarbamate
Fenuron	3-Phenyl-1,1-dimethylurea
Heptachlor	1,4,5,6,7,8,8-Heptachloro-3a,4,7a-tetrahydro-4,7-methanoindene
IPC	Isopropyl *N*-phenylcarbamate
Linuron	3- (3,4-Dichlorophenyl) -1-methoxy-1-methylurea
Malathion	S-[1,2-bis (ethoxycarbonyl) ethyl]*0,0*-dimethyl phosphorodithioate
MCPA	2-Methyl-4-chlorophenoxyacetic acid
Monuron	3- (*p*-Chlorophenyl) 1,1-dimethylurea
Parathion	0,0 Diethyl 0-*p*-nitrophenyl phosphorothioate
Phorate	*0,0*-Diethyl S-[(ethylthio) methyl] phosphorodithioate
Picloram	4-Amino-3,5,6-trichloropicolinic acid
PMA	Phenylmercuric acetate
Prometryne	2,4-Bis (isopropylamino) -6-methylmercapto-*s*-triazine
Propazine	2-Chloro-4,6-bis (isopropylamino) -*s*-triazine
Simazine	2-Chloro-4,6-bis (ethylamino) -*s*-triazine
2,4,5-T	2,4,5-Trichlorophenoxyacetic acid
2,3,6-TBA	2,3,6-Trichlorobenzoic acid
TCA	Trichloroacetic acid
Trifluralin	a,a,α-Trifluoro-2,6-dinitro-N,N-dipploplyipyl-*p*-toluidine

sented in Figures 3-1 and 3-2 are those resulting under normal agricultural conditions. However, some of the mechanisms responsible for loss or degradation of pesticides in soil are influenced by environmental and edaphic factors. As a result, the persistence of certain pesticides may vary considerably. For example, low moisture conditions have been shown to increase the persistence of atrazine, simazine, monuron and 2,3,6-TBA two to four times (4). On the other hand the degradation of BHC can be accomplished in about one-tenth the average time under flooded rather than well-aerated conditions (16).

Disappearance Curves

The loss of most pesticides from soils follows a first-order reaction (Fig. 3-3A). That is, at any time, the rate of pesticide loss is proportional to its concentration in the soil. Although the loss generally assumes the shape of a first-order reaction curve, several mechanisms may be acting on a soil residue at any given time.

Volatilization, photodecomposition, and mechanical removal by water and wind erosion are operative primarily on the surface before incorporation. Once the pesticide is incorporated (by leaching, cultivation, or translocation by plants and animals) other loss mechanisms (i.e. chemical decomposition, adsorption to soil colloids, microbial metabolism, and possibly plant uptake) become important. Oxidation, reduction, hydrolysis, and more recently, free radical reactions have been observed to decompose pesticides chemically (18). Adsorption to soil colloids results in inactivation of the pesticide, and may also contribute to its persistence. Adsorption may also cause chemical alteration or enhance catalysis of certain pesticides (2). Uptake by plants represents an avenue for the disappearance of some pesticides (7, 15). Although plant absorption may decrease soil residues, this route of dissipation may not be entirely beneficial. Plants may absorb a soil-pesticide residue through their roots and translocate the chemical to the economic portion of the plant. When biological metabolism is the primary avenue of disappearance, a different type of loss has been observed (Fig. 3-3D).

Regardless of which mechanism (except for biological metabolism) predominates in the overall disappearance of a pesticide, the loss usually follows that of a first-order reaction curve. Several examples of this type of loss behavior can be cited (5, 6, 13,15, 17, 20, 24).

Periodic application of degradable pesticides would yield the type of curves depicted in Figure 3-3B. Maximum and minimum

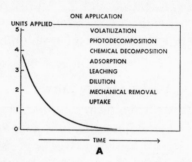

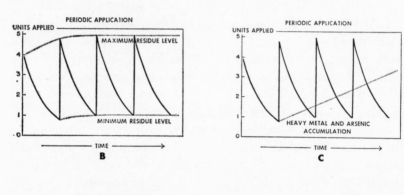

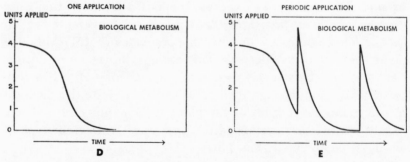

FIGURE 3-3. Loss of pesticides from soil following one or periodic applications.

residue levels would remain parallel to the base line or would not exhibit any progressive accumulation under actual field conditions, when the number of pesticide units lost in a given time equals the number of units applied. Therefore, Figure 3-3B represents a situation when 4 units are applied periodically and 20 per cent residue remains at the next application. Note that the minimum residue remains constant at 1 unit after about the fourth application. If the pesticide were degraded to only 50 per cent before the next application, the minimum residue would reach 6 to 8 units after eight applications. If 80 per cent of the pesticide remained at the next application, the minimum residue would reach 20 units after twenty applications. Similar curves have been developed by Sheets (20) and Foster *et al.* (7).

Periodic application of heavy metal and arsenic pesticides to soils should result in the situation shown in Figure 3-3C. This assumes that a portion of the pesticide molecule is not lost, but accumulates in direct proportion to its application. This is probably the case with the organic and inorganic arsenicals. Lead and calcium arsenate were used for many years in insect control in tobacco, cotton, orchards and other crops. Arsenic residues in the surface eight inches of orchard soils ranged for 23.2 to 580 ppm As_2O_3 (9) in twenty-one samples collected in the Northwest. In a more recent survey (26), a range of 18 to 2500 ppm arsenic, with average value of 304 ppm, was found in orchard soils last treated about twenty years ago. These samples were collected from the states of Washington, Idaho, Montana and New York. Subsequent crops often fail to survive on these abandoned orchard soils.

In recent years, the organic arsenicals have been used to control certain perennial weeds in cotton. Several of the more important herbicides are sodium cacodylate and sodium salts of methanearsonic acid. The monosodium methanearsonic acid (MSMA) is slowly metabolized by soil microorganisms to arsenate and CO_2 (23). In this case the organic portion of the molecule would presumably dissipate from soil similar to that shown in Figure 3-3D, while the inorganic portion of the molecule would accumulate as shown by the broken line in Figure 3-3C. Schweizer (19) has calculated that at normal rates of application of DSMA in Mississippi (equivalent to 6 kg per hectare (ha) per year) more than fifty years would elapse before arsenic residues would reach toxic levels in cotton. In contrast, rice is very susceptible to DSMA residues.

Arsenic reactions in soils are complex. Consequently, it would be erroneous to assume that equal applications to different soil types would yield essentially the same degree of residual phytotoxicity. Arsenic exists in four principal forms in soils, i.e. the water-soluble, calcium, aluminum and iron arsenates. Experiments are underway in this laboratory to assess the phytotoxicity of each of these four fractions. Incomplete data suggest that water-soluble arsenic and possibly calcium arsenic exhibit phytotoxicity (26).

Mercury residues from periodic application of certain agricultural fungicides may not necessarily follow the accumulation curves shown in Figure 3-3C. Metallic-mercury vapor and trace amounts of phenylmercury acetate (PMA) were present in air surrounding PMA-treated soils (14). About equal amounts of metallic-mercury vapors and a volatile ethylmercury compound were present in air when ethylmercury acetate was applied to soils. Likewise, with the methylmercury treated soils, methylmercury and trace amounts of mercury vapor were present in air. Increased soil moisture reduced the amount of organic mercury vaporization. Although the soil movement of volatile organo- and metallic-mercury was not investigated, this process probably accounted for some loss from the site of application. The status of mercury in the environment is the subject for discussion by other participants at this conference and therefore will not be considered in any further detail.

Biodegradable pesticides exhibit the type of behavior shown in Figures 3-3D and 3-3E (10, 11, 12, 25). A lag phase occurs after the first application in which relatively little pesticide is lost and then is followed by a rapid disappearance caused by soil microbial metabolism. Loss of dalapon, 2,4-D, and several of the phenylcarbamate herbicides follow this sigmoid pattern. After periodic applications, however, the soil microbial population has been adapted to metabolize the pesticide and the lag phase is not observed (Fig. 3-3E). Subsequently the minimum residue level would remain near zero.

DISCUSSION

CLARKSON: What is the mechanism of volatilization of mercury? Does the volatilization from cells depend on metabolism?

ISENSEE: I am not aware of specific mechanisms that regulate mercury volatilization from soil or plants. Some evidence exists that microorganisms can degrade certain organic mercury compounds. Compounds such as phenylmercury acetate are degraded to mercury and lost as mercury vapor while others are primarily volatilized as the organomercury compound.

CLARKSON: Do the microorganisms show some adaption to mercury so that mercury might not accumulate?

INSENSEE: Apparently this is possible, but the evidence is meager. Additional research is certainly needed to determine the extent of mercury loss.

WELCH: Are you aware of any organisms which can oxidize insecticides?

ISENSEE: Plants oxidize certain insecticides. There is some evidence from sterile vs. nonsterile soil studies that microorganisms play a role in the decomposition of certain insecticides. On the other hand, the chlorinated hydrocarbons are degraded at very slow rates.

WELCH: If, one could find an organism which might oxidize these compounds, it would be a way of getting rid of them from the soil since they would convert to more water-soluble compounds. Reducing converts them to more lipid-soluble compounds which would only enhance storage.

ISENSEE: Reference was made during the discussion of the last paper to the long-term study by Nash and Woolson on the persistence of chlorinated hydrocarbon insecticides. This was cited as evidence for a ten-to-twelve-year half life of DDT. I would like to point out that the purpose of this study was to determine the upper limit of persistence for insecticides applied at extremely high rates with minimal tillage, which would tend to maximize persistence. Therefore, one is not justified in claiming that the half-lives determined from this study are the same as those found under normal agricultural usage.

BERG: Dr. Isensee's presentation of the spectrum of agricultural chemicals gives us an excellent perspective on the ecological importance of organochlorines and arsenic on one hand, and practically everything else that is commercially used in agriculture on the other. The first two are persistent, the rest are not. I wonder, however, if persistence in soil is a good indication of persistence in the ecological system.

Dr. Isensee described nine mechanisms of removal: three of those altered the molecule chemically and perhaps even converted it into something that was not toxic, while the other six moved the molecule to somewhere else. Is it possible to distinguish degradation from transfer? Should we not think about possible accumulation elsewhere in the system?

ISENSEE: We know that several of these mechanisms are operative on the same compound at the same time. Also, degradation or metabolism could occur while a compound is in a state of transfer. Therefore, whether a compound could accumulate elsewhere in the system would be dependent in large part on the susceptibility of the compound to the loss mechanisms. For example, the organophosphates are quickly degraded while the chlorinated hydrocarbons may be moved about, bound, or made essentially unavailable in the soil.

BERG: When they are bound, are they really unavailable or is their transfer only slower? Can they equilibrate from one sink to another?

ISENSEE: It depends on the compound. Some of them are totally inactivated. For example, the herbicides paraquat and diquat, both used for aquatic weed control, are completely adsorbed to the colloidal fraction of soil.

RISEBROUGH: Are you responsible for recommending to farmers which herbicide to use?

ISENSEE: No, but we conduct research on which some of the uses are based. The objective of our research is to assure the safe and effective use of pesticides in the total environment. Our emphasis is on the behavior and fate of pesticides in soils. The primary recommendations are made by the manufacturers and later by State and Federal Extension Services.

RISEBROUGH: Do you think they are justified in still recommending the use of chlorinated hydrocarbons in soils?

ISENSEE: The United States Department of Agriculture is responsible for the registration of pesticides. It accomplishes this mission through cooperation with the Departments of Health Education & Welfare and Interior. The State Agricultural Extension Services in cooperation with the Federal Agricultural Extension Service advises farmers on safe and effective use. The registrant, which is usually the manufacturer, is responsible for the recommendations.

WURSTER: You did not mention in your list of mechanisms the property of codistillation, which is a very important mechanism of transfer from one place to another [Acree, F., *et al.*: *J Ag Food Chem, 11*:278, 1963]. I should also like to mention that the Pure Food and Drug laws apply to people, not to the remainder of the environment. The bulk of pesticide research has been directed either toward increasing crop yield or protecting people. I would suggest that we had better begin to worry more about the total world ecosystem, since these materials obviously are accumulating and have great biological activity.

REFERENCES

1. *Annotated Bibliography on Occurrence and Persistence of Pesticides in Soil and Water.* PHS Publ. No. 999-WP-17, 1964. 90 pp.
2. ARMSTRONG, D.E.; CHESTERS, G., and HARRIS, R.F.: *Soil Sci Soc Amer Proc, 31*:61, 1967.
3. BESS, H.A.; OTA, A.K., and KAWANISHI, C.: Persistence of soil insecticides for control of subterranean termites. *J Econ Entomol, 59*:911, 1966.
4. BURNSIDE, O.C.; WICKS, G.A., and FENSTER, C.R.: Herbicide longevity in Nebraska soils. *Weeds, 13*:277, 1965.
5. DECKER, G.C.; BRUCE, W.N., and BIGGER, J.H.: The accumulation and dissipation of residues resulting from the use of aldrin in soils. *Internatl. Congr. Entomol. Proc., 12th Cong.,* 1964, p. 558.
6. EDWARDS, C.A., and JEFFS, K.A.: The persistence of some insecticides in soil and their effects on soil animals. *Internatl. Contr. Entomol. Proc., 12th Cong.,* 1964, p. 559.
7. FOSTER, A.C.; BOSWELL, V.R.; CHISHOLM, R.D.; CARTER, R.H.; GILPIN,

GLADYS L.; PEPPER, B.B.; ANDERSON, W.S., and GIEGER, M.: Some Effects of Insecticide Spray Accumulations in Soil on Crop Plants. USDA Tech. Bull. No. 1149, 1956, 36 pp.

8. *Herbicide Handbook of the Weed Society of America,* 1st ed. Geneva, N.Y., W.F. Humphrey, 1967, 289 pp.

9. JONES, J.S., and HATCH, M.B.: The significance of inorganic spray residue accumulation in orchard soils. *Soil Sci, 44:*37, 1937.

10. KAUFMAN, D.D.: Microbial degradation of herbicide combinations: Amitrole and Dalapon. *Weeds, 14:*130, 1966.

11. KAUFMAN, D.D., and KEARNEY, P.C.: Microbial degradation of isopropyl-N-3-chloropheylcarbamate and 2-chloroethyl-N-3-chlorophenylcarbamate. *Appl Microbiol, 13:*443, 1965.

12. KAUFMAN, D.D.; KEARNEY, P.C., and SHEETS, T.J.: Microbiol degradation of simazine. *J Agr Food Chem, 13:*238, 1965.

13. KEARNEY, P.C.; SHEETS, T.J., and SMITH, J.W.: Volatility of seven s-triazines. *Weeds, 12:*83, 1964.

14. KIMURA, Y., and MILLER, V.L.: The degradation of organomercury fungicides in soil. *J Agr Food Chem, 12:*253, 1964.

15. LICHTENSTEIN, E.P.; MYRDAL, G.R., and SCHULZ, K.P.: Effect of formulation and mode of application of aldrin on the loss of aldrin and its epoxide from soils and their translocation into carrots. *J Econ Entomol, 57:*133, 1964.

16. MacRAE, I.C.; RAGHU, K., and CASTRO, T.F.: Persistence and biodegradation of four common isomers of benzene hexachloride in submerged soils. *J Agr Food Chem, 15:*911, 1967.

17. NASH, R.G., and WOOLSON, E.A.: Persistence of chlorinated hydrocarbon insecticides in soils. *Science, 157:*924, 1967.

18. PLIMMER, J.R.; KEARNEY, P.C.; KAUFMAN, D.D., and GUARDIA, F.S.: Amitrole decomposition by free radical-generating systems and by soils. *J Agr Food Chem, 15:*996, 1967.

19. SCHWEIZER, E.E.: Toxicity of DSMA soil residues to cotton and rotational crops. *Weeds, 15:*72, 1967.

20. SHEETS. T.J.: Review of disappearance of substituted urea herbicides from soil. *J Agr Food Chem, 12:*30, 1964.

21. SHEETS, T.J., and HARRIS, C.I.: Herbicide residues in soils and their phytotoxicities to crops grown in rotations. *Residue Rev, 11:*119, 1965.

22. UPCHURCH, R.P.: Behavior of herbicides in soils. *Residue Rev, 16:*46, 1966.

23. VON ENDT, D.W.; KEARNEY, P.C., and KAUFMAN, D.D.: Degradation of monosodium methanearsonic acid by soil microorganisms. *J Agr Food Chem, 16:*17, 1968.

24. WHEATLEY, G.A.: The persistence, accumulation and behavior of organochlorine insecticides in soil. *Internatl. Congr. Entomol. Proc., 12th Cong.,* 1964, p. 556.

25. WHITESIDE, J.S., and ALEXANDER, M.: Measurement of microbiological effects of herbicides. *Weeds, 8:*204, 1960.

26. WOOLSON, E.A.: Unpublished data, 1968.

Chapter 4

CONVERSION OF MERCURY COMPOUNDS

A. Jernelöv

ABSTRACT

It has been shown in Sweden that fish contain methylmercury in instances where the original discharge contains other types of mercury compounds. Mercury discharge in the form of metallic mercury, inorganic divalent mercury, phenylmercury and alkoxi-alkylmercury can all be converted in nature to methylmercury.

REPORT

Introduction

THROUGH THE WORK OF Westermark and his collaborators (9) we are beginning to get a clear picture in Sweden of the primary sources of mercury and the types of mercurial compounds in nature. The main types of discharges are the following:

1. As inorganic divalent mercury, Hg^{2+};
2. As metallic mercury, Hg^0;
3. As phenylmercury, $C_6H_5 Hg^+$;
4. As methylmercury, $CH_3 Hg^+$; and
5. As alkoxi-alkylmercury, $CH_3O\text{-}CH_2\text{-}CH_2\text{-}Hg^+$.

To understand the ecological effects of the different kinds of discharges and the risk factors involved, the transforming reactions between the different compounds of mercury in nature are of central significance. The consequences of these transforming reactions are particularly obvious when it concerns the deposits of mercury in the sediments of lakes and rivers, which can be mobilized through conversion to other, more hard-to-bind forms. These deposits are principally made up from phenylmercury in fiber banks and inorganic mercury (divalent and metallic) in bottom sediment.

Basic Transforming Reactions

1. The conversion of metallic mercury to divalent mercury.

$$(Hg^0 \rightarrow Hg^{2+})$$

An important quality about divalent inorganic mercury is its affinity for organic mud. This binding to organic mud is extremely strong, with an α-coefficient greater than 10^{21} in comparison to other complexes. (The α-coefficient is a measure of the binding strength of a complex.) According to Werner (10) under certain conditions the potential necessary for the oxidation of metallic mercury into divalent mercury ions can be written as follows:

$$E = 850 + 30 \log \frac{[Hg^{2+} \text{ total}]}{\alpha}.$$

Since the value of α has been shown to be greater than 10^{21} and since Hg^{2+}total $= 2$ ppm $\cong 10^{-5}M$ the potential necessary will be $E < + 80$ m V. If Hg^{2+}total is lower, the potential necessary for the oxidation is, of course, further decreased.

This illustrates how the oxidation of metallic mercury (Hg^0) to divalent mercury ions (Hg^{2+}) can occur under conditions present at the bottoms of lakes and rivers; it has also been shown to occur experimentally by Jernelöv and Werner (10, 8).
2. The conversion of divalent inorganic mercury to methylmercury.

$$(Hg^{2+} \rightarrow CH_3Hg^+ \leftrightarrow CH_3HgCH_3)$$

The idea that mercury is biologically methylated was first proposed by Japanese engineers to explain the Minamata catastrophy. However, during the investigation they found that it was possible to conclude that the formation of methylmercury had occurred within the factory. However, the plans for studying further this potential biological methylation seem to have been cancelled. Subsequently, Johnels and Westermark (9) and Westöö (12) have given new indications of a biological methylating of mercury, which Jensen and Jernelöv (2) later showed to occur in bottom sediment from aquaria. Further studies have shown that during certain conditions, for example, in connection with the decomposition of fish, the methylated mercury will evaporate as dimethylmercury (3). Also the methylated mercury in sediments and soil will in many cases depart as a volatilizable compound (7).

Investigations have been made on sediments from a large number of lakes and rivers regarding occurrence and rate of methylation of mercury. In all cases there have been present in the sediments, microorganisms capable of methylating mercury (2, 3, 7).

Chemical Fallout

It was previously stated, that divalent inorganic mercury has a great affinity for the binding site of organic sediment and that a transformation into mono- or dimethylmercury can contribute to a release of mercury from the sediments. Another way in which mercury, bound to the sediments, could come into the aquatic nutritional chain is by accumulation in the bottom fauna. Tubifex have been cultivated in aquaria in organic mud containing mercury. In these tests the factor of mercury accumulation in Tubifex was low (1.20 ± 0.26). Moreover, the amount of methylmercury vs. the total mercury was as low in worms as in bottom sediment (.5 to 5%). This indicates that the microbiological processes of methylating in bottom sediments are significant factors for the mobilization of the bound mercury.

To get a basis for judging the speed of this process in the cases where the mercury in the bottom sediments is an inorganic compound, a large number of samples of bottom sediment have been taken from such lakes and rivers and kept under natural conditions concerning light and temperature. The surface layer of the sediments was continuosly oxygenated during the experiments. Analyses have been made on the yield of mono- and dimethylmercury over periods lasting from 1 week to 2 months. Assuming that the rate of conversion of inorganic to methylmercury is the rate-determining step in removal of mercury from bottom sediments, it was calculated that this removal would take from 10 to 100 years.

The study of the methylating processes in mud anaerobic conditions seems far more complicated. The low solubility of mercuric sulfide (HgS) is well known. Hence, it is plausible to ask if inorganic divalent mercury is precipitated as a sulfide under anaerobic conditions in organic mud where sulphur is present. Experiments have shown that when mercury is added to mud as a sulfide (HgS), it is not methylized under permanent anaerobic conditions (5). However, if the mud is aerobic, methylation of the mercury occurs when the sulfide has been oxidized to sulfate (SO_4^{2-}).

It was shown by Jensen and Jernelöv (3) that large amounts of dimethylmercury was formed from either divalent inorganic mercury or from methylmercury during the anerobic putrefication of fish. Also in tests with natural sediments, mono- and dimethylmercury were formed under anaerobic conditions. That mercury bound to organic substances and subsequently subjected to anaerobic conditons could be methylated also in the presence of hydrogen sulfide (H_2S) was confirmed in later experiments (5). The speed of this methylating process under anaerobic conditions can sometimes be very high, with the end product being either mono- or dimethylmercury. Probably there are other compounds (e.g., CH_3-Hg-S-CH_3 and $(CH_3$-S$)_2$ Hg).

The presence of threshold values for biological-chemical processes is a classical item for debate. Threshold values imply a discontinuity in the relation of dose-response. The important relevant question is: Is the methylation of mercury faster over certain concentrations of inorganic mercury? The concentration of inorganic mercury in the methylating process is significant for at least two reasons: (a) It effects the transformation of metallic mercury to inorganic divalent mercury since it influences the factor (Hg^{2+}total). (b) If such a value exists, the dilution factor will be of great importance for the mobilization of mercury through methylation from the bottom sediments in nature.

Two experiments indicate the existence of a threshold value for the biological-chemical methylation of mercury. In tests with *Neurospora crassa,* no methylating of mercury was found to occur when the concentration of inorganic divalent mercury was less than 10 ppm in a minimum substrate, or when the concentration of mercury fell below 1 ppm in a substrate with a great surplus of methionine (6).

In experiments with the methylation of mercury in bottom sediment from aquaria, where the relationship between inorganic divalent mercury in the mud and the production of monomethylmercury was studied, a steep increase in the amount of monomethylmercury was observed as the concentration of added inorganic mercury reached 1 to 10 ppm (4). In this experiment, however, there are several possibilities other than a threshold value to explain the lack of a linear relationship between the concentration of (Hg^{++}) and the amount of former CH_3Hg^+: (a) competition between organisms with and without the ability to methylize mercury could give the same results and (b) synthesis primarily of dimethylmercury since a high concentration of inorganic divalent mercury would raise the probability for a reaction between these two compounds and the formation of monomethyl-mercury (2). Lastly, Westermark's (11) observation that sea eagle feathers from the 1840's contain the main part of mercury as methylmercury contradicts to some extent the hypothesis that a threshold value exists for the methylating of mercury.

3. The conversion of phenylmercury to methylmercury.

$$(C_6H_5Hg^+ \rightarrow CH_3Hg^+ \longleftrightarrow CH_3HgCH_3)$$

The conversion in bottom sediments of phenylmercury to mono- and dimethylmercury has been studied and shown to occur (7). However, it has been very difficult to repeat quantitatively the experiments. It seems plausible that the formation of mono- and dimethylmercury proceeds along more than one synthetic pathway. A further complication is that the different reactions interfere with one another. Because of the variability in the laboratory tests it is in no way possible to estimate the speed of mobilization of phenylmercury in fiber banks. Observations in nature repeatedly indicate that the discharge of phenylmercury has a stronger and faster effect of mercury concentration in the fish than the discharge of a similar amount of inorganic mercury (1).

4. The conversion of alkoxi-alkylmercury to inorganic divalent mercury.

$$(CH_3O - CH_2 - CH_2 - Hg^+ \rightarrow Hg^{2+})$$

This transformation is well known to occur.

In summary, some of the steps by which inorganic mercury and some mercury-containing compounds are converted in nature to methylmercury is the following:

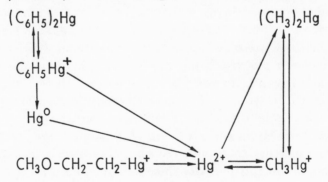

DISCUSSION

SUZUKI: What kind of bottom sediment is most effective in converting inorganic mercury to methylmercury? What ecological conditions are most important to such a conversion?

JERNELÖV: Generally the speed of the conversion is very much higher when the bottom sediment contains more organic substances. It is also influenced by oxygen tension and temperature.

SUZUKI: How much organic sediment is needed for the conversion?

JERNELÖV: We have not found any limit to it; the organic substances are there but that is not the only factor involved. Conversion rates can be very high in sewage treatment plants.

SUZUKI: How many different samples of bottom sediment did you test?

JERNELÖV: We have taken samples from more than one hundred rivers and lakes in Sweden, and in all the tested samples we found microorganisms capable of making methylmercury.

NORSETH: Have you tested the intestinal tracts of animals for conversion from inorganic to organic mercury?

JERNELÖV: Dr. Westöö has done such experiments.

WESTÖÖ: The rat does not convert injected inorganic mercury to organic mercury. But we have fed hens with inorganic mercury salts, phenylmercury salts, and methoxymethylmercury salts and in all cases observed some conversion to methylmercury.

REFERENCES

1. BOUVENG, H.: Personal communication.
2. JENSEN, S., and JERNELÖV, A.: *Nordforsk Biocindinformation*, No. 10, March 1967.
3. JENSEN, S., and JERNELÖV, A.: Nord*forsk Biocindinformation*, No. 14, Feb. 1968.
4. JENSEN, A.: Unpublished data.
5. JERNELÖV, A.: Unpublished data.
6. JERNELÖV, A., and LANDNER, L.: Unpublished data.
7. JERNELÖV, A., and RUDLING, R.: Personal communication.
8. JERNELÖV, A., and WERNER, J.: Personal communication.

9. JOHNELS, A., and WESTERMARK, T.: *Kvicksilverfrågan i Sverige* (Stockholm 1965) 25 (1964 års naturresursutredning, Jordbruksdep., Kvicksilverkonferensen 1965)
10. WERNER, J.: Svenska Kemistsamfundet, Analysdagarna i Lung, June 1967.
11. WESTERMARK, T.: Personal communication.
12. WESTÖÖ, G.: *Acta Chem Scand, 20:*2131, 1966.

Chapter 5

METHYLMERCURY COMPOUNDS IN ANIMAL FOODS

G. Westöö

ABSTRACT

A method for determining methylmercury compounds in fish by gas-liquid chromatography and thin-layer chromatography was developed in 1965. In 1966 and 1967 the procedure was improved and can now be used also for other animal foodstuffs. Methylmercury compounds were found, e.g., in eggs, meat, liver (average values $<$ 0.06 mg of Hg/kg), and fish. The substitution of methoxyethylmercury for methylmercury compound as a seed disinfectant in Sweden in 1966 caused a decrease in the methylmercury content of eggs, meat, and liver in 1967 to 1968 to approximately $\frac{1}{3}$ of the average values found earlier. Marine fish caught near the shore and freshwater fish often show high methylmercury levels. Frequently more than 0.4 mg of Hg/kg (very infrequently, as much as 5-10 mg/kg) of fish muscles has been found in Sweden. Levels above 1 mg of Hg/kg are usually caused by industrial discharge of mercury compounds. Regardless of the nature of the mercury pollutant, methylmercury has been found in the fish. When hens were fed with seed disinfected with different mercury compounds, part of the mercury in the eggs was in the methylated form.

REPORT

IN SWEDEN THE MERCURY problem was first noticed in connection with a study of birds. During the years 1956 to 1963 Borg *et al.* (1) observed that pheasants and several other seedeating species were poisoned, sometimes lethally, during or after the sowing periods. The viscera of the poisoned birds contained large amounts of mercury (up to 140 mg/kg in the liver and kidney of pheasants, and one-half to one-fourth of the liver level in the muscles). During the shooting season (October - December) the mercury content had decreased considerably (average value $<$ 1 mg of Hg/kg of muscle).

Methods

In 1960 to 1965 Sjöstrand (11), Westermark and Sjöstrand (18), and Christell *et al.* (2), presented methods for the determination of small amounts of mercury by activation analysis, which made it possible to analyze a large number of food samples for mercury. The sample (usually about 0.3 g) contained in a sealed quartz ampoule, is placed in a reactor and irradiated with neutrons. The mercury present in the sample is partly transformed into radioactive isotopes. To ensure radiochemical purity, mercury carrier (20 mg) is then added, the sample is subjected to wet combustion and distilled; the mercury is then electrolyzed onto a gold cathode, and weighed. Finally, the radiation emitted is measured with a γ-spectrometer using the [197]Hg line at about 70 keV. The radiation of the sample is compared with that of a mercury standard treated in the same way. Mercury levels of 0.001 mg of Hg/kg of food can be measured.

This method has been invaluable for dealing with the problem of mercury in food in Sweden. However, to obtain a real basis for a toxicological estimation of the hazards connected with the mercury content of foodstuffs, it was necessary to determine also the chemical binding of the mercury. In this laboratory five derivatives of the mercury compound in animal foodstuffs were analyzed by gas chromatography (electron capture detector) and thin-layer chromatography. The chromatograms showed that the mercury in fish, eggs, meat, and liver (Table 5-I) was present mainly as a methylmercury compound (10, 19, 20, 21, 28, 2) (Figs. 5-1, 5-2, 5-3). This was verified by mass spectrometry (4) (Fig. 5-4). In connection with the methylmercury poisoning of human beings by fish and shellfish, a similar gas-chromatographic procedure has been elaborated in Japan (7). In Sweden a method for the electropheretic separation of different mercury compounds has been developed (16), which has the advantage of simultaneously determining inorganic and organic mercury compounds.

The method for determining methylmercury by gas-liquid chromatography and thin-layer chromatography (19, 20, 22) is based on a benzene extraction of methylmercury from the aqueous

TABLE 5-I
SOME TOTAL MERCURY AND METHYLMERCURY CONTENTS OF
SWEDISH FOODS IN 1966

Foods	Total Mercury mg/kg	Methylmercury	
		mg of Hg/kg	% of total Hg
Meat (ox)	0.074	0.068	92
Meat (poultry)	0.023	0.017	74
Liver (pig)	0.130	0.095	73
Liver (pig)	0.096	0.075	78
Egg yolk	0.010	0.005	50
Egg yolk	0.010	0.009	90
Egg white	0.012	0.011	92
Egg White	0.025	0.024	96
Muscle tissue of perch	0.22	0.20	91
Muscle tissue of perch	3.25	2.99	92
Muscle tissue of pike	3.35	3.11	93
Muscle tissue of pike	0.56	0.55	98
Muscle tissue of cod	0.064	0.055	86
Muscle tissue of cod	0.026	0.022	85

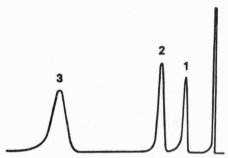

FIGURE 5-1. Gas chromatogram of methylmercury (1), ethylmercury (2) and methoxyethylmercury (3) compounds.

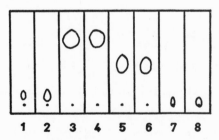

FIGURE 5-2. Thin-layer chromatogram on silica gel of authentic methylmercury chloride (1), iodide (3), bromide (5), and cyanide (7) together with the corresponding compounds prepared from methylmercury extracted from pike (2, 4, 6, 8).

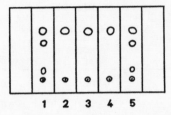

FIGURE 5-3. Thin-layer chromatogram on aluminum oxide of authentic methylmercury, phenylmercury, and mercury dithizonates (1, 5) together with dithizonates prepared from methylmercury extracted from pikes (2, 3, 4).

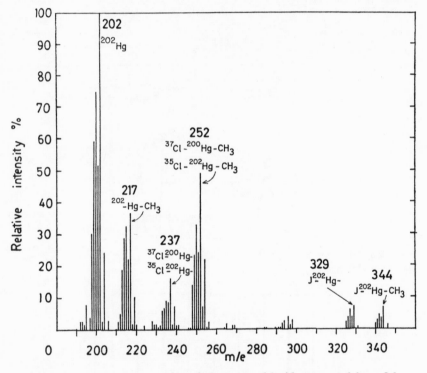

FIGURE 5-4. Mass spectrum of methylmercury chloride extracted from fish.

suspension of the food strongly acidified with hydrochloric acid, according to Gage (3). The methylmercury is purified by extraction with a cysteine acetate solution. With liver samples the proteins must be precipitated, e.g., with molybdic acid, for good results (22). In 1965 to 1966 ammonium hydroxide was used instead of cysteine; then, however, benzene-soluble compounds containing SH groups, especially in liver and egg yolk, had to be bound by excess mercuric ions, which were added to the benzene extract. Finally, the aqueous extract is strongly acidified with hydrochloric acid, and the methylmercury is reextracted in benzene (Fig. 5-5). The benzene solution is dried with anhydrous sodium sulfate and subjected to gas and thin-layer chromatography (Figs. 5-1 to 5-3). Polyethylene glycol (Carbowax 20 M) or phenyl diethanolamine succinate on Chromosorb W (acid-washed DMCS, 60/80 mesh) is used in the column. A combination of the mercuric ion and cysteine acetate procedures, with addition of the mercuric salt to the water suspension of the food, has also been used (22), though not for liver, which, in aqueous suspension, methylates to a slight extent the mercuric salt to methylmercury salt. With the ammonium hydroxide-mercuric chloride procedure the recovery of added methylmercury dicyandiamide is 91 ± 6 per cent (29 samples). With the cysteine acetate or cysteine acetate-mercuric chloride procedures recovery is slightly higher, e.g., for fish 98 ± 3 per cent (10 samples). Methylmercury levels of 0.001 mg of Hg/kg of food can be measured.

In the cysteine acetate procedure, no organic or inorganic mercury compounds interfere with the results. Dimethylmercury is not determined by this method. With the methods involving addition of mercuric chloride, dimethylmercury interferes, because methylmercury is formed according to the equation:

$$(CH_3)_2 Hg + Hg^{++} \rightarrow 2 CH_3Hg^+$$

Several hundred samples of foods, mainly fish and egg white, have been analyzed by both the cysteine acetate method and the mercuric chloride procedures. Within the methodical errors, identical results were obtained; therefore, it is reasonable to conclude that no dimethylmercury was present. Liver may, however, be an exception (*cf.* above).

Results

In 1963 Smart and Lloyd (12), and in 1964 Tejning and Vesterberg (15), reported a high mercury content ($\approx$ 10 ppm) in eggs from hens fed with seed containing about 6 or 14 ppm of

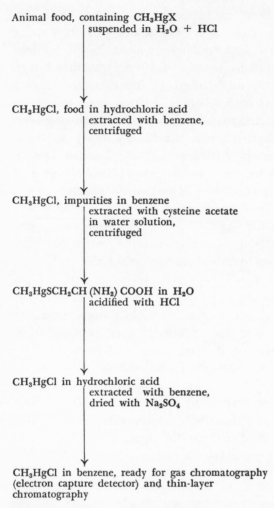

Animal food, containing CH_3HgX
  suspended in H_2O + HCl

CH_3HgCl, food in hydrochloric acid
  extracted with benzene,
  centrifuged

CH_3HgCl, impurities in benzene
  extracted with cysteine acetate
  in water solution,
  centrifuged

$CH_3HgSCH_2CH (NH_2) COOH$ in H_2O
  acidified with HCl

CH_3HgCl in hydrochloric acid
  extracted with benzene,
  dried with Na_2SO_4

CH_3HgCl in benzene, ready for gas chromatography
(electron capture detector) and thin-layer
chromatography

FIGURE 5-5. Extraction and purification of methylmercury in animal foods for chromatographic analysis.

methylmercury dicyandiamide. This compound was used as a seed disinfectant in Sweden from about 1940 through January 1966 and is still in use in the United States. However, of more than two hundred eggs bought on the open market in Sweden in 1964 to 1965 all contained less than 0.05 mg of Hg/kg, with an average of 0.029 mg/kg (23, 30). Eggs with mercury levels exceeding 0.1

mg/kg were found on a few small farms. On one farm eggs contained up to 1.5 mg of Hg/kg (30). This shows that some feeding of seed treated with methylmercury compound took place. Eggs from foreign countries were also analyzed. Except for the Norwegian eggs, all foreign eggs investigated contained less mercury than Swedish eggs (Table 5-II). In Norway as in Sweden methylmercury dicyandiamide was used as a seed disinfectant, whereas in Denmark methoxyethylmercury compounds were mainly used.

TABLE 5-II
MERCURY CONTENT OF EGGS ON THE OPEN MARKET

Period of Time	*Country*	*Number of Eggs*	*Mercury Content, mg/kg*	
			Limits	*Average*
March 1964-April 1966	Sweden	> 200	0.015-0.043	0.029
1964-1965	Norway	7	0.018-0.022	0.021
1964-1965	Other European Countries	48	0.004-0.013	0.007
April 1967-Sept. 1967	Sweden	> 200	0.005-0.020	0.009

From February 1, 1966 treatment of seed with methylmercury compounds was prohibited in Sweden. Instead, the use of methoxyethylmercury compounds was allowed. A few months earlier, in October 1965, the disinfection of seed had been strongly restricted. These two amendments, together with mass-media propaganda against the use of mercury compounds, caused a decrease in the mercury treatment of seed from about 80 per cent to about 12 per cent (spring sowing 1967). The effect was that the mercury poisoning of seedeating wild animals ceased. Furthermore, the mercury content of Swedish eggs decreased from an average of 0.029 mg/kg in 1965 to 0.009 mg/kg in 1967 (21), which is similar to the mercury content of foreign eggs (Table 5-II) (Fig. 5-6). This indicates that enough methylmercury compound migrated from the seed to the ripened grain to affect the eggs. That the methylmercury did migrate via this route was verified by a feeding experiment, which also indicated that grains, grown from seed treated with methoxyethylmercury salt, when fed to hens had no effect on the mercury content of the eggs (9) (Table 5-III).

Irrespective of the hens being fed with grain grown from untreated seed or from seed treated with methylmercury or methoxy-

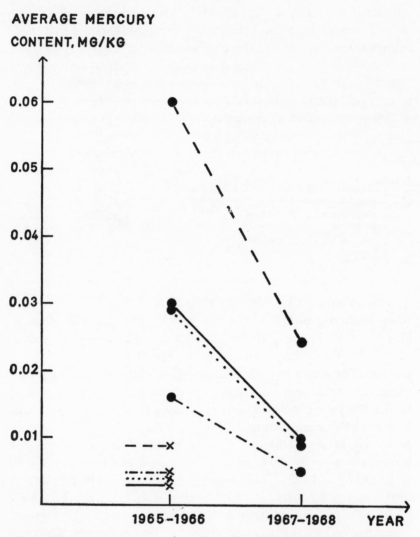

FIGURE 5-6. Decrease of mercury content of animal foods in Sweden, due to changes in seed disinfection.

● Sweden   x Denmark
———————— Pork chop   .... Egg
— — — — — — — Liver (pig)   -●-●-●-●-●-● Liver (ox)

ethylmercury compounds, methylmercury was found in the eggs
(9). As the grain was not the only feed given to the hens, the
methylmercury emanated, at least partly, from the rest of the feed,
which among other things contained a few per cent of fish meal.

TABLE 5-III
MERCURY CONTENT OF EGGS FROM FEEDING EXPERIMENTS

Grains Fed	Mercury Content, mg/kg	
	Limits	Average
Grown from methoxyethylmercury-treated seed	0.009-0.011	0.010
Grown from methylmercury-treated seed	0.022-0.029	0.027
Grown from untreated seed	0.008-0.015	0.012

Feeding experiments have shown that some formation of
methylmercury from other mercury compounds can take place in
hens (8). Grains were treated with phenylmercury, methoxyethylmercury or inorganic mercury compounds and fed to hens. The
mercury content increased in both egg white and yolk. In all the
experiments methylmercury was found in the egg white. In the
yolk, where the concentration of the mercury was up to five times
higher than in the white, only a small percentage of the mercury
was methylmercury. When, however, the corn feed was treated
with a methylmercury compound, all the mercury in the egg was
methylmercury and the concentration in the white was four to five
times higher than in the yolk.

Swedish meat and liver (poultry, ox, pig), except pig's liver,
contained on the average less than 0.05 mg of Hg/kg even before
1966, but, just as in eggs, the mercury content was higher than in
similar products from Denmark (26). The substitution in 1966, of
methoxyethylmercury for methylmercury as a seed-disinfectant in
connection with the decrease in the frequency of mercury-disinfec-
tion of seed, resulted also in a lower content of mercury in these
meats (Table 5-IV) (Fig. 5-6). However, after 1966 a high percentage of the mercury in meats was still in the methylated form
(Table 5-V) (27). The only exception among the foodstuffs analyzed so far was egg yolk, where only about one-third of the mercury was methylmercury.

The fact that Swedish eggs, meat, and liver, especially before
1967, contained more methylmercury than the corresponding

TABLE 5-IV
MERCURY CONTENT OF MEAT AND LIVER

Period of time	Country	Food	Number of Samples	Mercury content, mg/kg Limits	Average
Sept.-Nov. 1965	Sweden	Pork chop	35	0.016-0.130	0.030
October 1965	Denmark	Pork chop	6	0.002-0.007	0.003
March 1968	Sweden	Pork chop	6	0.006-0.016	0.011
November 1965	Sweden	Liver (pig)	26	0.014-0.183	0.060
October 1965	Denmark	Liver (pig)	6	0.005-0.020	0.009
January 1968	Sweden	Liver (pig)	10	0.011-0.049	0.024
Nov. 1965-April 1966	Sweden	Liver (ox)	23	0.005-0.046	0.016
October 1965	Denmark	Liver (ox)	6	0.003-0.007	0.005
March 1968	Sweden	Liver (ox)	6	0.003-0.008	0.005
Febr. 1965-March 1966	Sweden	Poultry meat	83	0.002-0.058	0.013
March 1965	Sweden	Poultry liver	31	0.016-0.062	0.030
Oct. 1965-April 1966	Sweden	Filet of beef	23	0.002-0.074	0.012
October 1965	Denmark	Filet of beef	6	0.002-0.004	0.003

TABLE 5-V
MERCURY AND METHYLMERCURY CONTENT OF SWEDISH MEAT,
LIVER, KIDNEY, EGGS IN 1967-1968

Period of Time	Food	Number of Samples	Total mercury mg/kg	Methylmercury % of Total Mercury Limit	Average
March 1968	Pork chop	6	0.006-0.016	80-100	88
January 1968	Liver (pig)	10	0.011-0.049	45-86	67
January 1968	Kidney (pig)	8	0.014-0.086	24-71	50
April 1967-Jan. 1968	Egg white	17	0.006-0.027	70-100	88
January 1968	Egg yolk	6	0.005-0.006	22-36	29

Danish foods has concerned many Swedes. A more important problem, however, is the mercury contamination of fish. As is well known, fish and shellfish accumulate water pollutants and are usually overlooked when tolerances are established for unintentional additives, e.g., metal salts. As early as 1934 Stock and Cucel

(13, 14) demonstrated that the mercury level in marine fish was 0.03 to 0.11 mg/kg and in freshwater fish 0.03 to 0.18 mg/kg. Westermark *et al.* (17) were the first to observe that Swedish fish often contain much more mercury than the fish analyzed by Stock. They also showed that the mercury content of fish increases with the age of the fish (*cf.* Johnels *et al.* (6) .

In Sweden today the mercury level in marine fish caught far from the shore is usually well below 0.1 mg/kg, as is the mercury level in freshwater fish caught in uncontaminated lakes. Marine fish caught near the shore, and especially freshwater fish, often contain 0.4 to 1.0, frequently 1 to 5, and very infrequently 5 to 10 mg/kg (5, 6, 24, 28) (Fig. 5-7) . For example, in the archipelago of Stockholm, samples of pike, perch and Baltic herring often contain more than 1 mg of Hg/kg. In Vänern, the largest lake in Sweden, the mercury content of pike in several creeks exceeds 1 mg/kg of fish muscles.

The high mercury content in fish is usually caused by industrial disposal of mercury into the water or the air (6, 10) . Downstream from pulp and paper mills, which were allowed to use phenylmercury compounds as a fungicide until January 1, 1966 and may still use sodium hydroxide and sulfuric acid contaminated with mercury compounds, the mercury content of fish is higher than that of fish caught upstream (6, 10) (Table 5-VI). Chlorine-alkali factories using mercury electrodes, and mercury rectifier factories have a similar effect on fish living downstream (Table 5-VI) (10) . Hospitals, dentists and laboratories are other examples of potential sources of mercury disposed into the air and water. Soot and smoke contain mercury (13) , and thus contribute to the mercury fallout, which must exert a widespread, though not strong, effect on the mercury level in fish. Soils and rocks containing mercury compounds also make some contribution. Today about forty water areas in Sweden are polluted by mercury to such an extent that sale of fish from these areas is prohibited. Regardless of the nature of the mercury pollutant, only methylmercury has been found in the fish (10, 19, 20, 28, 29) (Table 5-VI) , which indicates that a methylation of mercury compounds takes place.

Boiling or frying fish does not remove any methylmercury

(10, 25) . Since the fish loses water by these processes, a corresponding increase in the concentration of the methylmercury is observed
(10) (Table 5-VII) .

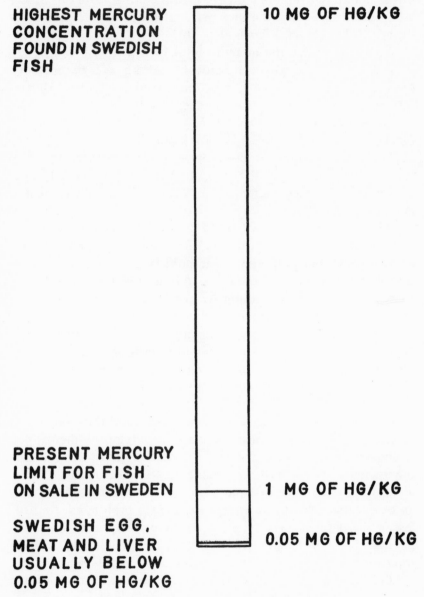

HIGHEST MERCURY
CONCENTRATION
FOUND IN SWEDISH
FISH

10 MG OF HG/KG

PRESENT MERCURY
LIMIT FOR FISH
ON SALE IN SWEDEN

1 MG OF HG/KG

SWEDISH EGG,
MEAT AND LIVER
USUALLY BELOW
0.05 MG OF HG/KG

0.05 MG OF HG/KG

Figure 5-7. Methylmercury levels in certain foods in Sweden.

TABLE 5-VI
METHYLMERCURY CONTENT OF FISH CAUGHT UPSTREAM AND
DOWNSTREAM FROM OUTLETS OF MERCURY COMPOUNDS

Lake, where the fish were caught	Species of fish	Total weight of fish, kg	Methylmercury Compounds mg of Hg/kg of fish muscles	% of total mercury
Viren, downstream from paper factory	Perch	0.12	1.91	~100
Viren, downstream from paper factory	Perch	0.12	2.18	99
Viren, downstream from paper factory	Perch	0.19	3.02	86
Viren, downstream from paper factory	Perch	0.41	2.81	91
Viren, downstream from paper factory	Pike	0.59	3.13	95
Viren, downstream from paper factory	Pike	0.91	3.48	92
Öjen, upstream from the same factory	Perch	0.064	0.18	100
Öjen, upstream from the same factory	Perch	0.071	0.20	91
Öjen, upstream from the same factory	Perch	0.10	0.70	93
Öjen, upstream from the same factory	Perch	0.14	0.42	~100
Öjen, upstream from the same factory	Pike	0.40	0.55	98
Övre Hillen, downstream from rectifier factory	Perch	0.015	0.83	94
Övre Hillen, downstream from rectifier factory	Perch	0.017	1.20	92
Övre Hillen, downstream from rectifier factory	Perch	0.21	2.48	86
Övre Hillen, downstream from rectifier factory	Pike	0.40	1.81	95
Övre Hillen, downstream from rectifier factory	Pike-perch	0.37	2.39	94
Övre Hillen, downstream from rectifier factory	Pike-perch	0.41	2.05	95
Övre Hillen, downstream from rectifier factory	Whitefish	0.035	1.40	100
Övre Hillen, downstream from rectifier factory	Whitefish	0.067	1.06	~100
Väsman, upstream from the same factory	Eelpout	0.24	0.35	90
Väsman, upstream from the same factory	Eelpout	0.32	0.50	93
Väsman, upstream from the same factory	Eelpout	0.32	0.70	95
Väsman, upstream from the same factory	Eelpout	0.36	0.37	100
Väsman, upstream from the same factory	Eelpout	0.42	0.53	79

TABLE 5-VII
METHYLMERCURY CONTENT OF PIKE BEFORE AND AFTER BOILING AND FRYING

Fish species	Preparation	Weight of sample, g before preparation	Weight of sample, g after preparation	Methylmercury content, mg of Hg/kg of fish muscle before preparation	Methylmercury content, mg of Hg/kg of fish muscle after preparation	Methylmercury content in prepared fish, recalculated to unprepared, mg/kg
Pike	boiling	50.0	35.5	0.52	0.76	0.54
Pike	frying	65.6	50.1	0.52	0.71	0.54

TABLE 5-VIII

Sample	Number of Samples With a Methylmercury Content, Calculated as mg of Hg/kg					
	0-0.050	0.051-0.200	0.201-1.00	1.01-2.00	2.01-4.00	> 4.00
Marine Fish	29	25	0	0	0	0
Freshwater Fish and Inshore Fish	37	36	857	141	38	10

Methylmercury in fish is today an important problem in food hygiene in Sweden (Table 5-VIII), and there are different opinions as to what amounts are tolerable. Regardless of which content will be judged acceptable, the fundamental consideration must be to stop, or essentially diminish, the mercury discharge.

REFERENCES

1. BORG, K.; WANNTORP, H.; ERNE, K., and HANKO, E.: *J Appl Ecol,* 3(suppl):171, 1966.
2. CHRISTELL, R.; ERWALL, L.G.; LJUNGGREN, K.; SJÖSTRAND, B., and WESTERMARK, T.: *Proc. Intern. Conf. Mod. Trends Activation Anal.,* College Station, Tex., 1965, p. 380.
3. GAGE, J.C.: *Analyst, 86:*457, 1961.
4. JOHANSSON, B.; RYHAGE, R., and WESTÖÖ, G.: To be published.
5. JOHNELS, A.G.; OLSSON, M., and WESTERMARK, T.: *Vår Föda, 19:*65, 1967.
6. JOHNELS, A.G.; WESTERMARK, T.; BERG, W.; PERSSON, P.I., and SJÖSTRAND, B.: *Oikos, 18:*323, 1967.
7. KITAMURA, S.; TSUKAMOTO, T.; HAYAKAWA, K.; SUMINO, K., and SHIBATA, T.: *Med Biol, 72:*274, 1966.
8. KIVIMÄE, A.; SWENSSON, Å.; ULFVARSON, U., and WESTÖÖ, G.: To be published.
9. LAGERVALL, M., and WESTÖÖ, G.: *Vår Föda* (to be published).
10. NORÉN, K., and WESTÖÖ, G.: *Ibid, 19:*13, 1967.
11. SJÖSTRAND, B.: *Anal Chem, 36:*814, 1964.
12. SMART, N.A., and LLOYD, M.K.: *J Sci Food Agr, 14:*734, 1963.
13. STOCK, A.: *Sv Kem Tidskr, 50:*242, 1938.
14. STOCK, A., and CUCUEL, F.: *Naturwiss enschaften, 22:*390, 1934.
15. TEJNING, S., and VESTERBERG, R.: *Poultry Sci, 43:*6, 1964.
16. VESTERMARK, A.; WESTERMARK, T.; LJUNGGREN, K., and HAGMAN, D.: Personal communication.
17. WESTERMARK, T., *et al.:* Kvicksilverfrågan i Sverige. Kvicksilverkonferensen 1965, Stockholm, 1965, p. 25.
18. WESTERMARK, T., and SJÖSTRAND, B.: *Int J Appl Radiat 9:*1, 1960.
19. WESTÖÖ, G.: *Acta Chem Scand, 20:*2131, 1966.
20. WESTÖÖ, G.: *Ibid, 21:*1790, 1967.
21. WESTÖÖ, G.: *Vår Föda, 19:*121, 1967.
22. WESTÖÖ, G.: *Acta Chem Scand* (to be published).
23. WESTÖÖ, G.: *Vår Föda, 18:*85, 1966.
24. WESTÖÖ, G.: *Ibid, 19:*1, 1967.
25. WESTÖÖ, G.: *Oikos, 9* (suppl.) :11, 1967.
26. WESTÖÖ, G.: *Vår Föda, 18:*88, 1966.
27. WESTÖÖ, G.: To be published.
28. WESTÖÖ, G., and NORÉN, K.: *Vår Föda, 19:*135, 1967.

29. WESTÖÖ, G., and RYDÄLV, M.: *Ibid* (to be published).
30. WESTÖÖ, G.; SJÖSTRAND, B., and WESTERMARK, T.: *Vår Föda, 17:5*, 1965.

DISCUSSION

RISEBROUGH: Why should mercury be a problem in the northern countries rather than here in the more temperate areas?

WESTÖÖ: I think the reason is that we have many paper mills and several chlorine-alkali factories that have discharged rather large amounts of mercury compounds which now pollute our water.

RUDD: Is this the major source of contamination?

WESTÖÖ: Yes.

RUDD: Sweden has been able to make what appears to be a proper restriction on the agricultural aspects of mercury contamination. Has any progress been made with regard to mercury contamination from industrial sources?

WESTÖÖ: The industrial sources to some extent have been stopped. Paper mills are not allowed to use mercury compounds anymore.

BERGLUND: I think one reason that this problem does not exist in the United States with mercury is that the levels are not known. Although there is a legal tolerance of zero for mercury compounds in foods in the United States, I feel, personally, that the problem also exists here as it does in other parts of the world but it is not recognized.

RISEBROUGH: I am sure that people are just not looking for it here. There is a related situation in this country which requires a Rachel Carson, and that is the distribution of lead from automobile exhausts in the ecosystem. Dr. Chow of the Scripps Institution of Oceanography (personal communication) has found that the concentration of lead in the marine air off La Jolla is about 0.5 $\mu g/m^3$ air. This is about ten times lower than the average found in cities but one hundred times higher than that found in the central Pacific. The movements of both mercury and lead through the ecosystem urgently require investigation.

CLARKSON: Are you speaking of tetraethyl lead or of lead carbonate from automobile exhaust?

RISEBROUGH: It is inorganic lead originating from tetraethyl lead. It has the same isotopic composition as lead in locally sold gasoline.

HUNT: I think that the reason we have not reported recently on mercury residues in our wildlife in this country is because we have been asleep and not looking for them. However, we have instances of birds dying after consuming mercury-treated cereal grains, and large amounts of mercury-treated grain are used in California each year. Do you find maximum levels depending on which time of year the sample is taken?

WESTÖÖ: We have taken pike and perch samples at different times of the year, and have not found any difference. In 1956 to 1963, when methylmercury dicyandiamide was used as a seed disinfectant in Sweden, Borg *et al.* (1) found the highest mercury levels in seedeating birds during and after the sowing periods.

STICKEL: Several years ago when the Swedes first began pointing up the mercury problem, the United States Fish and Wildlife Service had several samples analyzed in Sweden. These were eggs or livers of black ducks, gulls, or ospreys from various locations in the northeastern United States. All of them were found to be within normal limits, according to a Swedish specialist. Beyond all question, we should do a great deal more on the mercury problem in the United States, but at least these data do not suggest that we have a great bombshell ready to burst here.

CLARKSON: I understand that the mercury compounds have been used longer in Sweden than they have in other countries. When did you start using these in Sweden? In the 1930's, was it?

JOHNELS: The initial use of phenylmercury in industry was made about 1946, and increased afterwards, so it is not very old in Sweden. I think we copied it from the United States. Inorganic mercury compounds were used in agriculture in the twenties. During the thirties phenylmercury was used. And the use of alkylmercury compounds started in the early forties. It is important to know which species of birds were used in the United States tests, because it is possible to miss the evidence of contamination. Pure land birds accumulate mercury in quite another way than water birds, as I will show later. Mammalian herbivores do not accumulate mercury to high levels: we see levels below 10 ng/g in cow, moose and deer.

BERGLUND: I think two factors diminish the role of mercury along the coast of the United States: (a) the absence of an archipelago that closes off large amounts of water as in parts of Japan and Sweden; (b) a very strong tide with an appreciable movement of water.

BERLIN: We are a little ahead in Sweden because we were forced to get better analytical methods. We now have neutron activation analysis and an excellent method for methylmercury, and we can detect mercury in foodstuffs with these tools. We expect that these methods will show high concentrations of methylmercury in fish any place in the world where mercury is used in industry.

NORSETH: How fast is your analytical method?

WESTÖÖ: One person can make the extraction and cleanup of six to ten samples a day. The injection on the gas chromatograph is done by another person who can serve four or five people preparing extracts.

NORSETH: Is inorganic mercury found in the liver as a result of breakdown of the methylmercury, or is it accumulated with time because of exposure to inorganic mercury?

WESTÖÖ: Perhaps both. If the animal receives food contaminated with inorganic mercury some of it will be present in the liver. But if there is change of the methylmercury to inorganic mercury it will probably also be found in the liver. I am not sure that the difference between the total mercury and methylmercury contents must be inorganic mercury salts, but probably it is. As to other organic mercurials, phenylmercury is not used in Sweden by now, and methoxyethyl mercury is used as a seed-disinfectant, but is very unstable and is degraded to salts of inorganic mercury.

CLARKSON: I understand that the evidence that it is inorganic mercury is the fact that it is not extracted by benzene in hydrochloric acid. Is that correct?

WESTÖÖ: If phenylmercury were present in the extracts I would not have seen it in the ordinary gas chromatograms. Therefore I analyzed samples for phenylmercury on a shorter column with higher temperature, and found no phenylmercury. This was confirmed by thin-layer chromatography.

BUTLER: Is there any localization of the mercury residues in the animal bodies comparable to DDT? When fish are prepared for eating most of the DDT residues are discarded in processing.

WESTÖÖ: We analyzed only fish muscles and fish liver, and liver usually contains lower levels than muscle.

JERNELÖV: Uptake is more rapid in the liver and slower in the muscles. When the mercury content in the water is very high and the amount of mercury in the fish is increasing, you'll find very high levels in the liver. With stable environmental conditions the amount of mercury in the liver is about 20 per cent of that in the muscles.

NORSETH: Is this higher amount in the muscle an actual transport of mercury from liver to muscle, or is it a retention of the mercury in muscle compared to liver?

WESTÖÖ: I don't know.

CLARKSON: The experiment where you fried fish and found no loss of methylmercury was most impressive. It would suggest a very strong binding.

VOSTAL: Is there a good correlation between the age of the fish and the amount of the accumulated mercury?

WESTÖÖ: Yes. It indicates that the binding of the mercury must be very strong to create such an accumulation during the whole life of the fish.

FASSETT: How is methylmercury attached to proteins?

WESTÖÖ: Probably like this (structural formula R-S-Hg CH_3). Adding excess hydrochloric acid converts the proteins and other RSHg CH_3 compounds to the RSH-form and releases free methylmercuric chloride. Methylmercury is also displaced by an excess of mercuric ions. The protein itself is not hydrolyzed.

FASSETT: If you put a fish in uncontaminated water, is methylmercury removed from the protein?

JERNELÖV: We had fish with quite a high level of methylmercury and put them in clean water. It took a very long time to remove some of the mercury. The decrease was very small, really. I don't think it is something you can count on. The fish that have mercury will have to die and be replaced by a new generation.

SESSION II
IMPACT ON NATURAL POPULATIONS
OTHER THAN MAN

Chapter 6

THE TOXICITY PROBLEM—COMMENTS BY AN APPLIED INSECT ECOLOGIST

R. van den Bosch

ABSTRACT

The applied insect ecologist's concept of toxicity is discussed and contrasted with the concepts of persons concerned with narrower aspects of toxicity. The organic insecticide revolution is reviewed with emphasis on the lack of ecological consideration in the development of modern pesticides. Impact of chemicals on entomophagous arthropods is particularly stressed and it is pointed out how this has contributed to past and present toxicity problems. The problems associated with Azodrin, an organophosphate newly introduced for cotton pest control in California, are discussed in considerable detail. The future of pest control is treated speculatively.

REPORT

Introduction

WHEN I FIRST GLANCED over the letter of invitation to this conference I seriously questioned whether the organizers had invited the right person. The conference was supposed to deal with persistent pesticides, particularly organochlorines, materials with which I haven't worked much lately. But as I re-read Dr. Rothstein's letter, I discovered a neat loophole, the option, within reason, to choose my own subject. I have seized upon this option and have chosen to discuss the toxicity problem from the viewpoint of the applied insect ecologist. I am quite excited over the opportunity to participate in this conference for it gives me a chance to discuss a basic flaw in pest control that has been largely overlooked in the recurrent dialogue concerning toxicity. It is my conviction that until this flaw is generally recognized and the situation corrected, toxicity problems of a serious nature will continue to plague us, no matter what families of pesticides we evolve.

In the following discussion I shall try to illustrate why this is so, but first I should like to emphasize how one's status as an applied insect ecologist colors his concept of toxicity. This is important because this concept is extremely broad and perhaps embraces aspects with which some people may not be familiar.

An applied insect ecologist deals with agro-ecosystems, not just elements within them. This means that his concern over toxicity covers a multitude of things; e.g., pests, potential pests, predators, parasites, pathogens, pollinators, wildlife, livestock, plants, people and even people's pocket books. Thus, when the applied insect ecologist is charged with the control of a particular pest he makes decisions which can have almost infinite ecological ramifications. What makes his work so terribly difficult is that in the face of manifold pressure to kill pests, save fish, protect pollinators, preserve song birds, safeguard human health, ad infinitum, he does not have the ecologically sophisticated chemical tools with which to do all these things. And so his decisions frequently involve compromise, and because of this they are not always satisfactory from the standpoint of the overall toxicity problem. Needless to say, with all the furor that such decisions engender, the applied insect ecologist is one of the major victims of toxicity.

The Modern Pesticide

It is perhaps symbolic and at the same time ironic that among the modern Nobel laureates, a chemist, Paul Mueller, comes nearest to being an entomologist. Mueller was honored for his discovery of the insecticidal properties of DDT and the saving of countless human lives from disease and starvation. His discovery also triggered the organic insecticide revolution and its avalanche of synthetic materials. This is where irony comes in, for as avalanches often do, this one has created havoc. This derives from the fact that the insecticide revolution occurred so quickly and with such overwhelming impact that ecological consideration (and the ecologist) were almost completely swept aside. Riding in the wake of the avalanche were the chemist, toxicologist and sales executive who poured into the vacuum as the prime movers in pest control, employing toxicology and economics as their major criteria.

The resultant chaos is a matter of record. However, what is

disturbing is the possibility that if it had not been for the sensation caused by Rachel Carson's *Silent Spring* (4) very little attention might be given to the pesticide problem even today. The fact that the problem is recognized and that something is being done about it is hardly basis for comfort, for the efforts aimed at its alleviation still have not reached the core of the problem. By this I mean that there is as yet no coordinated effort to invoke ecological criteria at the basic level of pesticide development where they are most critically needed. Too often we become concerned about toxic materials only after they have been developed and especially after they have had some traumatic effect in the environment. In other words, except for some effort to reduce the hazard to warm-blooded animals and wildlife, chemistry and economics still dominate pesticide development and as a result problems continue to proliferate. This is a major flaw in pest control and until it is corrected, our problems of pollution and toxicology will not abate.

Pest Control in the United States of America

In discussing American pest control I will speak principally from my own experiences in California. This parochial stand should not seriously limit me, for California agriculturists are the largest utilizers of insecticides in the United States. Consequently the state has a substantial array of pesticide-associated problems and this certainly holds true for the crops in which I work, particularly cotton.

In introducing the subject of pest control, I will follow the example of Brian Beirne (2) and compare it with human medicine, since in many ways the two are similar. Thus, in the pest-crop relationship we deal with a complex living entity, the crop, which at times becomes ill or is threatened with illness through such factors as malnutrition, physical stress, microbial infection, arthropod infestation, etc. In essence, then, the crop can and often does become "sick" and just as with diseased or threatened humans, means have been developed to cure or prevent these illnesses. In their own way, crop protection researchers have developed immunological techniques (e.g., pest-resistant plant varieties), surgery (e.g., roguing of diseased plants, pruning of diseased plant parts) , physical therapy (e.g., heat treatment of seeds, bulbs, etc.) , isolation techniques (e.g., growth of seed crops in disease-free areas, timing of plantings to avoid infection or infestation), chemotherapy (e.g., use of fumigants and tropical and systemic pesticides) and other techniques as well.

The research that has gone into the development of these techniques and materials has been of a high order and it has contributed significantly to America's leadership in agriculture. Plant protection technology at the research and development level has had a proud and sophisticated history that is not totally paled by that of human medicine. But if there are similarities between the two, there are also differences and of the latter none is more striking than that which exists at the level where technique and material are put into use. Whereas in medicine there exists a highly trained, technologically skilled corps of practitioners to diagnose human illness and effect measures for its prevention or cure, literally nothing of the sort exists in plant protection. It is difficult to conceive of a highly complex technology without a profession to implement it and yet this is true of plant protection. Needless to say this is very much at the heart of the chaos in modern pest control.

At this point it is perhaps best to describe how this rather incredible situation came about. It relates in large measure to the fact that plant protection technology got off to a relatively late start, especially in the area of effective chemical control. Before World War II pest control relied to a great extent on cultural or other crop manipulation methods and although chemicals were used, they were limited in variety, mode of action and effectiveness. In other words, in the period before World War II chemical control was not very efficient and there was very little that could be done to control a great many pests through chemotherapy. Thus, there was hardly room for the professional practitioner who, though he might diagnose problems, did not have the efficient "drugs" and "medicines" with which to treat them. In the field of human medicine this situation never occurred because the therapeutic and prophylactic drugs and medicines and the quality and training of the professional practitioners evolved more or less simultaneously. Even so, serious problems and questionable practices have occurred, but they appear to have been minor compared with those in plant protection.

In plant protection the breakthrough in the chemical area came all at once and it caught the field without the technocracy to implement it. Literally overnight a welter of unbelievably effective pest killers was released into the hands of people concerned with pest control; farmers, salesmen, entomologists and farm advisers. For the most part, these people were unprepared to utilize the new materials and especially to comprehend the genetic and ecological implications of their use. Unfortunately under this aura of ignorance a very great demand developed for the proliferating variety of organic insecticides and production skyrocketed. The snowball was on its way, careening downhill pellmell and it is still rumbling along at frightening speed.

Thus, pest control burst into its modern age in a flash and it arrived with hardly more than a handful of professional practitioners to implement it. In other words a great vacuum existed and into it rushed the chemical producers and agricultural service companies with their salesmen and technical representatives; men oriented essentially to sales quotas and "bug" killing and literally devoid of ecological training or perception. This created a unique (and disturbing) situation, wherein a highly developed technology utilizing complex techniques and highly toxic ecologically disruptive materials came to rely in great measure on salesmen for its implementation. It takes little imagination to envisage the chaotic state of medicine, were the diagnosis of illness and its treatment particularly through the use of drugs and medicines the responsibility of the pharmaceutical-house salesman. Yet this is the case in plant protection.

The description I have just given is not exaggerated. No doubt some will argue that the agricultural experiment stations and agricultural extension service (farm advisers, extension entomologists) largely dominate pest control practice in the various states, but this is a myth. For example, a recent study in Iowa (1) showed that farmers in that state place greater reliance on the chemical dealer than on the agricultural extension service for information on agricultural chemicals. I am certain that in the much more complex agricultural economy of California, even greater influence is exercised by the chemical and agro-service company salesmen and field representatives. To illustrate this point I should like to discuss the recent history of the organophosphate insecticide, Azodrin, in California.

The Azodrin Story in California

Azodrin, a product of the Shell Chemical Company, was first registered for use against cotton insects in California by the State Department of Agriculture in 1965. It saw limited use that year. In 1966 a production setback greatly restricted its use but in 1967 the material was applied to almost one million acres of cotton. This is one of the most striking use-acceptances of a new insecticide that has ever occurred in California (3). I will discuss the reasons for this later.

Now, during the period 1964 through 1967, we in the Agricultural Experiment Station tested Azodrin exhaustively in the field. Quite significantly, after four years of study and a number of experiments we found that we could not recommend the use of this material to the California cotton growers (7). There were several reasons for this. For one thing we found Azodrin to be one of the most highly destructive materials to a broad spectrum of entomophagous arthropods (predators and parasites) of any that we had ever tested (5, 6). Thus although it was a highly effective killer of *Lygus hesperus* we decided not to recommend it for control of that pest because its impact on natural enemies of such pests as cabbage looper (*Tricoplusia ni*), beet armyworm (*Spodoptera exigua*), and bollworm (*Heliothis zea*) might leave the cotton fields open to severe attack by these species. When we tested Azodrin directly upon bollworm we again recorded heavy destruction of predators and parasites, and in two experiments bollworm infestations increased because of predator elimination. In several other experiments inconclusive results were obtained, and only in a single test did we record a rather substantial increase in cotton yield through use of the material.

In summary, our experiments showed that at times, use of

Azodrin aggravated the bollworm problem; on other occasions it had indifferent effects, while in one case its use resulted in a substantial increase in yield. We know Azodrin will kill bollworm, but we have not yet found a way to consistently use it economically and with minimum ecological impact. This is hardly a basis for recommending the material, and in fact Azodrin's adverse characteristics have caused us to warn the growers of the possible hazards, ecological and economic, that its use might entail (8) .

How then did this material, which is not recommended for use against cotton bollworm by the University of California Agricultural Experiment Station and Agricultural Extension Service, come to be so widely accepted by California cotton growers? This came about in large measure through an aggressive sales program by the Shell Chemical Company, Azodrin's producer. Shell was aware of our research findings through publications and a seminar which I was invited to give at their Modesto, California laboratory in January, 1965. Nevertheless the company decided to promote their material for use against bollworm in the San Joaquin Valley. The California cotton growers, plagued by an increasingly serious bollworm problem and desperately in need of a replacement for the largely outlawed organochlorine insecticides, apparently eagerly accepted Azodrin along with an unprecedented multiple treatment program. There can be little doubt that Shell's aggressive salesmanship in this period of grower concern over the bollworm problem played a major role in the spectacular use-increase of Azodrin in 1967.

It is difficult to assess the economic impact of the intensive Azodrin program in California cotton. In doing this, one has to take into account the cost of the insecticide and the price of its application, which together approximated several dollars per acre or several millions of dollars for the San Joaquin Valley during 1967. But cost is only part of the formula; economic gain is the other. In other words, the San Joaquin Valley cotton industry had to regain its multimillion dollar investment and an additional sum through increased cotton yield in order to justify its expenditure. Whether or not it did this is and will remain a moot question. However, the bulk of evidence would seem to indicate that no outstanding benefit occurred, since the 1967 cotton yield in the San Joaquin Valley was one of the lowest of the past decade. In addition our own experiments indicate considerable odds against significant economic gain resulting from the use of Azodrin.

But the matter of the economic benefit of the material's wide-scale use is only one of several interesting aspects of the Azodrin story. The complete breakdown in the Agricultural Extension Service's communication of the research findings on Azodrin to the growers is especially disturbing to me. The inability of the official research and extension establishment to reach the California cotton grower and influence his decisions in the face of an aggressive sales program by a major chemical producer is a striking example of the waning role of the traditional extension agency. In this connection it should be stressed that more was involved than just the use of a nonrecommended material, for the prophylactic manner (i.e. multiple treatments on a calendar basis) in which it was widely used was also a departure from recommended practice. Prophylactic use of insecticides in cotton has long been recognized as bad practice by California's research and extension personnel, since it entails treatment of uninfested acreages, tends to hasten pest resistance to pesticides, is costly, and contributes excessively to environmental pollution. Similar programs in Peru and Columbia have had such disastrous results that they have led to strong governmental controls over the use of organic insecticides and in Nicaragua a state of chaos now exists in cotton for the same reason. Yet in California, perhaps the most advanced agro-economy in the world, prophylactic use of Azodrin occurred literally overnight despite the efforts of the Agricultural Experiment Station and Extension Service to prevent it. Such a development seems almost incredible after all the public furor, political hand wringing and legislation engendered by the insecticide problem in recent years. But such is the case and I have still told only part of the Azodrin story. I have done this deliberately because as an insect ecologist I wanted first to stress those problems of insecticide use which relate to arthropods, agro-economy systems and grower economics. Too often public concern transcends these aspects of the pest control problem and directs itself only to matters of public health and livestock and wildlife safety.

I have tried to show that major pesticide-associated problems occur in plant protection long before we ever start poisoning babies, bunnies or beef. Now I'll carry the Azodrin story a step further and allude to another problem. This concerns the considerable impact Azodrin appears to have on wildlife (3). What makes this so disturbing is that such impact was unforeseen and its nature is still not understood. One would have thought that we had learned our lessons years ago with organochlorine poisoned ospreys, grebes and salmon, but apparently we haven't. The extent of the Azodrin impact on wildlife is unknown largely because the California Department of Fish and Game with its limited manpower has not been able to mount extensive surveys. However, in 1967 fish and game personnel received thirty-five reports of wildlife losses due to Azodrin. Twenty-two of these were investigated and in nineteen instances dead or affected wildlife was verified (3). Wildlife involved in losses were pheasants, doves, quail, jackrabbits, horned larks, meadowlarks, blackbirds, kill-deer and sparrows. Reports were also received of dead feeder geese, while several ranchers reported losses of peacocks and a hunter reported that his dog died after retrieving birds from a treated cotton field. No livestock losses were reported. It is emphasized that the Fish and Game Department figures do not reflect the actual extent of losses as only the edges of treated fields were searched for carcasses. This was so because (a) carcass counts were not practical in dense cotton, particularly in fields under irrigation, and (b) because the hazard to humans was not known and fish and game personnel were requested to stay out of the treated areas for three days after application. In other words, wildlife kills were probably much greater than the record shows and indeed may well have been massive. In this connection it is of interest that crop dusters reported that shortly after they had sprayed cotton, they clearly observed sick pheasants. In Arizona where heavy wildlife kills were also recorded observers reported that the most impressive factor was not the number of carcasses counted (which in one case involved 365 dead birds of seventeen species found along $12\frac{1}{4}$

miles of field edge) but "the lack of living birds in treated areas where bird life had formerly been abundant" (3).

These are highly disturbing statistics and observations but as I said earlier what is perhaps even more disturbing is that it is not known how Azodrin affects wildlife. However, speculation and some experimentation indicate that the animals ingest the insecticide through feed (both plant and animal) and through water. The latter possibility is especially disturbing since Azodrin hydrolizes very slowly in water and thus may remain in puddles in the fields or in other pools of standing water for considerable periods. The implication here hardly requires elaboration except for the comment that a highly toxic material scattered about in standing water certainly poses a hazard to public health as well as wildlife.

So here with the insecticide Azodrin we have a case history that entails literally everything that is disturbing about modern chemical pest control; a broadly toxic, highly hazardous material with known adverse ecological characteristics, developed by chemists, exploited aggressively by sales personnel and in the end having serious unforeseen ecological side effects. One would have thought this to be impossible after all the furor of the past decade — and so it comes as a shock that a case as disturbing in its implications as many that occurred in the "pre-Carsonian era," could occur in 1967. Indeed, we seem to have gained little ground in our efforts to attain scientific pest control. But then why shouldn't this be? Chemists remain the prime movers in pesticide development and they are still guided by toxicological and economic criteria while merchandising continues to be a primary factor in insecticide use. Salesmen and chemical company field representatives play a major role as pest diagnosticians and pesticide dispensers, and they still are valued more by their companies for their competence as salesmen than as ecologists. In this connection I should like to quote from a recent address by R. H. Wellman of the Union Carbide Corporation to the Western Agricultural Chemicals Association on one aspect of the distributor's responsibility in the Manufacturer-Distributor relationship (9). Wellman states; "The distributor must maintain and continually improve an aggressive knowledgeable sales force who SELL products. The order takers and price salesmen are not the type on which the distributor can rely for both profit and volume. Nor do they make up the kind of sales force for which the manufacturer is justified in providing adequate suggested margins on his products." The message comes through clearly!

In the past decade we have seen the evolution of a considerable series of regulations and tests and tolerances and labeling and registration requirements for insecticides — all of which have been aimed at reducing the hazard to human health. But essentially nothing has been done to invoke ecological criteria at the basic level of insecticide development. Lip service is given to the need for selective materials, but distressingly few have appeared on the scene. Registration of chemicals for use in crops is still a slipshod affair, the main criteria being, (a) Do the materials cause reasonable pest kills? (b) What hazards do the materials pose to warm-blooded animals? Salesmen, the diagnosticians of pest problems and dispensers of nostrums, are still simply salesmen — they aren't even licensed and of course need not have technical skills.

In short, chemical pest control is approaching a state of chaos, there is no other way to describe the situation.

Conclusions

But what's to be done about this unhappy situation. This is like asking, how do we get out of Vietnam? As with Vietnam there is no quick and easy answer; the extrication process is going to be messy. First, certain almost predictable events will have to occur. For one thing matters are going to have to get a lot worse, and in fact approach the point of disaster. In other words, something very bad is going to have to happen to people, livestock or wildlife so as to cause great public concern and therefore severel legal restrictions on the use of pesticides, or, economic disaster will force the grower to alter his attitudes and demand changes. It has been my experience in developing integrated control programs and in analyzing those developed elsewhere, that only after the grower has been backed to the wall by economic disaster does he seek alternatives to the traditional use of insecticides. Until he is desperate he wants only the cheapest most broadly toxic types of insecticides and he is under constant pressure from the insecticide industry to use such products. At the present time we in research and extension simply cannot reach him without ideas of a more sophisticated approach to pest control.

Assuming that a chain of events leads to an overwhelming public and grower clamor for major changes in chemical pest control — what then should transpire?

In my opinion the following events would be most desirable:

1. The disappearance of the insecticide company salesman from the pest diagnosis and control advisement scene and his replacement by the independently practicing agro-technologist who should be licensed and required to qualify for licensing through examination.

2. The development of insecticides with limited toxicity spectra within the *Arthropoda* as well as for warm-blooded animals (i.e. ecologically selective materials).

3. A reduced effort by research entomologists in the area of unilateral pesticide use, and the substitution of more sophisticated programs which blend a variety of techniques, concepts and materials into pest management schemes. This portends the development of research teams which will include persons (e.g., agronomists, economists, biometricians) outside entomology.

4. The wide-scale reeducation of entomologists, chemical company personnel, growers and the general public as to the nature and complexity of pest control. This will be the function of the universities which must largely abandon traditional concepts in training economic entomologists and emphasize instead the ecological nature of pest control. Part of the educational responsibility will also fall on the Agricultural Extension Service which must concentrate on inculcating the ecological concept into the grower and citizen in general. The Extension Service must also come to realize that its role in "on the farm" pest control advisement will be taken over by the agro-technologists and that increasingly its function will be to transmit information to these professional plant protection specialists.

When will this all come about? Anyone's guess is as good as mine, but I'm sure that change is coming because chaos will bring it, and chaos is near at hand.

REFERENCES

1. BEAL, G.M.; BOHLEN, J.M., and LINGREN, H.G.: Agr. and Home Econ. Expt. Sta., Iowa State U. of Sci. and Technology, Special Report 49, 1966.
2. BEIRNE, B.P.: Pest Management. CRC Press International Scientific Series, 1967.
3. Calif. Dept. Fish and Game: Azodrin wildlife investigations in California. FWIR Pesticides Investigation Project Dec. 1967. (Mimeo.)
4. CARSON, R.: *Silent Spring.* Boston, Houghton, 1962.
5. FALCON, L.A., et al.: *Calif Agr, 19* (7) :13, 1965.
6. Univ. of Calif.: Pest and disease control program for cotton. Calif. Agr. Expt. Sta. and Ext. Service Leaflet 83, 1965.
7. *Ibid,* 1966.
8. *Ibid,* 1967.
9. *Ibid,* 1968 (in press) .

DISCUSSION

RUDD: You are predicting catastrophe, and I happen to believe you, but people say if we have all the ingredients of catastrophe, why has it not come about?

VAN DEN BOSCH: Of course it has, in certain areas, but this does not convince people. It is our job to delineate the crisis and prevent total disaster. But, the grower and the dealer are members of the same community — church, Rotary, and country club — and many growers think that everybody from the University community (especially Berkeley) is some kind of radical kook. The credibility of the University Experimental Station is pretty low, but we are making progress. For example, one of the largest growers in California gave us 2500 acres for experimental purposes this year, and that is three quarters of a million dollars worth of cotton. He wants us to prove to him that our program works.

I still feel, that until near or total chaos has come about, it will be very hard to get acceptance for the integrated control approach to pest control. The more the situation worsens the greater will be the demand for the integrated (ecological) approach — a sort of feedback situation.

RUDD: Are ecologists very much involved in this field?

VAN DEN BOSCH: Most of the basic ecologists are not pragmatic enough to deal with this very complex situation in the hard area of economics. This is why I use the term "applied insect ecologist" because we are selling a program, an ecologically sophisticated program to a cotton grower; the ultimate question for him is "am I going to make money — more money than I did with my previous program?" So, it is understandable that the transition is going to be slow, and I am afraid that cotton in California may verge on the threshold of economic collapse before a rational pest control program is attained. There is no question about it.

The pink bollworm is invading the San Joaquin Valley. The valley is a two-bale cotton area. They have got to produce $1\frac{3}{4}$ bales to break even; in other words they've got about $50 to $75 margin and as they go into $30 and $40 insecticide programs they are just not going to stay in business. As an ecologist one has to keep economics in mind, before recommending a scheme that depends, say, on putting parasitic wasps out in the fields.

HERMAN: In view of the clear evidence indicating hazard to wildlife from Azodrin, what has been the recommendation of the Department of Fish and Game in California with respect to its widespread use?

HUNT: There are two ways a chemical may be withdrawn in the State of California: these are through, (a) the chemical company's withdrawing it of their own volition, and this takes some pretty damaging information, or (b) a whole series of hearings before the Department of Agriculture, which is the regulating agency in the State of California. The easiest, most effective way is either to get the company to withdraw or to ask the Department of Agriculture for a policy statement, that a specific pesticide should not be used on a specific crop.

In this situation, the information presented by the Department of Fish and Games was insufficient from the standpoint of a State Department of Agriculture, to prohibit the use of Azodrin on cotton. The manufacturer had considerable information, gathered by a University in another state, indicating the insecticide was relatively harmless to wildlife. We had done some experimenting with this chemical in an area where they also used parathion and other organophosphates. We found some wildlife losses, but there was no specific analytical method for Azodrin only a cholinesterase test. We were not able to demonstrate specifically that Azodrin killed these birds.

The University of California's role is to recommend or not to recommend the use of certain pesticides but I am not sure how many of the University's recommendations the farmers actually use. The decisive question in the State of California is whether the material is registered by the State Department of Agriculture; once it is registered, the farmers may or may not use it, as they see fit.

VAN DEN BOSCH: When an insecticide is once registered however, it's almost impossible to get it withdrawn. How does it get registered? This is the point of my talk, and the only reason for choosing Azodrin as an example. The College of Agricultural Experiment Station and Extension Service is the official research and recommendation agency in the State of California. We worked for four years on this material and could not recommend it, yet it was registered and caused disruption of agricultural ecosystems. The flaw of the registration system is this: all the company has to show is reasonable pest kill, and that it is reasonably safe to humans. As insect ecologists, we have criteria which are basic to this whole problem, and they are completely ignored.

BERG: It seems that agriculture is following the mistakes of medicine — three quarters of a century later and with more powerful tools. Doctors used to dispense calomel and jalop the way chemical salesmen dispense organomercurials and organochlorines. Would you say that there is a positive feedback inherent in the marketing system, so that whether a chemical works or does not work one uses more of it next time?

VAN DEN BOSCH: Well, yes. The grower's philosophy has traditionally been "if five pounds will do a pretty good job, then ten pounds ought to really kill them." Furthermore, from the standpoint of merchandising, the pesticide salesmen who are diagnosing problems and advising growers on pest control are out to move their product and make a profit. A force of perhaps some five hundred salesmen, and a budget of many millions of dollars is involved in promoting and selling pesticides to the growers of the San Joaquin Valley of California. The companies have to recoup these costs and make a profit. As matters now stand, it is the growers, and ultimately, the consumer who pay for this.

So, merchandising is a fundamental premise on which pest control is conducted today in California, the most advanced agrotechnology in the United States, if not the world.

BERG: Essentially the same problems are found in medicine, and it has taken a century of work to start controlling promotion and to eliminate some of the conflicts of interests. The worst of the positive feedback is seen when it pays to market a preparation that does not work, because then more can be sold — only promotion is limiting. These abuses have occurred in the history of drugs, and a great many controls have been incorporated into the law as a result. Perhaps something is to be learned, not from the achievements of chemical therapy, but from the controls that made the achievements possible at a reasonable cost in lives.

VAN DEN BOSCH: I might add that Azodrin used against bollworm is an example of a poor insecticide. It has to be used repetitively to effect reduction of a bollworm population. But repetitive use entails volume, and this is profitable from the seller's standpoint.

Chapter 7

ORGANOCHLORINE INSECTICIDES AND BIRD POPULATIONS IN BRITAIN

J. ROBINSON

ABSTRACT

The relationship between the use of organochlorine insecticides and changes in the population status of birds in Britain is discussed. Two independent sources of information have been used: one is concerned with the assessment of changes in bird populations during the period of use of these compounds; the other is based on an assessment of the biological significance of residues of these compounds found in the tissues and eggs of birds. A critical evaluation of the two types of information shows that there are a number of important deficiencies. In some cases it is difficult to assess the representative nature of the samples studied; in other cases the analytical results are of doubtful accuracy. Attention is also drawn to some of the difficulties in the evaluation of the results; a particular difficulty is that a set of observations may be used to generate a number of implications and these have been submitted to an incomplete process of explication in some cases. Notwithstanding the difficulties of interpretation of the published results it is concluded on the basis of convergent probabilities that (a) grain-eating birds were poisoned by cyclodiene insecticides as a result of one particular usage of these compounds between 1955 and 1962; (b) whilst the possibility cannot be excluded, the declines of certain predatory birds in the 1950's cannot be related to organochlorine insecticides with any degree of practical certainty; (c) there is no evidence of serious effects on the population status of predatory birds by these compounds since 1962; and (d) common farmland species of birds have suffered no inimical effects attributable to the organochlorine insecticides.

REPORT

Introduction

IN THE LAST TEN YEARS or so, it has become apparent that residues of organochlorine insecticides, particularly DDT and its derivatives or congeners and some of the cyclodiene insecticides, occur in the tissues and eggs of a wide range of bird species. Thus

113

residues of these compounds have been reported in more than eighty species of British birds by Moore (43), and another series of surveys has indicated the presence of residues in 118 species (16). The phenomenon is not, of course, confined to Britain. An indication of the potentially worldwide distribution of very small residues is given by reports of the detection of these compounds in the atmosphere (2), rainwater (19), and Anarctic fauna (23, 75, 80). These observations prompt a number of questions, the more important being (a) What is the biological significance of the residues found? (b) Is the contamination of the environment increasing with continued usage of these compounds? (c) What are the mechanisms of dispersion of the residues from the point of application? An attempt is made in this chapter to review the evidence available in Britain in a critical and, as necessary, detailed manner with a view to indicating whether any answers, even if tentative, can be given to these three questions. A considerable amount of information is available in regard to the first question, but the assessment of the toxicological and ecological significance of the residues in wild birds is still a matter of some difficulty. In this chapter, two types of evidence on this aspect are summarized: changes in the population dynamics of birds in Britain since the introduction of organochlorine insecticides and a comparison of the concentrations of DDT, etc. in tissues of experimentally poisoned birds with those found in wild birds. The second question, that of the relationship of environmental contamination to time, can only be answered in a fragmentary and tentative manner at present, and this comment also applies to the third question.

Population Dynamics of Birds

The fecundity of many birds is such that unless some mechanism is operating which limits their numbers it is difficult to see why gross overpopulation has not occurred, whereas it is generally found that the numbers of a particular species remain fairly constant. The main theories suggested to explain the regulation of animal numbers are associated with Nicholson (49), Andrewartha and Birch (4), and Wynne-Edwards (91). In the case of birds, Lack (37, 38) has espoused Nicholson's theory of density-dependent mortality and suggested that the food supply was the major cause of regulation of numbers; Wynne-Edwards also supports a density-dependent mechanism but he suggests an additional mechanism whereby the increase is checked well before the available food supply is fully utilized. This mechanism includes emigration and a social control of the rate of breeding. The controversy on the mechanisms of population control of birds will not be discussed, but the divergence of views on such a fundamental topic indicates the care required in assessing the causal relationship between pesticides and changes in bird populations.

For the purpose of this chapter, it is postulated that the population of a bird species, within its particular ecological niche, changes from time to time such that the energy flow through the ecosystem tends to a maximum; this theory is based upon the work of Odum (51), MacFadyen (42), and others. Random variations in the edaphic factors are reflected in random variations in the population numbers about a mean value which is a function of time. Thus the introduction of a new species into an ecosystem in which a suitable unoccupied niche is available, results in an initial rapid increase in the numbers of that species, an increase which approximates to an expotential function of time. An outstanding recent example of this type in Britain is presented by the collared dove *(Streptopelia decaocto)*. This bird reached Britain between 1952 and 1955 and since 1955 the numbers have increased a thousandfold (30) as shown graphically in Figure 7-1. The exponential increase of a population is not maintained indefinitely and the rate of increase begins to decline and eventually becomes zero. At this point in time the recruitment by production is balanced by the losses. Lack (37, 38) quotes the population data for the heron *(Ardea cinerea)* in Britain in support of this concept of a steady state of population numbers. The numbers of a particular species of birds, on this simplified and idealized basis, can be represented as a function of time since entry into the niche, as in Figure 7-2. The number of birds in the population of a particular species is a function of a number of variables, including time:

$$N = f\ (\alpha_a \relbar\joinrel\relbar x_j\ t)$$

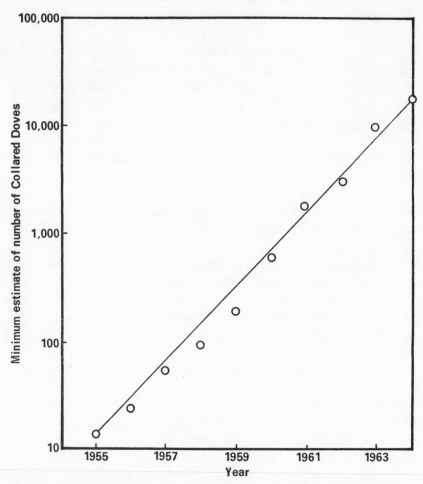

FIGURE 7-1. Minimum estimates of the numbers of collared doves in Britain and Ireland, based on the data summarized by Hudson (30).

The effects of the introduction of a new variable may be tested by comparing the rates of increase of the population before and during the exposure to the new factor (α_i) ; if the factor is terminated, then the rate of increase following the cessation of the factor may also be compared with that before and during the exposure. Difficulties arise in practice in that the information available may not be sufficiently precise to determine the phase of the population growth at which the new factor (α_i) came into operation. The analysis may be simplified by transforming the animal population estimates to their logarithms, the relationship from $T_1 - T_3$ for example would then approximate to a straight line (as in Fig. 7-1).

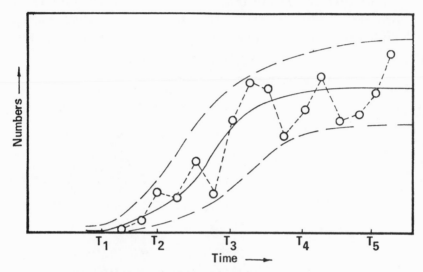

FIGURE 7-2. Population growth. Animal numbers represented 0, mean growth curve ———, limits of variation of increase ----.

An investigation of some actual population counts for birds in Britain is given below. Before considering these cases, however, it is relevant to discuss some of the factors which may affect the population dynamics of birds. According to Lack (37) bird populations are limited by the food supply, predation is an important factor in some species and diseases in others. Diseases do not seem to be an important factor in regulating the numbers of most wild birds but they may be responsible for large random fluctuations in numbers. Climatic variation may have serious effects and a good example was given by the dramatic decreases of many bird species in Britain in the severe winter (for that country) of 1962-63 (18). The effects of this winter will be considered further below. A major and sometimes drastic factor in the population dynamics of many bird species is man. It would be a mistake to assume that species of birds have declined in numbers or even become extinct as a result only of man's activity. Until about 10,000 years ago man was merely a food gatherer and hunter; his impact upon animals and plants was probably no greater than that of other predatory species. During this period before the beginning of settled agriculture, namely, the Pleistocene epoch which started about one million years ago, there was a series of climatic changes, including four very severe periods of glaciation. These climatic

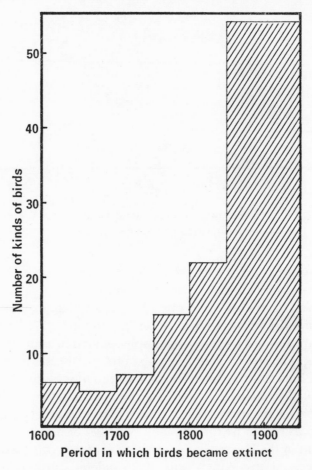

FIGURE 7-3. Periods during which kinds of birds (species and subspecies) became extinct.

changes had catastrophic effects upon flowering plants for example (24). Presumably birds were also seriously affected and Brodkorb (9) has estimated that reductions in avifauna of up to 25 per cent may have occurred.

During the last 10,000 years there is no doubt that man's activities have had an increasing impact upon birds, both as regards species survival and numbers. These effects have probably been most serious in recent centuries and it is estimated that some 160 kinds of birds (species and subspecies) have become extinct since 1600 (86). The dates of extinction and geographical location are shown in Figures 7-3 and 7-4 respectively. Some sixty kinds of birds are considered to be in danger of extinction at present, and the geographical distribution of all birds that are giving rise to concern is also shown in Figure 7-4. Of the birds in danger of extinction it is reported that some 60 per cent of the species or subspecies have declined as a result of man's activities. Some slight reassurance is given by the small proportion of birds, either extinct or in danger, which were (or are) in geographical

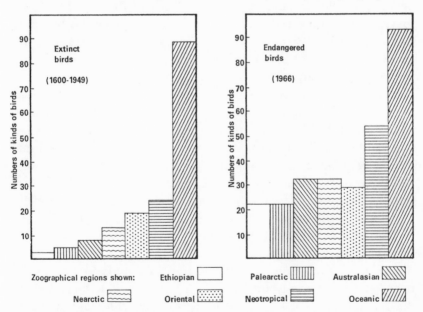

FIGURE 7-4. Geographical distribution of kinds of birds that have become extinct or are endangered.

areas with large human populations. Birds in oceanic islands, particularly those with specialized habits, have been and are in the greatest potential danger. It is also pertinent to point out that of the species in danger there is only one (86), the American condor (*Gymnogyps californianus*), for which it is suggested that pesticides may be involved. A recent example is that of the Bermuda petrel. The decline of this extremely rare bird in recent years has been linked with the presence of DDT-type residues in their eggs (90).

Bird Populations in Britain

Information concerning the numbers of various types of birds in Britain is derived from a variety of sources. In some cases there have been studies of a particular species for a number of years. Summaries of these types of investigation have been published by Alexander and Lack (3) who reviewed the changes in population status of 207 species of birds during the 100 years up to about 1940, and by Parslow (52) who has published a series of papers on the post-1940 population status of birds in Britain. The latter survey is not quite complete, but a comparison of the breeding status of birds in Britain, as shown by these two surveys, is of interest as none of the organochlorine insecticides was in use during the first survey (up to 1940). Consequently, a comparison of the two surveys might provide information on the general effects of these compounds upon bird populations. Such a comparison, which can only be of a qualitative nature, is given in Table 7-I. This table includes only those species for which sufficient information is available to draw conclusions on population status, and it will be seen that there is no plausible reason to believe that a new inimical factor has affected the majority of bird species since 1940. In about 16 per cent of the species, the population status is less favorable since 1940 and presumably various detrimental effects have been occurring. In some cases the investigations have given indications of the causes of the decline in population status between the two periods, but in only a few cases is it suggested, in Parslow's review, that organochlorine insecticides are a major or contributory cause of the declines of particular species since

1940. Particular concern is expressed in the papers summarized by Parslow about the relationship of organochlorine insecticides to the declines in predatory birds (hawks, falcons and owls) ; these birds will be discussed after reviewing the information from other investigations on nonraptorial species.

TABLE 7-1
COMPARISON OF THE POPULATION STATUS OF SPECIES
OF BRITISH BIRDS BEFORE AND AFTER 1940

Population Status Pre-1940	Post-1940	Number of Species
Increasing	Decreasing	4)
No evidence of change	Decreasing	11)
Increasing	No evidence of change	13) } 28
Decreasing	Decreasing	24)
No evidence of change	No evidence of change	39)
Doubtful change or change geographically variable	Doubtful change or change geographically variable	23) } 86
Increasing	Increasing	44)
Decreasing	No evidence of change	3)
No evidence of change	Increasing	6) } 60
Decreasing	Increasing	7)

The pre-1940 estimates are based on Alexander and Lack (3), the post-1940 ones on Panslaw (52).

A scheme which has given accurate information on the population status of farmland birds in Britain is the Common Bird Census. This was started in 1961 with the aim of collecting information on changes in the populations of the commoner British birds. Details concerning the methods used have been given by Williamson and Homes (89) ; statistical analyses of the results have been given by Taylor (81) and Bailey (6). The census methods used are negatively biased in some cases (76) but provided the biases are consistent from year to year the results of the Common Bird Census can be used to measure annual changes. Bailey (6) has summarized the results of the Scheme for the period 1962 to 1966. Of the forty-one species studied, twelve decreased significantly in the period 1962 to 1963, two increased significantly, and the remainder showed no significant change. Some of the changes in the species that declined were very large, thus the mistle thrush *(Turdus viscivorus)*, wren *(Troglodytes troglodytes)*, and the pied wagtail *(Motacilla alba)* declined by 89 per cent, 85 per cent, and 83 per cent respectively. The significant decreases are attributed to the winter of 1962-63; that winter was one of the most severe in Britain for about 200 years. Since 1963, the numbers of the majority of species have been increasing and by 1966 had returned to approximately the levels found in the 1962 survey. There are two outstanding exceptions to this general recovery, the common partridge *(Perdix perdix)* and the lapwing *(Vanellus vanellus)*. The decline of the partridge is a continuation of one that has been occurring for some years, and the reasons for this decline, together with the failure of the lapwing to recover, are not understood. It appears to be a combination of inclement weather during the spring periods of several years and loss of suitable habitat. The overall conclusion that may be drawn from the results of the Common Bird Census is that no serious adverse effect (such as that arising from large scale poisoning by toxic chemicals) was occurring from 1963 to 1966 and the changes during 1962 to 1963 appear to be mainly, if not wholly, the result of the very cold winter. This conclusion is very important since it refers to farmland species, those species which would be expected to be amongst those at maximum risk from the use of pesticides in agriculture. It must be appreciated that the results cannot be interpreted as showing the absence of any environmental effects since the necessary control populations are not available. However, the population changes in different geographical regions, indicated by the Common Bird Census, were consistent and any such changes must therefore be occurring equally in the different areas. Since the usage of insecticides varies from area to area, it seems plausible to conclude that the effects of insecticides are either nil or are too small to be detected.

A more detailed investigation of the population dynamics of the blackbird *(Turdus merula)* by Snow (77), using the results of bird ringing and recoveries, gives general support to this conclusion. He found that the annual mortality of adult blackbirds had tended to decline from 1951 to 1961, the decline being rather more pronounced in the southern part of Britain. The adult mortality rates increased in 1962 and 1963, presumably as a result of the effect of the two winters, and returned to the 1961 level in 1964. Once again we cannot conclude that the declining adult mortality indicates an absence of deleterious effects by some factor such as pesticides unless we have a control population. Fortunately, there are two other estimates of mortality of adult blackbirds, one covering the period 1932-33 to 1953-54 (13) and the other being for the period prior to 1940 (39). It is noteworthy that the average annual mortalities for these earlier periods were higher than the annual mortalities for most of the years since 1952. Consequently, we are justified in concluding that it is unlikely that the mortality of blackbirds in Britain was significantly affected by pesticides from about 1951 onwards. Snow (76, 77) also investigated the breeding success of the blackbird, the investigation being based in this case on the Nest Record Scheme. For the period 1951 to 1965 no trend in the clutch size was detected, hatching success was nearly constant, but there was a slight decline in the numbers of nestlings surviving the age of 9 days. This latter observation may be pertinent in relation to the potential effects of organochlorine insecticides (see section on the sublethal effects of organochlorine insecticides). The breeding statistics of another nine common species have been examined (17): no marked alterations in clutch size, fertility, or fledging success were found and in only one species, the chaffinch (*Fringilla coelebs*), was a small but significant change found, namely, in the hatching success for the years 1959 to 1961.

Attention has been drawn to the care required in interpreting the data considered so far, as a control population is often lacking. It is of interest, therefore, to see if there is any information that can be used retrospectively in an attempt to assess the population status of birds before the introduction of organochlorine insecticides. Unfortunately, little information is available prior to 1945, i.e. it is difficult to obtain the population status before the introduction of BHC and DDT. However, the comparison given above of the changes in bird populations before and since 1940, although of a qualitative nature, is suggestive of a general lack of inimical effects from these compounds, and the detailed study of the blackbird again indicates the absence of such effects. The cyclodiene insecticides (aldrin, dieldrin and heptachlor) did not come into general use in Britain until 1955 and the succeeding years. The Nest Record Scheme, which has been reviewed by Burton and Mayer-Gross (10) , and the Ringing (Banding) Scheme however, have been operating for some thirty years and fifty years respectively. Since about 1950, in particular, the amount of information has been steadily increasing and it is desirable that the annual records of these schemes should be examined to ascertain their general usefulness as population indices in addition to the detailed use of information from the schemes of the type described by Snow. It would first be necessary to establish the validity of using the annual records as indices of population status, and one method of assessing this would be by comparison of the results of the Common Bird Census with the Annual Records for these two schemes since 1962. An investigation of this type is in hand and the preliminary results indicate that the results of the three schemes are correlated. Whether the significant correlations indicate that the three schemes are independent measures of population status requires closer investigation of the changes in the numbers of participants in the schemes from year to year and the consistency of their observations.

This summary of the population status of the commoner birds indicates that there are no significant changes amongst these birds that can be plausibly attributed to organochlorine insecticides; individual birds of some species have been poisoned but the total mortalities of the various species do not seem to have increased. These conclusions refer to the commoner birds, and evidence for raptors requires separate consideration.

Population Status of Birds of Prey

The question we are asking is, Are there any indications of change in the population dynamics of British birds of prey following the introduction of organochlorine insecticides into agriculture in Britain? To answer this question, we require reliable data on the number birds of prey, their breeding success, etc., before and since 1945. As the introduction of organochlorine insecticides is not the only change that may have occurred, a control population is required in order to assess the significance of any changes that are found in the raptor populations since 1945. Obviously a true control population is not available, but comparisons between different geographical areas, and between different time periods, may be very helpful in interpreting the causal link, if any, between changes in population status and the effects of organochlorine insecticides. The difficulty in giving a valid answer to the question is increased by the paucity of reliable and consistent information on the population status of birds of prey from year to year.

General reviews of the status of the rarer birds of prey up to about 1956 were prepared by a number of authors, including Nicholson (50) who discussed the status of the golden eagle (*Aquila chrysaetos*) and Ferguson-Lees (21) who discussed that of the peregrine falcon (*Falco peregrinus*). A detailed review of the status of the buzzard (*Buteo buteo*) in 1954, consequent upon outbreaks of myxomatosis in Britain, was published by Moore (44). A decline in the numbers of peregrine falcons in Cornwall in the period 1955 to 1960 was reported by Treleaven (82); he was unable to find an obvious explanation for the decline but suggested that the excess of females in the population might be an important factor, and that disturbance by man and increased predation of the eggs and young were also possible contributory causes. Ash (5) studied the numbers of birds of prey (10 species) in a game preserve in southern England between 1952 to 1959. He considered that the kestrel (*Falco tinnunculus*) and buzzard showed marked fluctuations in numbers from year to year and that the sparrow hawk had declined greatly, the cause being given as persecution by game keepers. The merlin (*Falco columbarius*) and peregrine falcon were also said to have declined in numbers but no explanation of the decline was offered. Hen harriers (*Circus pygargus*) increased in spite of persecution. The results of Ash's observation on sparrow hawks and peregrine falcons are of particular interest: the numbers of sparrow hawks seen per 100 hours of observations declined from 12.28 in 1952 to 2.70 in 1959; and the decline is significantly correlated with time since 1952 ($r_s = 0.77$, P <0.05) ; the decline in numbers of peregrine falcons observed per 100 hours is also significantly correlated with time since 1952 ($r_s = -0.88$, P <0.01) . The possibility that the declines of these two species were the result of poisoning by organochlorine insecticides was not discussed by Ash and it is worthwhile examining his observations in this regard. The decline in sightings of sparrow hawks was almost continuous from 1952, *i.e.*, the factor(s) operating were in existence by or before 1952. In the peregrine the onset of the decline is less clear-cut, one may say either that it occurred during 1953/54, with a partial recovery in 1955, or that it occurred during 1955/56. DDT and lindane have been used in Britain since about 1945/46, the cyclodiene insecticides since 1955 and consequently it may be argued that the decline of these sparrow hawks was brought about by DDT and/or lindane, and that the peregrine decline was caused either by the cyclodienes (assuming 1955 as the onset of decline) or a combination of DDT and cyclodienes (assuming 1953 as the onset) . The occurrence of high residues in pigeons, arising from eating dressed grain, is consistent with the postulated involvement of organochlorine insecticides in the decline of the Peregrine in this particular area of southern England. However, proof of the hypothesis requires evidence that contaminated pigeons were available in that area and that the level of contamination was sufficient enough to affect the peregrines. This additional evidence is not available in this case, and no firm conclusions can be drawn concerning the influence of organochlorine insecticides.

Changes in the breeding status of diurnal birds of prey in Europe were summarized by Ferguson-Lees (22) . Of the thirty-seven species of birds of prey in Europe, twenty were said to have decreased markedly, a further five to have probably declined markedly, nine to be showing no significant change of status, and three to have increased. The grounds for these conclusions are not discussed. A list of various causes of the decline and increases was suggested, man being directly or indirectly responsible for all of them. Ferguson-Lees gives them as shooting and trapping, poisoning, modern hygiene, habitat destruction, adaptability, climate, and interspecific competition. In this chapter we are only concerned with one of these classes of causes, namely poisoning. No evidence is produced by Ferguson-Lees for implicating agricultural chemicals in the declines, but reference is made to an article by Cramp (14). This latter paper is mainly concerned with residues of toxic chemicals. However, Cramp compared the numbers of nest record cards for the kestrel in South and East Britain with those for the rest of Britain. The trends in the annual numbers of nest records for the two areas tend to diverge, but it is difficult to test the statistical significance of the divergence of the two sets of data in the form in which it is published. However, the decline in the South and East is not statistically significant, and the apparent increase in the remainder of the country is of doubtful significance $(P = 0.1)$. Even if there is a significant difference between the two areas there is no necessary implication (as suggested by Cramp) that the difference arises from differences in the usage of organochlorine insecticides in the South and East as compared with the remainder of Britain — the changes may be a reflection of changes in numbers of field voles in the two areas rather than changes in the total population of kestrels.

The status of the peregrine falcon in Britain has been studied intensively since 1961 by Ratcliffe (55, 56, 57). It is perhaps ironical that the initial survey in 1961 to 1962 was done as a result of claims that peregrines were preying upon racing pigeons. The resultant information on population of this raptor is now proving most valuable, however, in another field of enquiry: the relationship between the use of organochlorine insecticides and the status of the peregrine. In the context of this paper the following conclusions by Ratcliffe are particularly important. Firstly, the population of peregrines increased rapidly after 1945 following heavy persecution during the period 1939 to 1945. Secondly, a rapid decline, beginning about 1955, occurred in southern England and this decline spread northwards. By 1961 the population was only about 60 per cent of the pre-1939 population. Thirdly, breakage and infertility of eggs, and sterility of the birds, often preceded the actual disappearance of peregrines. Fourthly, the decline continued from 1962 to 1963, but some slight improvement was apparent in 1964 as compared with 1963. Fifthly, no appreciable change in the level of the breeding populations occurred during the period 1964 to 1966. Sixthly, failure of young peregrines from successful eyries to build up the breeding population of depleted areas is probably a reflection primarily of the lack of improvement in the adverse factor which initiated the decline. Finally, circumstantial evidence pointed strongly to agricultural toxic chemicals as the adverse factor and cause of the decline.

The evidence given by Ratcliffe does not seem sufficient to warrant the firm conclusions that he draws. For example, he cites Ferguson-Lees (21) in support of the statement that "several southern districts depleted by wartime persecutions showed only a partial recovery to the prewar level of population." Ferguson-Lees gives only one example of partial recovery, namely that eight to twelve pairs were occupying 16 miles of cliffs in Sussex prewar but only five to six pairs were in occupation in 1954. Ratcliffe then asserts that by 1961 "there were clear signs of decrease in nesting populations in southern England," he cites only Treleaven (82), whose observations were summarized above, in support of this assertion. Ratcliffe's conclusions regarding the population status in southern England may be correct, but the evidence cited is insufficient to establish their general validity. Again, Ratcliffe states that "egg-eating, which I believe to be a first symptom of decline, was first noticed in 1951 and has been a common occurrence ever since." However, Ratcliffe, in an earlier paper (58) reported broken eggs in an eyrie on 18th April, 1949. He gives the rate of egg breakage in 1949-1956 as the following:

1949-50 one eyrie with broken eggs out of thirty-five examined,
1951   four eyries with broken eggs out of nine examined,
1951-56 thirteen eyries out of fifty-nine

He also quotes several examples of this phenomenon in other species of birds. Ratcliffe stresses that the "peculiar incidents" are a recent phenomenon. No evidence in support of this conclusion is given, but Harper Hall (26) reported observing broken eggs in a peregrine eyrie in Montreal in 1942, and an apparent loss of eleven out of fifty eggs from 1940 to 1952 inclusive. Ratcliffe (59) reported breakage and disappearance of eggs from eight sparrow hawk nests out of forty-eight in the period 1943 to 1959 and he also reports similar phenomenon for golden eagles (5 cases between 1941 and 1958). Other examples of egg-breaking in sparrow hawks are reported by Campbell (11), one of the cases being in 1927, the other in 1953. It is obvious that egg-breaking cannot be regarded as a new phenomenon restricted to the post-1945 era, and it is difficult to ascertain whether the rate of egg breaking has increased in recent years as claimed by Ratcliffe. These detailed examples indicate that there are weaknesses in the arguments used by Ratcliffe to support at least some of his conclusions. The other type of evidence used by Ratcliffe concerns the presence of residues in the tissues and eggs of peregrines and this is considered in a later section together with the reported declines in the weight and thickness of eggs of raptors.

Prestt has published the results of an enquiry into the breeding success of some of the British birds of prey, the buzzard, merlin, sparrow hawk, kestrel, barn owl and tawny owl, in the period 1953 to 1963 (53). On the basis of the questionnaires completed by the participants in the enquiry, Prestt reached the following conclusions: firstly, the sparrow hawk and barn owl had decreased widely, the buzzard, merlin and kestrel less so, and the tawny owl only locally; secondly, the distribution of the sparrow hawk, buzzard and kestrel had changed in half or more of the areas for which information was given; thirdly, the decreases had mostly occurred in the 1950s; and fourthly, the decreases were thought by most of the participants to be due to toxic chemicals except in the case of the buzzard and merlin.

The validity of these conclusions is dependent on (a) the accuracy of the data, and (b) the correctness of the inferences drawn from the data. It is difficult to assess the accuracy of the data, but the replies are based upon personal observations by the 238 participants and others they consulted, and upon local bird records. The competence of the participants as observers can be taken for granted, but it is less certain that their estimates of changes in bird numbers during the period can be regarded as either accurate or consistent in the absence of published data. With regard to the inferences drawn from the observations there are good reasons for regarding them as, at best, tentative. For example, Prestt states that the enquiry has "its limitations, as it is necessarily largely based on subjective impressions." Bearing these points in mind two of the conclusions will be discussed with particular regard to the sparrow hawk, kestrel and barn owl. It is stated that the decreases mostly occurred in the late 1950's, and the estimated dates of onset of declines in various areas are given by Prestt in the form of histograms. For these it appears that the rates of geographical spread of the declines in the three species are similar and that the onset of the declines of the sparrow hawk occurred about one year before that of the kestrel and barn owl. It is plausible to suggest a common cause for these declines, and in fact the declines of all three of these species are attributed predominantly to organochlorine insecticides by the participants, but the evidence in support of this conclusion is not given, and cannot, therefore, be assessed. However, a comparison of the declines of the sparrow hawks, as assessed by the participants in this enquiry, with the sightings per 100 hours observations given by Ash (5) indicates either a lack of concordance between different observers or a difference in the time of onset of the declines between Hampshire (the area in which Ash made his observations) and the rest of Britain. Ash also reported the numbers of sparrow hawks killed each year in the observation area from 1949 to 1959. It is noteworthy that these indicate 1953 as the date of onset of the decline in the numbers killed per year. This estimate of the date of onset of the declines is consistent with the number of sparrow hawks sighted per one hundred hours (see above). On the other hand, the nest records for the sparrow hawk in southeast England are not significantly correlated with the numbers of sparrow hawks seen per 100 hours by Ash in Hampshire, nor with the numbers killed in that area, whereas the sightings and numbers killed are correlated ($r_s = 0.88$, $P < 0.01$). With regard to the kestrel the relationship between the numbers sighted by Ash and the numbers of nest record cards for southeast England is not significant.

A study of buzzards in the New Forest area has been made by Tubbs since 1962 (83). At least thirty-three or thirty-four pairs were constantly present during the 5 years 1962 to 1966; the sizes of the clutches were small and the output of young was low, with failure to hatch and egg-breaking common. No correlation was found between breeding success and the known factors which might adversely affect it.

This review of the evidence regarding the population and breeding status of birds of prey has purposely been very critical: the consistency between the evidence produced by various authors and the conclusions they have drawn have been investigated, and the consistency of the various types of evidence with one another has also been examined. It is highly probable that significant decreases in the numbers of sparrow hawks and peregrine falcons have occurred in southern and eastern Britain during the last fifteen years or so. The significance of the relationship between these decreases and the periods of usage of organochlorine insecticides cannot be tested rigorously, as the published information on populations is either too fragmentary or imprecise, and in some cases the various reports are contradictory. The population status of the kestrel is more difficult to assess in view of the apparent cyclical fluctuations in the records for this species (64), and the difficulty is increased by our lack of knowledge of the nature of the fluctuations, e.g., do they represent real cyclical changes in numbers or are they a reflection of temporary changes in the geographical distribution of a mobile species? (I am indebted to Dr. D. W. Snow for this suggestion.) Finally, the causes of the fluctuations, whatever their nature, are not known. The annual records for the barn owl and tawny owl show similar cyclical fluctuations which are approximately synchronous with those for the kestrel (64) and these fluctuations make it difficult to assess whether these birds also have shown a significant trend in population status over the last fifteen years.

In view of the gaps in our knowledge on the population status of raptors, and the ambiguities in the evidence, it is necessary to consider another and independent type of information, namely, residues of organochlorine insecticides in raptors and other birds, with a view to assessing the possibility that these compounds have adversely affected birds in general and raptors in particular.

Residues of Organochlorine Insecticides in Birds and Their Eggs

General reviews of the occurrence of residues in British birds and their eggs have been given by Moore (43) and Robinson (64).

It is obvious that the value of the published analytical results is dependent upon the reliability of the analytical methods. Some discussion of the problems encountered in the determination of residues of organochlorine insecticides in animal tissues is therefore appropriate at this point. All the results summarized in Tables 7-II to 7-VI were obtained by analytical methods based upon gas-liquid chromatography in combination with an electron capture detector (25); a subtractive type clean-up procedure (20, 63) being used. There are a number of difficulties in the various stages of the analytical method, which may give rise to negative or positive biases in the results (62, 63). The accuracy of an analysis depends upon two major factors, firstly, the response of the final detection step must correspond solely to the compound being determined, i.e., we require either a detector which is specific for the particular compound, or a procedure which separates the compound of interest from all other compounds, so that the response of a nonspecific detector corresponds to only one compound. The uniqueness of the answer given by present techniques cannot be established *a priori* and a number of techniques have been used to confirm the identity of the component detected by gas-liquid chromatography in an attempt to overcome this difficulty. Whether these other techniques increase the information on identity requires more consideration than is usually given to the interpretation of the experimental results: the amount of information given by a combination of techniques is not necessarily a simple additive function of the information given by the individual techniques, and the independence of the various types of information is important in this regard (70). A combination of gas-liquid chromatographic retention times using polar and nonpolar stationary phases, R_F values found by thin-layer chroma-

TABLE 7-II

CONCENTRATIONS OF ORGANOCHLORINE INSECTICIDES IN TISSUES OF DEAD KESTRELS

Period of Collection	Tissue	Statistic	Concentrations of Organochlorine Insecticides, ppm				Other Organochlorine Insecticides, and Other Remarks
			BHC	Heptachlor Epoxide	HEOD	Total DDT-type Compounds	
Spring, 1962 (85)	Liver	N	2	2	2	NA	Residues of DDT-type compounds not reported; HEOD concentrations may be too high (analytical confusion with pp'-DDE)
		GM	1.7	37.5	14.5	NA	
		AM	14.5	38.0	23.5	NA	
		Range	nd-28.8	32.0-44.0	5.0-42.0	NA	
Sept. 1962 - July, 1963. (15, 16)	Liver/ (Viscera)	N	6	6	6	6	0.1 ppm HHDN reported in one specimen, (c), (d).
		GM	0.015	0.21	1.40[a]	2.15	
		CL	nd-0.04	0.02-2.37	0.22-9.05[a]	0.21-21.88	
		AM	0.025	0.87	4.08[a]	6.22	
		Range	nd-0.1	nd-3.0	0.2-14.0[a]	nd-14.0	
August, 1963 - July, 1964 (16)	Viscera/ (Liver)	N	18	18	18	18	Endrin not detected in any specimen, (b), (c).
		GM	0.04	0.21	0.47[a]	0.71	
		CL	0.02-0.08	0.08-0.55	0.20-1.07[a]	0.34-1.50	
		AM	0.10	1.48	1.06[a]	1.59	
		Range	nd-0.4	nd-14.0	nd-4.4[a]	nd-7.6	
1963 (87)	Liver	N	16	16	16	16	Published data not suitable for calculation of geometric means, (b), (c), (d).
		AM	NA	3.0	4.3	9.2	
		Range	NA	nd-40.0	0.2-15.0	nd-45.0	
1964 (87)	Liver	N	6	6	6	6	Published data not suitable for calculation of geometric means, (b), (c), (d).
		AM	NA	1.4	3.5	6.7	
		Range	NA	0.1-7.0	0.3-17.0	nd-26.0	

TABLE 7-III

CONCENTRATIONS OF ORGANOCHLORINE INSECTICIDES IN TISSUES OF
DEAD SPARROW HAWKS AND PEREGRINES

Species and Period of Collection	Tissue	Statistic	Concentrations of Organochlorine Insecticides, ppm				Other Organochlorine Insecticides, and Other Remarks
			BHC	Heptachlor Epoxide	HEOD	Total DDT-type Compounds	
Sparrow Hawk, Sept. 1962 - July, 1963 (15, 16)	Liver/ (Muscle)	N	5	5	5	5	(b), (c), (d).
		GM	0.02	0.02	0.64[a]	3.40	
		CL	0.01-0.07	0.01-0.07	0.22-1.91[a]	0.66-17.51	
		AM	0.03	0.03	0.86[a]	6.58	
		Range	nd-0.1	nd-0.1	0.2-2.1[a]	0.8-21.2	
Sparrow Hawk, 1963 (87)	Liver	N	8	8	8	8	Published data not suitable for calculation of geometric means, (b), (c), (d).
		AM	NA	1.0	2.9	17.4	
		Range	NA	nd-6.0	0.4-8.6	0.8-48.0	
Sparrow Hawk, August, 1963 - July, 1964 (16)	Viscera	N	7	7	7	7	Endrin not detected in any specimen (b), (c).
		GM	0.05	0.45	2.09	3.99	
		CL	0.01-0.27	0.09-2.26	0.94-4.64	0.64-24.92	
		AM	0.20	1.07	2.74	9.02	
		Range	nd-0.80	nd-2.80	0.56-5.50	nd-29.5	
Sparrow Hawk, 1964 (87)	Liver	N	2	2	2	2	Published data not suitable for calculation of geometric means, (b), (c), (d).
		AM	NA	0.2	1.4	5.8	
		Range	NA	nd-0.4	1.4-1.4	nd-9.1	
Peregrine Falcon		N	4	4	4	4	(b), (c), (d).
		GM	0.04	0.55	1.78	20.1	
		CL	nd-0.5	0.04-7.1	0.5-6.3	2.7-148	
		AM	0.10	1.20	2.2	33.4	
		Range	nd-0.3	0.1-1.5	0.6-4.0	6.3-7.0	

TABLE 7-IV

CONCENTRATIONS OF ORGANOCHLORINE INSECTICIDES IN TISSUES OF
DEAD EAGLES, BUZZARDS AND MERLINS

Species and Period of Collection	Tissue	Statistic	Concentrations of Organochlorine Insecticides, ppm				Other Organochlorine Insecticides, and Other Remarks
			BHC	Heptachlor Epoxide	HEOD	Total DDT-type Compounds	
Golden Eagle, Sept. 1962 - July, 1963 (15)	Liver	N	2	2	2	2	(b), (c), (d).
		GM	0.014	nd	0.04	0.17	
		AM	0.015	nd	0.05	0.18	
		Range	nd-0.02	nd	nd-0.08	0.1-0.2	
Buzzard Sept. 1962 - July, 1963 (15)	Liver	N	1	1	1	1	(b), (c), (d).
		AM	nd	12.0	4.0	nd	
Buzzard August, 1963 - July, 1964 (16)	Liver	N	3	3	3	3	0.34 ppm endrine in one specimen (b), (c).
		GM	0.03	0.04	0.19	0.24	
		CL	nd-0.3	nd-1.2	nd-24	nd-5.4	
		AM	0.04	0.08	0.4	0.4	
		Range	nd-0.1	nd-0.2	nd-0.7	nd-0.7	
Merlin (16)	Liver	N	1	1	1	1	(b), (c), (d).
		AM	15.0	nd	nd	5.0	
Merlin (16)	Liver	N	1	1	1	1	(b), (c), (d).
		AM	0.47	5.8	3.2	2.3	

TABLE 7-V

CONCENTRATIONS OF ORGANOCHLORINE INSECTICIDES IN TISSUES OF
DEAD BARN OWLS AND TAWNY OWLS

| Species and Period of Collection | Tissue | Statistic | Concentrations of Organochlorine Insecticides, ppm | | | | Remarks |
			BHC	Heptachlor Epoxide	HEOD	Total DDT-type Compounds	
Barn Owl, Sept. 1962 - July, 1963 (15)	Liver/(Viscera)	N	10	12	10	10	Residues of 0.1 HHDN and 0.2 ppm heptachlor reported for one specimen, (d).
		GM	nd	0.17	1.41	1.13	
		CL	nd	0.05-0.6	0.44-4.5	0.3-4.2	
		AM	nd	0.45	2.5	3.1	
		Range	nd	nd-1.1	nd-7.0	nd-14.0	
Barn Owl, August 1963 - July, 1964 (16)	Liver/(Viscera)	N	12	12	12	12	No residues of endrin detected (b), (c), (d).
		GM	0.04	0.18	0.73	1.56	
		CL	0.02-0.11	0.05-0.69	0.19-2.78	0.36-6.77	
		AM	0.10	0.87	2.47	12.54	
		Range	nd-0.3	nd-4.5	nd-7.7	nd-113	
Tawny Owl, Sept. 1962 - July, 1963 (15,16)	Liver/(Viscera)	N	10	10	10	10	(b), (c), (d).
		GM	0.02	0.06	0.09	1.28	
		CL	nd-0.04	0.02-0.14	0.04-0.22	0.27-6.06	
		AM	0.05	0.12	0.16	5.43	
		Range	nd-0.4	nd-0.5	nd-0.5	nd-15.0	
Tawny Owl, August 1963 - July, 1964 (16)	Liver/(Viscera)	N	8	8	8	8	(b), (c), (d).
		GM	0.05	0.17	0.21[a]	1.44	
		CL	nd-0.20	0.03-1.10	0.05-0.92[a]	0.12-17.0	
		AM	0.22	2.87	0.64[a]	16.0	
		Range	nd-1.4	nd-22.0	nd-3.2	nd-88.6	

TABLE 7-VI

CONCENTRATIONS OF ORGANOCHLORINE INSECTICIDES IN TISSUES OF
DEAD LITTLE OWLS, LONG-EARED OWLS AND SHORT-EARED OWLS

Species and Period of Collection	Tissue	Statistic	Concentrations of Organochlorine Insecticides, ppm				Remarks
			BHC	Heptachlor Epoxide	HEOD	Total DDT-type Compounds	
Little Owl, Sept. 1962 - July, 1963 (15, 16)	Liver/ (Muscle)	N	6	6	6	6	2.8 ppm heptachlor reported in one specimen, (b), (d).
		GM	nd	0.02	0.04	1.44	
		CL	nd	0.02-0.03	nd-0.15	0.32-6.52	
		AM	nd	0.02	0.58	2.91	
		Range	nd	nd-0.03	nd-0.58	0.3-7.9	
Little Owl, August 1963 - July, 1964 (16)	Liver	N	2	2	2	2	(b), (c), (d).
		GM	nd	0.03	0.09	0.55	
		AM	nd	0.04	0.20	1.42	
		Range	nd	nd-0.05	nd-0.37	0.1-2.7	
Long-Eared Owl, August 1963 - July, 1964 (16)	Viscera	N	2	2	2	2	(b), (c), (d).
		GM	0.06	1.12	1.90	0.67	
		AM	0.17	7.05	3.26	2.32	
		Range	nd-0.3	0.09-14.0	0.61-5.9	0.1-4.5	
Short-Eared Owl, Sept. 1962 - July, 1963 (16)	Liver/ (Muscle)	N	2	2	2	2	(b), (c), (d).
		GM	0.045	nd	0.25	3.12	
		AM	0.06	nd	0.35	3.10	
		Range	nd-0.1	nd	0.10-0.60	2.8-3.4	
Short-Eared Owl, August 1963 - July, 1964 (16)	Viscera	N	3	3	3	3	(b), (c), (d).
		GM	0.06	4.95	1.06	0.69	
		AM	0.12	5.37	7.24	0.72	
		Range	nd-0.3	3.7-8.6	nd-18.5	0.5-1.0	

tography, and one or more appropriate chemical reactions, appears
to provide a satisfactory compromise between an unequivocal
proof of identity and a set of conveniently determined experimental parameters. The second requirement of an accurate analytical method is that the amount of the compound used in the
detection device must be an accurately known fraction of the total
amount originally present in the sample. This requires that the
whole of the compound in the sample be completely extracted
and that no significant unmeasured losses occur during the "cleanup" procedures. Examples of the difficulties experienced in extraction and cleanup procedures have been given by Richardson (62)
and Robinson (63).

An illustration of the difficulties faced by the analysts is given
by the reports of the presence of unidentified organohalogen compounds in wildlife specimens (28, 69, 74). Some, at least, of these
compounds may be polychlorinated biphenyls (29, 34), some of
which have gas-liquid chromatographic retention times similar to
those of various organochlorine insecticides (29, 63), particularly
those of the DDT-group. False values for the concentrations of
pp'-DDT and pp'-TDE may easily arise unless the analyst is aware
of the interference arising from polychlorinated biphenyls also
present in the sample. The determination of HEOD is also potentially open to intereference by these compounds unless the
pre-GLC cleanup procedure has incorporated a step for the
separation of the polar and nonpolar organochlorine insecticides
(63).

A considerable increase in the certainty of identification is
obtained if the infrared spectrum and/or mass spectrum of the
component can be studied. In the following discussion it is assumed that the concentrations reported in the various surveys are
analytically reliable. Further, in view of the information available
on the population status of the commoner British birds it is only
considered necessary to discuss the residues in two classes of birds,
namely, grain-eating birds and raptors, hawks, falcons and owls,
since specimens of these have been found to contain the highest
residues of cyclodiene insecticides and since it has been suggested

the information on population status indicates the possibility of new inimical factors in the environment.

The effects of cyclodiene compounds, used as seed-dressing agents, upon granivorous birds began to raise concern in the late 1950's and as a consequence of investigations by Turtle *et al.* (84) and Murton and Vizoso (48) it was agreed by the parties concerned that the use of these compounds as seed-dressing agents should be confined to seed sown in winter. The particular circumstances which occur in Britain in the Spring, namely, shortage of food and the consequent increased probability that seed sown by farmers will be eaten by pigeons, etc., and the proximity of many roosts to wheat growing areas combined to make grain seed dressed with organochlorine insecticides a particular hazard to granivorous birds. The conclusion that a causal link exists between the use of cyclodiene insecticides as seed-dressing agents and the death of considerable numbers of grain-eating birds, particularly wood pigeon (*Columba palumbus*), pheasant (*Phasianus colchicus*) and partridge (*Perdix perdix*), is supported by the following observations. Firstly, incidents involving granivorous birds are correlated in time and place with the usage of cyclodiene insecticides as seed-dressing agents in Britain; secondly, the number of incidents have declined dramatically since the cessation of the use of seed, dressed with these compounds, in the Spring; and thirdly, the concentrations of HEOD, heptachlor epoxide and lindane in the tissues of many dead granivorous birds are similar to those in the tissues of experimentally poisoned birds. The latter point is illustrated in Figure 8-5 which is based upon the results of Turtle *et al.* (85). The concentrations of HEOD and heptachlor epoxide in about 30 per cent of the granivorous birds were similar to those found in experimentally poisoned pigeons (66). The concentrations in the remaining 70 per cent of the birds were similar to those found in birds that had been shot, and it is probable that these birds died from other causes.

The possibility that birds preying upon pigeons and other grain-eating birds were ingesting relatively large amounts of these compounds, and the reports of declines in the numbers of birds of prey, particularly sparrow hawks and peregrine falcons, has resulted in an intensive investigation of the residues in raptorial species. The results of the various surveys are summarized in Tables 7-II, 7-III and 7-IV (diurnal raptors) and Tables 7-V and 7-VI (nocturnal raptors). The latter prey upon mammals rather than birds, but the concept of biological amplification of residues through food chains is still applicable. The toxicological significance of the residues in birds is a controversial one (73) and the evaluation of the concept of toxic concentrations in tissues as indices of death by poisoning, requires discussion.

Concentrations of Organochlorine Insecticides in Tissues as Criteria of Death by Poisoning

Only a small amount of work on the concentrations of lindane and heptachlor epoxide in the tissues of experimentally poisoned birds has been published (84), but more detailed investigations have been made on the residues found in birds that have died after exposure to DDT or dieldrin.

The concentrations of DDT-type material in the tissues of adult house sparrows (*Passer domesticus*) that had been poisoned by feeding with diets containing various quantities of DDT were determined by Bernard (7). The material incorporated in the diets of the birds was not specified in detail, and it is assumed that technical DDT was used; technical DDT consists mainly of pp'-DDT with up to 30% op'-DDT. The Schechter-Haller method was used to analyze the tissues, a method which is not specific for pp'-DDT since a number of compounds form nitration products which, in the presence of alkali, have spectrophotometric absorption spectra similar to those of pp'-DDT. Interference may arise from pp'-DDD, pp'-DDE and op'-DDT. Consequently the concentrations of DDT reported by Bernard may correspond to a mixture of pp'-DDT, pp'-DDD, pp'-DDE and op'-DDT. This lack of specificity is a considerable drawback to using these results as indicators of toxic concentrations in tissues. The range of concentrations at death in a particular tissue, found by Bernard, is quite large and this scatter of results poses problems in interpretation. Boykins has reported the results obtained in similar trials in which DDT was fed to house sparrows (8). Boykins' results are lower than those of Bernard. Another trial, in which cowbirds (*Molothrus ater*) were fed a diet containing 500 ppm pp'-DDT, has been reported by Stickel *et al.* (79). These workers used gas-liquid chromatography in the analysis of the tissues of poisoned birds, and determined the concentrations of pp'-DDT, pp'-DDD and pp'-DDE in brain and liver. They concluded that the concentration of pp'-DDT + pp'-DDD in the brain of a number of species were quite similar at death, but as the range of results was quite wide, a single concentration indicative of death could not be established. However, when the concentration of pp'-DDT + pp'-DDD in the brain of a dead animal was in the range found in the experimentally poisoned animals, it was plausible, they suggested, to conclude that these animals had been at least seriously endangered by poisoning. A tentative lower limit of 30 ppm pp'-DDT + pp'-DDD in the brain was proposed as a criterion for assessing the possibility of poisoning. Stickel *et al.* (79) suggested that residues in the liver are less suitable than those in the brain for interpreting causes of death since they found that the residues appeared to be more variable than those in the brain, and are also dependent upon the duration of the exposure.

Concentrations of HEOD in the tissues of experimentally poisoned birds have been studied by Turtle *et al.* (84), and Robinson *et al.* (66), and some preliminary studies of the effects of mixtures of organochlorine insecticides have also been made (67). As regards the concentrations of HEOD in tissues (liver and brain) at death, three bird species, quail, pigeons and house sparows, show no significant interspecies differences (66, 72). This conclusion in the case of HEOD is similar to that of Stickel *et al.* (78, 79) in the case of residues of DDT-type compounds in the brain.

The results obtained in these experiments are given in Tables 7-VII, 7-VIII, and 7-IX. An attempt has been made to provide a rational basis for the use of tissue concentrations as criteria of death by poisoning (66) and the proposed model is shown diagrammatically in Figure 7-5. A lower limit for the concentrations of HEOD in the liver that may be used in the diagnosis of poisoning by dieldrin has been proposed — namely 20 ppm (66).

TABLE 7-VII

CONCENTRATIONS OF DDT-TYPE COMPOUNDS IN THE TISSUES
OF EXPERIMENTALLY POISONED BIRDS

Species	Treatment	Compounds Determined	Statistic	Concentration of DDT-type compounds in tissues, ppm		
				Liver	Muscle	Brain
House Sparrows (7)	Dietary intake of DDT	DDT-type compounds[e]	N	20	16	55
			GM	261	181	97
			CL	211-324	147-222	81-116
			AM	289	199	107
			Range	96-576	97-357	25-200
House Sparrows (8)	Dietary intake of DDT	DDT-type compounds[e]	N	37	—	37
			GM	42.8	—	27.9
			CL	37.1-49.5	—	24.7-31.5
			AM	46.4	—	29.6
			Range	12-79	—	9-51
Cow Birds (79)	Dietary intake of DDT	pp'-DDT[f] pp'-DDD	N	17		17
			AM	415		66
			Range	(154-751)[a]	8	35-99
Robins (7)	Dietary intake of DDT	DDT-type compounds[e]	N	8	8	8
			GM	173	85.4	80.2
			CL	123-243	55-133	68-95
			AM	186	97.5	81.7
			Range	104-313	50-207	61-116
Bald Eagles (78)	Dietary intake of DDT	DDT-type compounds[e]	N	3	5	5
			GM	428	143	73
			CL	132-1390	76-270	59-92
			AM	463	159	74
			Range	280-718	73-291	58-86

TABLE 7-VIII

CONCENTRATIONS OF HEOD IN THE TISSUES OF
EXPERIMENTALLY POISONED BIRDS

Species	Treatment	Analytical Method	Statistic	Concentration of HEOD in tissues		
				Liver	Muscle	Brain
Feral Pigeons (84)	Oral dosing with aldrin or dieldrin	h	N	—	30	—
			GM	—	18.3	—
			CL	—	15.3-21.9	—
			AM	—	20.4	—
			Range	—	6.2-39.0	—
Domestic Pigeon (66)	Dietary intake or oral dosing with HEOD	f	N	20	—	19
			GM	45.6	—	20.0
			CL	37.5-55.5	—	18.1-22.0
			AM	45.6	—	19.3
			Range	26.2-71.9	—	10.1-26.9
Quail (66)	Dietary intake or oral dosing with HEOD	f	N	36	—	65
			GM	40.0	—	17.4
			CL	34.7-46.2	—	15.9-19.0
			AM	43.4	—	18.3
			Range	18.1-112.7	—	3.1-37.0
House Sparrows (66)	Dietary intake of HEOD	f	N	19	19	19
			GM	44.7	14.7	20.0
			CL	38.4-52.2	11.9-18.1	17.6-22.7
			AM	—	—	—
			Range	20.8-74.3	6.5-36.3	8.9-28.9

TABLE 7-IX

CONCENTRATIONS OF ORGANOCHLORINE INSECTICIDES IN THE TISSUES OF
EXPERIMENTALLY POISONED QUAIL AFTER SIMULTANEOUS TREATMENT WITH
HEOD + ONE OR MORE ORGANOCHLORINE INSECTICIDE

(all analyses by gas-liquid chromatography) (67)

Treatment	Tissue	Statistic	Concentrations of Organochlorine Insecticides in Tissues, ppm					
			HEOD	Lindane	Heptachlor Epoxide	pp'-DDT	pp'-DDD	pp'-DDE
Oral doses of HEOD + γ-BHC	Liver	N	17	17	—	—	—	—
		GM	58.0	0.34				
		CL	46.1-73.0	0.17-0.68				
	Brain	N	17	17	—	—	—	—
		GM	12.1	0.84				
		CL	9.1-16.0	0.30-2.35				
Oral doses of HEOD + heptachlor	Liver	N	19	—	19	—	—	—
		GM	42.1		9.45			
		CL	36.3-48.9		6.9-13.0			
	Brain	N	19	—	19	—	—	—
		GM	11.52		2.5			
		CL	10.1-13.2		1.9-3.3			
Oral doses of HEOD + pp'-DDT + pp'-DDE	Liver	N	17	—	—	17	17	17
		GM	76.6			4.6	20.5	58.1
		CL	54.9-106.8			1.8-11.9	14.9-28.0	40.8-82.8
	Brain	N	17	—	—	17	17	17
		GM	16.4			4.1	0.6	4.0
		CL	13.7-19.8			3.1-5.3	0.4-0.8	2.4-6.7
Oral doses of HEOD + heptachlor + pp'-DDT + pp'-DDE	Liver	N	28		28	28	28	28
		GM	63.2		9.2	0.4	14.2	31.6
		CL	55-73		7.4-11.4	0.2-1.0	11.5-17.4	25.1-39.7

The concentrations of HEOD in the tissues of Quail that died after treatment with mixtures of two or more insecticides (see Table 7-IX) are particularly interesting. The purpose of these particular experiments was a restricted one, to study the variations in the concentration of HEOD at death when sublethal amounts of lindane, heptachlor or DDT-type compounds were administered simultaneously with HEOD. No attempt has been made to investigate mixtures containing different ratios of the various organochlorine insecticides with a view to determining isoboles. In the case of the residues in the liver, the concentration of HEOD at death is not reduced by the presence of sublethal concentrations of other organochlorine insecticides, if anything there is a tendency for the concentrations of HEOD in tissues at death to be increased by simultaneous exposure to DDT-type compounds. The residues of HEOD in the brain, however, appear to be reduced when there is a simultaneous exposure to other organochlorine insecticides. The ratios of the concentrations of the various organochlorine insecticides in the brain and liver are interesting; the concentration of lindane in the brain is greater than that in the liver but the reverse is found with the other compounds. Certain aspects of the relevance of the results obtained in laboratory experiments as criteria for diagnosing death by poisoning in the case of wild birds have been discussed elsewhere (64). The environmental conditions of laboratory birds are quite different from those of wild birds and the possibility of interaction between toxicants and such variables as temperature, nutritional status or disease, must be borne in mind, particularly interactions of the additive or synergistic type. Interactions of the latter type may result at death in concentrations in tissues which are considerably lower than those found in birds dying as a result of an exposure of healthy birds to only one toxicant under laboratory conditions. However, as shown in Figure 7-6, there is a good concordance between the concentrations of HEOD + heptachlor epoxide in the tissues of many granivorous birds which had died in circumstances indicative of poisoning by these compounds, and the concentrations of HEOD in experimentally poisoned birds. A similar conclusion is reached if we compare the results obtained

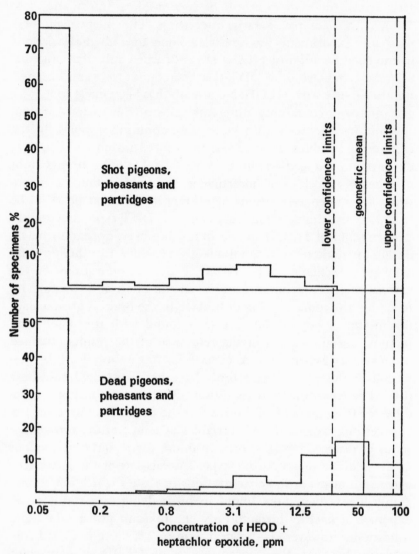

FIGURE 7-5. Concentrations of HEOD + heptachlor epoxide in the livers of shot birds and birds found dead (P=0.95) during spring 1961 to spring 1964 (85).

by Bernard (7) and Boykins (8) for dead wild birds, collected in circumstances indicative of poisoning by DDT, with those for experimentally poisoned birds (see Fig. 7-7).

The residues in the majority of species of birds in Great Britain are small compared with those found in experimentally poisoned birds (64) and in view of the conclusions drawn from the

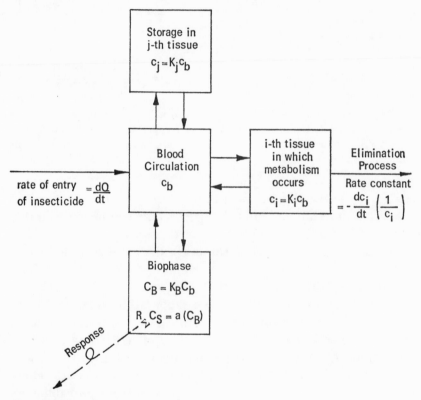

$R\Sigma$ = Receptor site
c_b = concentration of insecticide in the circulating blood
c_j = concentration in a storage tissue
c_i = concentration in a tissue in which there are metabolizing enzymes
C_B = concentration in biophase in which receptor sites are situated
C_S = concentration of insecticide at the receptor site, and is a function of C_B

FIGURE 7-6. Pharmacodynamic model of the relationship between the concentration of HEOD in tissues and toxicological effects.

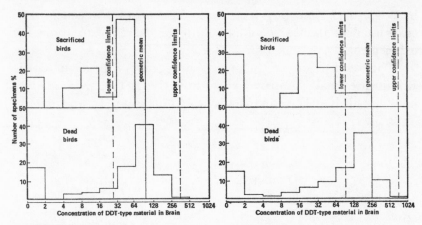

FIGURE 7-7. Concentration of DDT-type material in the brains and livers of dead birds (7, 8). Residues in the tissues of experimentally poisoned birds are shown. Geometric mean, ————, upper and lower confidence limits of samples: (P=0.95) -----.

studies of the population status of the commoner birds, a detailed discussion of residues of raptorial species *only* will be given in this paper, (residues in granivorous birds having been discussed above). The results of various surveys of the residues in the tissues of dead raptors are summarized in Tables 7-II through 7-VI. The residues of pp'-DDT and pp'-DDD are generally only a small proportion of the total for DDT-type compounds. Consequently, the assessment of their toxicological significance, using the value of 30 ppm, which is suggested by Stickel, *et al.* (78, 79) as the concentration to be used in the diagnosis of poisoning by DDT, will overestimate the toxicological significance of the residues. However, even in comparison with this criterion it seems probable that poisoning by DDT-type compounds is not a significant cause of death. The residues of lindane are also very small and are considered to be insignificant. Only the residues of HEOD and heptachlor epoxide are sufficient, *prima facie,* to warrant detailed evaluation. The toxicological significance of the residues can only be assessed in the light of presently available evidence on toxic concentrations in the tissues of experimentally poisoned birds. These are available only for house sparrows, robins and cowbirds (DDT-type compounds) , house sparrow, quail and pigeons (HEOD residues) . Is the use of criteria based on these species a valid procedure in the case of raptors? Obviously the criteria can only be defended by analogical arguments and these can, at the best, be regarded only as suggestive and not conclusive.

It has been suggested (33) that the concentrations of HEOD, indicative of the death by poisoning of hawks and falcons are significantly lower than those for pigeons, etc. This conclusion is based upon the results of analyses of two lanners (*Falco biarmicus*) and five peregrine falcons. The two lanners were captive birds being trained for falconry; four of the peregrines were found dead in Britain (one appeared to be a bird that had escaped from captivity) , and the fifth peregrine had died in captivity in Holland. In the postmortem examination of the two lanners no evidence of disease was found: there were no gross lesions indicative of virus, bacterial or parasitic infections. It is not clear from the report whether any attempt was made to detect bacterial infection by culturing. Small amounts of BHC residues were found, but the residues consisted predominantly of pp'-DDE, HEOD and heptachlor epoxide. The results (ppm in the liver) are summarized below:

	pp'-DDE	*HEOD*	*Heptachlor Epoxide*
Lanner 1	2.2	3.4	1.9
Lanner 2	3.8	6.1	1.4

Postmortem examination of one of the dead peregrines showed the presence of *Escherichia coli* and large round worms (*Serratospiculum sp.*) in the abdominal air sacs. This bird also had an ulcerated oesophagus. A second peregrine was also found to have air sac worms (*Serratospiculum sp.*) in the body cavity, and amorphous yellow nodules, of unknown nature, were found in the peritoneum. The postmortem of the third specimen showed that it was thin and no body fat was found (it is not clear whether a pathological examination was made) . No postmortem details are given for the fourth British bird. Parasitological and pathological examination findings in the Dutch bird were negative, and this bird had died in convulsions. The residues (ppm) found in the livers of the peregrines were as follows:

	pp'-DDE	HEOD	Heptachlor Epoxide	Lindane	pp'-DDD
Peregrine 1[a]	6.3	0.6	0.2	—	—
Peregrine 2[a]	50.0	2.2	3.0	0.3	—
Peregrine 3[b]	70.0	4.0	1.5	—	—
Peregrine 4[c]	7.3	1.9	0.1	—	—
Peregrine 5[d] (Holland)	60.0	9.3	—	2.3	4.4

[a]positive postmortem findings of bacteriological and/or parasitological infection.
[b]no postmortem findings available.
[c]may be a bird that had escaped from captivity.
[d]postmortem findings negative.

Jefferies and Prestt suggest that the concentrations of pp'-DDE in these birds were not toxicologically significant. This conclusion is probably correct but the basis for it is interesting. Their argument is an analogical one: namely the low oral toxicity of pp'-DDE in experimental birds (quail, etc.) and the high residues of pp'-DDT + pp'-DDE in the liver of 1 experimental bird (Bengalese finches, (*Lonchura striata*), i.e. they assume no interspecies difference in susceptibility to pp'-DDE between falcons and the experimental bird species. They further conclude that peregrines 1 and 4 died of causes other than poisoning (peritonitus and lack of food respectively) whereas the deaths of peregrines 2, 3 and 5 were only explicable (sic) on the grounds of poisoning. Consequently they further concluded that the concentrations of HEOD, or HEOD + heptachlor epoxide, in the liver of falcons at death following poisoning are in the ranges 2.2 to 9.3 and 5.2 to 9.3 respectively, and that large falcons are four times more sensitive to organochlorine insecticides than pigeons. This last inference is based on the comparison of the residues found in these falcons, presumed to have been poisoned, with those in the tissues (breast muscle) of experimentally poisoned pigeons (84). In the case of pp'-DDE they assume that the residues in livers of poisoned falcons are similar to those found in experimentally poisoned birds, whereas because the residues of HEOD in the falcons are only about one quarter of those found in experimentally poisoned pigeons they conclude that falcons are more sensitive to the cyclodiene type insecticides. Now one may argue that the absence of interspecies differences in residues of pp'-DDE in tissues following poisoning does not prove that there is a corresponding lack of difference in the case of HEOD, but the evidence they produce does not prove the case either way. A comparison of the results for these five falcons with those found in Dutch falcons is instructive.

Koeman and van Genderen (36) have published the residues found in a few raptors in Holland, and by combining these with other results from Holland (35), it is found that the mean (geometric) concentration of HEOD in the livers of these dead raptors (28 in number) was 13.2 ppm (confidence limits P = 0.95, 9.4 to 18.6). Some of the birds included had died from other causes (e.g., shooting) so that the geometric mean is not representative of poisoned birds only, i.e. the mean concentration in the livers of poisoned raptors in Holland is greater than 13.2 ppm. This does not prove that there is no significant difference between the concentrations of HEOD in the tissues of poisoned raptors and those in the tissues of experimentally poisoned birds of other species, but the geometric mean for the Dutch raptors is significantly higher than that for the five hawks discussed by Jefferies and Prestt (33). This implies that either there is a geographical variation in the concentrations of HEOD in tissues at death (which seems very improbable) or that a second factor, absent in the Dutch birds, was interacting synergistically with HEOD in the British birds, or that the deaths of the five British birds were not caused by poisoning by HEOD. The last alternative seems more likely.

A comparison of the concentrations of HEOD in the tissues of the various raptors examined in the various surveys indicates that significant differences occur, not only between species but also within species. Consequently, if all the birds analyzed had died from poisoning by HEOD we must postulate that the mean sensitivity within a species may vary by a factor of 80 in the kestrel (or, if the highest results given in Table 7-II are rejected as unrepresentative, by a factor of 8), and by a factor of 6 in the sparrow hawk, and that the between species mean sensitivity may vary by a factor of 2600 (800 if the high kestrel results are rejected). Variations in sensitivity of this magnitude seem highly improbable to say the least and it is concluded that it is unlikely that all the

hawks and falcons in Tables 7-II, 7-III and 7-IV died from poisoning by organochlorine insecticides. The same conclusion is
reached when the residues in owls are considered (although the
variations between and within species are less extreme than those
found with hawks and falcons). An attempt to assess the proportion of raptors that had been poisoned has been made by comparing the residues of HEOD + heptachlor epoxide in the tissues of
dead birds with those found in experimentally poisoned birds (of
other species), and with the residues found in the Dutch falcons.
The comparisons are shown graphically in Figure 7-8.

The graphical comparison indicates that the proportion of
British raptors that died from poisoning by HEOD is between 5
and 20 per cent, the balance of the evidence indicating that the
proportion is nearer to 5 per cent than 20 per cent.

The discussion of the residues in British raptors has been confined so far to those found in the livers of the birds. Jefferies and
Prestt (33) also give the residues found in the brain tissues of
four of the falcons which they considered had died from poisoning
by HEOD. In view of the concentrations of HEOD + heptachlor
epoxide in the brains of experimentally poisoned birds (see Table
7-IX) it was considered that the toxicological significance of residues in the falcons' brains should also be evaluated. The concentrations found in the brains were between 2.0 and 7.8 ppm HEOD
and 3.0 to 7.8 ppm HEOD + heptachlor epoxide. It will be seen
that these are about 1/3 of the concentrations found in the experimentally poisoned birds, and it appears that either the falcons are
more sensitive to HEOD than the experimental birds (by a factor
of about 3) or the birds had died from other causes. The latter
conclusion is the more probable in view of the discussion on the
residues in the livers. Effects on reproductive success, of course,
may also be occurring, and it is necessary to consider the residues
found in the eggs of raptors with a view to assessing their implications in relation to reproduction. These residues are discussed in
the following section.

The conclusions reached above are only applicable of course to
the period from 1962 onwards. There is no information on the
residues in the tissues or eggs of raptors prior to 1962. This is a

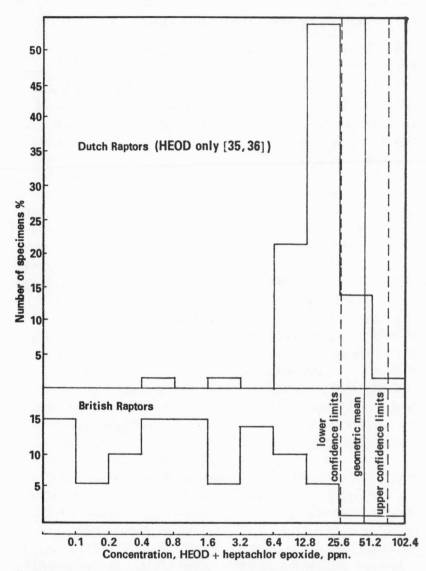

FIGURE 7-8. Concentrations of HEOD + heptachlor epoxide in the tissues (mainly livers) of dead raptors. Mean (geometric) concentration of HEOD in the livers of experimentally poisoned birds (66) are shown ————, upper and lower confidence limits of samples of experimental birds -----.

major gap in our knowledge as we can draw no conclusions on the effects of such residues as were present prior to 1962.

Effects of Sublethal Concentrations of Organochlorine Insecticides

The possible sublethal effects of these compounds has been given considerable attention, as is apparent from the general reviews by Moore (43, 45, 47) for example. The main sublethal effects that have been considered are reduced reproductive success, and behavioral changes that are inimical to the survival of the individual or the species. The effects of organochlorine insecticides upon fertility, numbers of eggs laid, hatching success of the eggs, and post-hatching survival can all be studied in experimental birds, but the study of behavioral changes is much more difficult.

A detailed review of residues of organochlorine insecticides in the eggs of British birds is in hand, and from a comparison of these residues with those in the eggs of experimental birds fed diets containing HEOD, it appears that the residues of HEOD found in the eggs of the majority of bird species are unlikely to be of any significance. The most sensitive of the parameters measured in the experimental birds appears to be the post-hatching survival of the chicks. These conclusions appear to be consistent with the evidence on the population status of most birds. The latter evidence would also appear to indicate that behavioral changes, if any have occurred, have not be inimical to species survival to date. The possibility that one group of birds, the raptors, have undergone a behavioral change as a result of ingestion of organochlorine insecticides has received particular attention in view of the finding of broken eggs in or near their nests (see above), and Ratcliffe (60) has suggested that this is related to the decrease in eggshell weight of certain birds of prey during the period 1945 to 1947. No further change appears to have occurred since 1948 and it would appear that the factor involved was not operating before 1945 and that no additional change has occurred since 1948. If organochlorine insecticides are the causal agents involved, then the date of onset of the decrease in eggshell weight indicates that only DDT or BHC could be involved. Laboratory trials should indicate whether this is a plausible explanation, although the problem of possible interspecific sensitivity may again complicate the interpretation of the results. The possibility that some other environmental factor is involved also requires further investigation. The delays in ovulation which have been observed by Jefferies (32) in birds fed diets containing pp'-DDT are a further indication of potential hazard to birds containing high residues of DDT-type compounds. A change in male wood pigeons, which has been related on circumstantial grounds to organochlorine insecticides, was the marked reduction in gonad development in April, 1961 (41). This association of cause and effect is one which should be susceptible to experimental verification in laboratory birds.

In relation to the possible induction of behavioral changes and/or decline in reproduction in raptors by organochlorine insecticides, it is worthwhile considering the residues in the eggs of peregrine falcons. The results of analyses of such eggs have been reported by Ratcliffe (57). He classified the eyries from which eggs were collected according to their geographical location and their success in rearing young. (Unsuccessful eyries were those whose eggs were broken or failed to hatch, successful eyries reared at least one young). Assuming that the difference between the two types of eyries is the result of pesticide residues (and this is implicit in Ratcliffe's discussion of the residues in eggs) one would expect significant differences in the residues. Ratcliffe states that the difference between the total residues of organochlorine insecticides is suggestive but not statistically significant. Ratcliffe's analysis of the data appears to be deficient in several respects, thus the pp'-DDE content constitutes about 88 per cent of the total, and comparison of total residues only is not appropriate unless it is considered that the residues of pp'-DDE make a significant contribution to the difference in success between the eggs from the two classes of eyries. Secondly, the distribution of the HEOD and heptachlor epoxide residues tend to be positively skew, and thirdly, there appear to be differences between geographical areas. The residues for the individual pesticides in eggs from the two geographical areas have therefore been analyzed statistically, after transformation to their corresponding logarithms, with the further subdivision into successful and unsuccessful categories for one of the geographical areas (all the eyries were classified as successful in the other areas). The concentrations of organochlorine insecticides in the three categories of eggs are given in Table 7-X.

There are no significant differences between the residues in the eggs from successful and unsuccessful eyries in the area (northern England and southern Scotland) in which both types are found. The only significant differences are those between different geographical areas; the residues of pp'-DDE in eggs from the central Scottish Highlands are significantly lower (P = <0.05) than those from the other area. A significant difference (P = <0.05) is also found for total residues but this is merely a reflection of the pp'-DDE residues which constitutes such a large proportion of the total residues. As there were no significant differences between the concentrations of organochlorine insecticides in the eggs from successful and unsuccessful eyries within a geographical area it does not seem plausible to attribute the differences in success to these compounds.

The results of other surveys of the concentrations of organochlorine insecticides in the eggs of hawks and falcons are summarized in Table 7-XI. The significance of these residues can be assessed in two ways, both of which are open to the criticism that interspecific variations in sensitivity cannot be taken into account. The first is the comparison between the residues in these eggs and those of other wild species whose reproductive success seems to have been only slightly affected, if at all, and whose eggs contain larger residues of organochlorine insecticides. Two such species are the shag (*Phalacrocorax aristotelis*) and the heron (*Ardea cinerea*). Residues in eggs of shags have been discussed by Robinson *et al.* (69), and there are a number of reports of relatively high residues in herons' eggs, e.g., the mean (geometric) concentration of HEOD in the eggs of one heronry in eastern England was 3.47 ppm (54). From Table 7-XI it is apparent that there is a mixture of insecticide residue in most of the raptors' eggs, but only the residues of HEOD are, *prima facie,* of potential signifi-

TABLE 7-X

CONCENTRATIONS OF ORGANOCHLORINE INSECTICIDES IN
THE EGGS OF PEREGRINE FALCONS

Organochlorine insecticide	Geographical area in which eggs were collected	Mean (geometric) concentration of insecticides in the two types of eyries, ppm	
		Successful	Unsuccessful
BHC-isomers	Central Scottish Highlands	0.025 (0.006-0.101)	0.119 (0.057-0.247)
	N. England & S. Scotland	0.100 (0.021-0.474)	
pp′-DDT	Central Scottish Highlands	0.072 (0.017-0.318)	0.050 (0.023-0.110)
	N. England & S. Scotland	0.038 (0.007-0.197)	
pp′-DDD	Central Scottish Highlands	0.036 (0.001-0.138)	0.024 (0.012-0.048)
	N. England & S. Scotland	0.021 (0.0051-0.095)	
pp′-DDE	Central Scottish Highlands	4.93* (2.75-8.85)	13.24 (9.73-18.02)
	N. England & S. Scotland	13.65 (7.10-26.22)	
Heptachlor Epoxide	Central Scottish Highlands	0.26 (0.08-0.72)	0.49 (0.27-1.14)
	N. England & S. Scotland	0.79 (0.23-2.78)	
HEOD	Central Scottish Highlands	0.43 (0.19-0.99)	0.46 (0.30-1.39)
	N. England & S. Scotland	0.60 (0.24-1.54)	
Total Residues	Central Scottish Highlands	6.19* (3.49-10.98)	15.24 (11.27-20.61)
	N. England & S. Scotland	15.67 (8.26-29.72)	

Based on results published by Ratcliffe (57).

*Significant difference (P <0.05) between geographical areas.

TABLE 7-XI

CONCENTRATIONS OF ORGANOCHLORINE INSECTICIDES IN THE EGGS OF HAWKS AND FALCONS

Species	Number of Eggs	Statistic	Concentrations of Organochlorine Insecticides, ppm			
			BHC	Heptachlor Epoxide	HEOD	Total DDT-type Compounds
Various (15)	4	GM	0.02	0.14	0.29	5.0
		CL	nd-0.29	nd-6.46	0.02-5.56	1.1-21.5
		AM	0.07	0.84	0.61	6.4
		Range	nd-0.25	nd-0.30	nd-1.0	1.5-10.7
Various (16)	9	GM	0.03	0.15	0.23	1.7
		CL	nd-0.10	0.06-0.36	0.05-1.01	0.7-4.3
		AM	0.11	0.27	1.17	3.2
		Range	nd-0.50	nd-1.30	nd-8.00	0.2-11.8
Various (61)	10	GM	0.08	0.07	0.46	0.71
		CL	0.06-0.10	0.05-0.09	0.17-1.25	0.27-1.89
		AM	0.08	0.08	1.03	1.60
		Range	nd-0.10	nd-0.20	nd-4.20	0.1-7.8
Golden Eagle (40)	10	GM	0.06	0.004	0.59	0.51
		CL	nd-0.11	0.002-0.008	0.21-1.65	0.21-1.26
		AM	0.09	0.006	1.34	0.99
		Range	0.03-0.27	0.001-0.020	0.04-6.90	0.11-3.35
Buzzard (83)	10	GM	—	0.03	0.36	0.82
		CL	—	0.02-0.07	0.13-1.02	0.45-1.52
		AM	—	0.03	0.74	1.08
		Range	nd	0.20	0.05-4.4	0.2-2.5
Buzzard (87)	9	AM	nd	nd	1.2	0.7
		Range	nd	nd	0.04-4.2	0.1-2.3
Buzzard (87)	4	AM	nd	nd	0.4	0.5
		Range	nd	nd	0.1-0.7	0.2-0.7
Kestrel (87)	5	AM	nd	nd	0.1	0.2
		Range	nd	nd	nd-0.1	nd-0.7
Sparrow Hawk (87)	3	AM	nd	nd	1.2	8.5
		Range	nd	nd	1.1-1.3	3.8-12.4
Sparrow Hawk (87)	13	AM	nd	nd	1.9	9.5
		Range	nd	nd	0.2-7.2	2.4-15.0

cance. The residues of HEOD in the eggs of these raptors are smaller than those in the eggs of shags (see Fig. 7-11) and considerably smaller than those in the eggs of herons; it is plausible to conclude that significant increases in chick mortality (the most sensitive parameter) are not very probable. The second method of evaluating the significance of the residues summarized in Table 8-XI is by comparing them with the residues in eggs of birds fed diets containing HEOD. From trials in progress at Tunstall Laboratory it appears that there is no significant increase in the mortality of chicks of quail, pigeons or domestic fowl hatched from eggs containing up to 15 ppm HEOD. This estimate of a no-effect level in eggs must be regarded as tentative until the trials are completed. However, it seems improbable (unless there are large differences in species sensitivity) that residues of 5 ppm HEOD or less are of any significance, and it is provisionally concluded that the residues in Table 7-XI are insufficient to be potentially inimical to reproductive success. Once again, of course, we are unable to make any statements regarding the concentrations in raptors eggs before 1962 or 1963, but if there is a "failure of young peregrines to build up the breeding population (57)," and if this failure cannot be plausibly attributed to residues or organochlorine insecticides, then it is reasonable to consider the possibility that some other factor has been operating since 1962/63. This other factor may also have been operating prior to 1962 and may have been responsible for the reported declines of peregrine falcons.

Dispersal of Organochlorine Insecticides in the Environment

It is obvious from the results quoted above and from general surveys of the concentrations of organochlorine insecticides in wildlife (43, 64) and man (12, 19, 31, 71) in Britain, that organochlorine insecticides are not confined to the point of application, and the reports of residues in Antarctic fauna (23, 75, 80) given an even greater emphasis to this point. Various mechanisms are involved in the dispersion of an insecticide from its point of application; these mechanisms are not confined to organochlorine insecticides since any chemical is potentially capable of dispersion from the point of application. The mechanisms can be divided into two major classes: one type of movement depends upon the inherent properties of the substance, namely its vapor pressure and molecular weight, since the rate of evaporation in a still atmosphere is dependent upon vapor pressure and molecular diffusion. A still atmosphere exists in the immediate vicinity of a surface, but this local atmosphere is very thin. At greater distances from the surface there is relative lateral movement of air which transfers the molecules more rapidly than does molecular diffusion (27). This phenomenon is an example of the second class of mechanisms of dispersion, those involving other agents. The other agents can be atmospheric turbulence (as just described), movement in water either as a solution or as suspended particles, dispersal of dust (containing the chemical) by wind, translocation in plants (e.g., from the soil, via the roots, to the aerial portion of the plant), and transfer in animals or birds that ingest the chemical and then move from the point of contact. These mechanisms are shown diagrammatically in Figures 7-9 and 7-10. The entry into plants and animals constitutes the beginning of a series of complex trans-

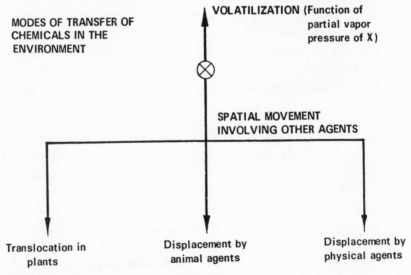

FIGURE 7-9. Mechanisms of dispersion of organochlorine insecticides in the environment.

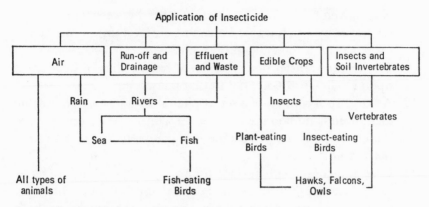

FIGURE 7-10. Transfers of residues of organochlorine insecticides in the abiotic and biotic components of the biosphere.

fers that can occur in ecological systems by means of food chains or webs. The potential capacity of organochlorine insecticides to be transferred along food chains, and hence between trophic levels, is one which is arousing considerable concern and it is necessary to consider this topic in some detail. A simple model of an ecosystem, defined as a set with trophic levels as subsets, has been proposed (65) and the following is an extension of that model:

1. An ecosystem consists of a number (approaching infinity) of phases.
2. A phase is a distinct part of the ecosystem, separated from other phases by a definite bounding surface.
3. A component is a chemical entity.
4. The chemical potential of a component is uniform throughout a phase.
5. The bounding surface of a phase may be the interface between immiscible physical phases or a biological membrane.
6. The bounding surface may be either permeable or nonpermeable to a component.
7. Any change of the chemical potential of a component at a point in a phase is followed by rapid equilibration of the chemical potential throughout the phase.

8. Net movement of a component occurs between two phases, separated by a permeable bounding surface, unless the chemical potential is the same in both phases.

9. The rate of transfer of a component from a phase is dependent upon its chemical potential in that phase.

10. An organism consists of a set of phases separated by permeable membranes and bounded by a heterogeneous envelope, part of the envelope is nonpermeable to a component and the rest is permeable.

11. Loss of a component from a living organism occurs by secretion or excretion through the permeable parts of the envelope, by chemical conversion within one or more phases, or by a temporary breakdown of a portion of the nonpermeable membrane.

12. The rates of secretion, excretion and chemical conversion of a component are dependent upon its chemical potentials in the particular phases involved.

13. Transfer of a component into a living organism is dependent upon its chemical potential in a phase external to the envelope of the organism but contiguous with a permeable portion of the envelope.

14. Transfer of a component from an organism also occurs when the envelope is damaged irreversibly, a process which precedes or is simultaneous with the death of the organism.

The model has not been defined in a completely rigorous manner (e.g., some of the terms are undefined, although they have their commonly accepted meaning). Nevertheless it can be used to investigate a number of simple cases, cases which have qualitative applications at least to a real system. Thus, if we have two phases, separated by a permeable membrane, in which the chemical potentials of a particular component are unequal at a particular instant, we can write down the mass balance and rate equations, and these can then be solved. Another example is that in which a component is entering a phase at a constant rate, and the component is undergoing chemical conversion in the phase at a rate proportioned to its chemical potential in that phase. The concentration of the component in the phase is obtained by integration of a differential equation of the form

$$\frac{dc}{dt} = \alpha - kc$$

and a particular solution obtained, allowing for the boundary conditions, is

$$C_t = C \infty \ (1\text{-}e^{-kt})$$

This type of problem can be generalized to say, n-1 phases in parallel with an nth phase in which conversion is occurring, the n-1 phases being separated from the nth phase by permeable membranes, but other contiguous phases being separated by either permeable or nonpermeable membranes. This model has a general solution involving n exponential terms:

$$C = A - \sum_{i=1}^{i=n} B_i \ e^{-\alpha_i t}$$

The interesting point about these solutions is that they approach finite asymptotes as t approaches infinity. Asymptotic solutions are found for models in which the rate of conversion in a phase is a simple algebraic function (e.g., $f(c) = c^n$) of the concentration in that phase.

The importance of these rather abstract models is that they enable us to make rational predictions about the probable future distribution of residues in the environment. Thus we would expect that the concentrations of a chemical in the environment will not increase in arithmetic manner with time, as use of it continues. Investigations of the pharmacodynamics of organochlorine insecticides have given results consistent with the above model (65) and surveys of residues of these compounds in the tissues of the general population of the United Kingdom and United States (man being regarded as a part of an ecosystem) appear to be consistent with it. The trend in the concentrations of HEOD in the eggs of Shags (68) living in a colony off the northeast coast of England (see Fig. 7-11) can also be explained on the basis of the proposed model.

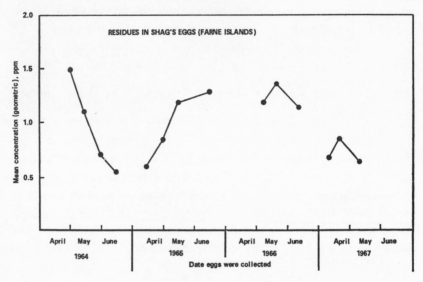

FIGURE 7-11. Concentrations of HEOD in the eggs of Shags collected 1964 to 1967.

The chemical conversions considered so far in the proposed model have all been occurring in living organisms. However, the detection of residues of organochlorine insecticides in air and rainwater, and in marine animals, indicates the importance of considering movements of insecticides in the inorganic components of the biosphere, and of degradation processes that may be occurring there. Degradation of pesticides by physico-chemical processes, particularly in the atmosphere, may be of the greatest importance and it is desirable that investigations into the existence of such processes should be undertaken.

General Discussion and Conclusions

The published information on the population status of various species of birds in Britain, and on the residues of organochlorine insecticides in the eggs and tissues of certain species of British birds, has been discussed critically and in detail as required. The discussion has been purposely critical in an attempt to define the areas of knowledge in which there are marked deficiencies. Such deficiencies in information, with consequent uncertainty regarding the probable influence or lack of influence of the continued use of organochlorine insecticides, are unsatisfactory both for the conservationists and the users of these compounds.

The conclusions drawn in this paper may be summarized as follows:

1. General comparisons of the status of various species of birds in Britain, before and after 1940, do not reveal the existence of a new and generally inimical factor in the period since 1940, the period in which organochlorine insecticides were introduced.

2. The Common Bird Census, which includes forty-one species of the common farmland birds, has shown that no serious adverse effects, attributable to organochlorine insecticides, have occurred over the period for which results are available, 1962 to 1966.

3. The annual mortality of the blackbird declined from 1951 to 1961, increased in 1962 and 1963 (the increase in mortality being attributed to cold winters) and decreased again in 1964.

4. The breeding statistics of the blackbird and nine other common species show no marked alteration or trends in clutch size, fertility or fledgling success, except in the chaffinch in which some decreases in hatching success occurred in 1959 to 1961.

5. The preliminary results of an attempt to assess population trends in other nonraptorial species from about 1950 are also consistent with the more precise conclusions drawn from the Common Bird Census and from the studies of the breeding success of the blackbird and 9 other common species.

6. Significant decreases in the sparrow hawk and peregrine falcon appear to have occurred in the 1950s. The population status of the kestrel, barn owl and tawny owl is more difficult to assess in view of cyclical oscillation in the available records.

7. The significance of the relationship between the decline of sparrow hawks and peregrines, and the usage of organochlorine insecticides, cannot be tested rigorously as the published information is too imprecise.

8. The results of studies of the concentrations of organochlorine insecticides in the tissues of experimentally poisoned birds have been summarized and the relevance of these concentrations in the diagnosis of death by poisoning, particularly in regard to birds of prey, has been discussed. It is proposed that concentrations in the livers of dead birds of pp'-DDT = pp'-DDD greater than 30 ppm, or 20 ppm HEOD, are a presumptive indication of poisoning of a bird by a single toxicant.

9. On this basis it is concluded that about 5 per cent of the dead raptors that have been examined since 1962 probably died from poisoning by HEOD or HEOD + heptachlor expoxide.

10. Sublethal effects of organochlorine insecticides do not appear to be occurring to any significant extent in most common bird species. The evidence for such effects in raptors since 1962 is rather tenuous.

11. No information is available on residues in raptors prior to 1962 and no conclusion as to the significance of such residues can be drawn.

12. The mechanisms of dispersal of organochlorine insecticides have been discussed in relation to the dynamics of organochlorine insecticides in ecosystems.

REFERENCES

1. ABBOTT, D.C.; HARRISON, R.B.; TATTON, J.O'G., and THOMSON, J.: *Nature 211*:259, 1966.
2. ABBOTT, D.C.; HARRISON, R.B.; TATTON, J.O'G., and THOMSON, J.: *Nature, 208*:1317, 1965.
3. ALEXANDER, W.B., and LACK, D.: *Brit Birds, 38*:42,62,82, 1944.
4. ANDREWARTHA, H.G., and BIRCH, L.C.: *The Distribution and Abundance of Animals.* Chicago, 1954.
5. ASH, J.S.: *Brit Birds, 53*:285, 1960.
6. BAILEY, R.S.: *Bird Study, 14*:195, 1967.
7. BERNARD, R.F.: Mich. State University, Publ. of Mono. Biol. Ser. 2, 1963, p. 155.
8. BOYKINS, E.A.: *Bioscience, 17*:37, 1967.
9. BRODKORB, P.: *Bull Fla State Museum Biol Sci, 5*:41, 1960.
10. BURTON, J.F., and MAYER-GROSS, H.: *Bird Study, 12*:100, 1965.
11. CAMPBELL, B.: *Brit Birds, 53*:221, 1960.
12. CASSIDY, W.; FISHER, A.J.; PEDEN, J.D., and PARRY-JONES, A.: *Mth Bull Min Hlth Lab Serv, 26*:2, 1967.

13. COULSON, J.C.: *J Anim Ecol, 29:*251, 1960.

14. CRAMP, S.: *Brit Birds, 56:*124, 1963

15. CRAMP, S.; CONDER, P.J., and ASH, J.S.: 4th Rpt. Brit. Trust Ornithology and Roy. Soc. Protection Birds on Toxic Chemicals, 1964.

16. CRAMP, S., and CONDER, P.J.: 5th Rpt. Joint Committee Brit. Trust Ornithology and Roy. Soc. Protection Birds on Toxic Chemicals, 1965.

17. CRAMP, S., and OLNEY, P.J.S.: 6th Rpt. Joint Committee Brit. Trust Ornithology and Roy. Soc. Protection Birds on Toxic Chemicals, 1967.

18. DOBINSON, H.M., and RICHARDS, A.J.; *Brit Birds, 57:*373, 1964.

19. EGAN, H.; GOULDING, R.; ROBURN, J., and TATTON, J.O'G.: *Brit Med J, 2:* 66,1965.

20. DE FAUBERT MAUNDER, M.J.; EGAN, H.; GODLEY, E.W.; HAMMOND, E.W.; ROBURN, J., and THOMSON, J.: *Analyst, 89:*168, 1964.

21. FERGUSON-LEES, I.J.: *Brit Birds, 50:*149, 1957.

22. FERGUSON-LEES, I.J.: *Brit Birds, 56:*140, 1963.

23. GEORGE, J.L., and FREAR, D.E.H.: *J Appl Ecol, 3* (suppl.) :155, 1966.

24. GOOD, R.: *The Geography of Flowering Plants,* 3rd ed. London, Longmans, 1964.

25. GOODWIN, E.S.; GOULDEN, R., and REYNOLDS, J.C.: *Analyst, 86:*697, 1961.

26. HARPER-HALL, G.: *Brit Birds, 51:*151, 1958.

27. HARTLEY, G.S.: Paper presented at the 153rd Meeting of the Amer. Chem. Soc., Miami, April 1967.

28. HOLDEN, A.V.: *Nature, 216:*1274, 1967.

29. HOLMES, D.C.: SIMMONS, J.M., and TATTON, J.O'G.: *Nature, 216:*229, 1967.

30. HUDSON, R.: *Brit Birds, 58:*105, 1965.

31. HUNTER, C.G.; ROBINSON, J., and RICHARDSON, A.: *Brit Med J, 1:*221, 1963.

32. JEFFERIES, D.J.: *Ibis, 109:*266, 1967.

33. JEFFERIES, D.J., and PRESTT, I.: *Brit Birds, 59:*49, 1966.

34. JENSEN, S.: *New Scientist, 32:*612, 1966.

35. KOEMAN, J.H.: private communication.

36. KOEMAN, J.H., and VAN GENDEREN, H.J.: *J Appl Ecol, 3* (suppl.) :99, 1966.

37. LACK, D.: *The Natural Regulation of Animal Numbers.* London, Oxford U.P., 1954.

38. LACK, D.: *Population Studies of Birds.* London, Oxford U.P., 1966.

39. LACK, D.: *Brit Birds, 36:*166, 1943.

40. LOCKIE, J.D., and RATCLIFFE, D.A.: *Brit Birds, 57:*89, 1964.

41. LOFTS, B., and MURTON, R.K.: *Brit Birds, 59:*261, 1966.

42. MACFAYDEN, A.: Symposium of the British Ecol. Soc., Bangor, 1962.

43. MOORE, N.W.: *Bird Study, 12:*222, 1965.

44. MOORE, N.W.: *Brit Birds, 50:*173, 1957.

45. MOORE, N.W.: *Proc Roy Soc [Biol] 167:*128, 1967.

46. MOORE, N.W.: Proc. 4th Brit. Insecticide Fungicide Conf., 1967, vol. 2, p. 493.

47. MOORE, N.W.: In Cragg, J.B. (Ed.) : *Advances in Ecology Research.* London, Academic, 1967, vol. 4, p. 75.
48. MURTON, R.L., and VIZOSO, M.: *Ann Appl Biol, 52:*503, 1963.
49. NICHOLSON, A.J.: *J Anim Ecol, 2:*132, 1933.
50. NICHOLSON, E.M.: *Brit Birds, 50:*131, 1957.
51. ODUM, E.P.: *Fundamentals of Ecology,* 2nd ed. Philadelphia, Saunders, 1959.
52. PARSLOW, J.L.F.: *Brit Birds, 60:*2,97,177,261,396,493 (1967), and *61:*49, 1968.
53. PRESTT, I.: *Bird Study, 12:*66, 1965.
54. PRESTT, I.; ROBINSON, J., and CRABTREE, A.N.: Upublished work.
55. RATCLIFFE, D.A.: *Bird Study, 10:*56, 1963.
56. RATCLIFFE, D.A.: *Bird Study, 12:*66, 1965.
57. RATCLIFFE, D.A.: *Bird Study, 14:*238, 1967.
58. RATCLIFFE, D.A.: *Brit Birds, 51:*23, 1958.
59. RATCLIFFE, D.A.: *Brit Birds, 53:*128, 1960.
60. RATCLIFFE, D.A.: *Nature, 215:*208, 1967.
61. RATCLIFFE, D.A.: *Brit Birds, 58:*65, 1965.
62. RICHARDSON, A.: Paper presented at the 6th Internat. Congress Plant Protection, Vienna, 1967.
63. ROBINSON, J.: Proc. 4th Brit. Insecticide and Fungicide Conference, 1967, vol. 1, p. 36.
64. ROBINSON, J.: *Chem Industr,* 1967, p. 1974.
65. ROBINSON, J.: *Nature, 215:*33, 1967.
66. ROBINSON, J.; BROWN, V.K.H.; RICHARDSON, A., and ROBERTS, M.: *Life Sci, 6:*1207, 1967.
67. ROBINSON, J.; BROWN, V.K.H., and ROBERTS, M.: Unpublished work.
68. ROBINSON, J.; CRABTREE, A.N.; COULSON, J.C., and POTTS, G.R.: Unpublished work.
69. ROBINSON, J.; RICHARDSON, A.; CRABTREE, A.N.; COULSON, J.C., and POTTS, G.R.: *Nature, 214:*1307, 1967.
70. ROBINSON, J.; RICHARDSON, A., and ELGAR, K.E.: Paper presented at the 152nd Meeting Amer. Chem. Soc., New York, 1966.
71. ROBINSON, J.; RICHARDSON, A.; HUNTER, C.G.; CRABTREE, A.N., and REES, H.J.: *Brit J Industr Med. 22:*220, 1965.
72. ROBINSON, J.; WRIGHT, E.N., and JONES, F.J.S.: Unpublished work.
73. ROBINSON, J.: *New Scientist,* 1966, p. 159 and subsequent correspondence pp. 232,359,432,499,645.
74. ROBURN, J.: *Analyst, 90:*467, 1965.
75. SLADEN, W.J.L.; MENZIE, C.M., and REICHEL, W.L.: *Nature, 210:*670, 1966.
76. SNOW, D.W.: *Bird Study, 12:*287, 1965.
77. SNOW, D.W.: *Nature, 211:*1231, 1966.
78. STICKEL, L.F.; CHURA, N.J.; STEWART, P.A.; MENZIE, C.M.; PROUTY, R.M.,

and REICHEL, W.L.: Trans. 31st N. Amer. Wildlife and Natural Resources Conf., March 1966.

79. STICKEL, L.F.; STICKEL, W.H., and CHRISTENSEN, R.: *Science, 151:*1549, 1966.
80. TATTON, J.O'G., and RUZICKA, J.H.A.: *Nature, 215:*346, 1967.
81. TAYLOR, S.M.: *Bird Study, 12:*268, 1965.
82. TRELEAVEN, R.B.: *Brit Birds, 54:*136, 1961.
83. TUBBS, C.R.: *Brit Birds, 60:*381, 1967.
84. TURTLE, E.E.; TAYLOR, A.; WRIGHT, E.N.; THEARLE, R.J.P.; EGAN, H.; EVANS, W.H., and SOUTAR, N.M.: *J Sci Food Agric, 14:*567, 1963.
85. TURTLE, E.E., *et al.*: Report of the Infestation Control Laboratory 1962-64. London, H.M.S.O., 1965.
86. VINCENT, J.: *Red Data Book.* Aves. Intern. Union for Conservation of Nature and Resources, Lausanne, 1966, vol. 2.
87. WALKER, C.H.; HAMILTON, G.A., and HARRISON, R.B.: *J Sci Food Agric, 18:*123, 1967.
88. WHEATLEY, G.A., and HARDMAN, J.A.: *Nature, 207:*486, 1965.
89. WILLIAMSON, K., and HOMES, R.C.: *Bird Study, 11:*240, 1964.
90. WURSTER, C.F., and WINGATE, D.B.: *Science, 159:*979, 1968.
91. WYNNE-EDWARDS, V.C.: *Animal Dispersion.* Edinburgh, Oliver & Boyd, 1962.

DISCUSSION

WURSTER: Were the population data on number of birds based on ringing and banding records?

ROBINSON: Yes.

WURSTER: Does that include the peregrine and sparrow hawk records?

ROBINSON: Yes.

WURSTER: Your data disagree with the publications of other ornithologists in Britain [e.g., Ratcliffe, D.A.: *Bird Study, 14:*238, 1967; *Ibid., Brit Birds, 58:*65, 1965; and Cramp, S.: *Brit Birds, 56:*124, 1963]. I wonder if they do not instead reflect an increase in the number of people putting rings on birds?

ROBINSON: Except that in the case of the sparrow hawk and the peregrine falcon there has been an apparent but very small recovery from about 1960, in a study that was pressed by ornithologists since 1950 to 1952. But you may be quite correct, and this is one of the points I make in my paper. These ringing records have to be looked at from the point of view of the consistency of the observers.

HUNT: Why do you chose to analyze liver tissue?

ROBINSON: Partly convenience, plus the fact that other people in Britain have been using it.

RADOMSKI: In your data on experimentally poisoned birds, were these minimally poisoned birds ?

ROBINSON: These were birds which were poisoned by various rations, either short-term dosing or long-term dosing, up to three months in some cases.

Well, if we just revert for a moment to the conclusions of Pennwillington in 1966. He could see no evidence in those particular species of birds of any detrimental effect within the environment such as could arise from toxic chemicals.

WURSTER: Many of the birds you have studied may be irrelevant to the problem. You have studied birds that are primarily herbivores, such as the chaffinch which is primarily a seedeater.

ROBINSON: There is one group of those two species of birds in Britain in one small area that is dying of DDT poisoning. The residues in their brains are of the order of 100 to 200 ppm total DDT, most of which is DDE; the residues in the liver are of the order of 490 ppm (ref.: yours and Ministry of Agriculture) .

HERMAN: For the past twenty years there has been a fairly well-defined controversy about the effects of pesticides on wildlife. This is easily extendable to human beings. In my view (and I'm simplifying) this controversy has been between industry and industrial scientists on one hand, and the conservationists and some ecologists on the other. Dr. Robinson's paper represents to me industry's position, and I wish to comment on this position.

We conservationists have not always been able to get a reasonable hearing. I'd like to give one example. In 1965 the University of Wisconsin sponsored a conference to consider the reasons for the widespread decline of the peregrine falcon. This conference considered all kinds of evidence: disease, human persecution, etc. One session was devoted to the consideration of pesticides and Dr. Ratcliffe showed a post-correlation between the decline of peregrines and the introduction and use of organochlorines. Dr.

Spencer was present; he was at that time Chief of the Animal Biology Staff of the Pesticide Regulation Division whose responsibility it was to review registration of use of pesticides in the field from the standpoint of environmental and wildlife hazard. Dr. Spencer expressed genuine concern about the peregrine falcon, and said he was prepared to investigate the situation with regard to the use of pesticides if some concrete evidence could be supplied of a correlation between the use of pesticides and the decline of the peregrine falcon. What, however, will be accepted as proof and what action will be taken on it?

RUDD: With regard to your question about what constitutes acceptable proof, I think more convincing evidence can be offered today than in 1965. However, the coming problems due to the misuse of pesticides were obvious to biologists before 1965. We contested the commercial practice, but we did not get a hearing. This is an uphill struggle, and we must face the fact that it is a confrontation of opposing interests. The peregrine falcon is symbolic of an issue which concerns the future of the entire food-producing ecosystem.

RISEBROUGH: It is clear now that contamination of the environment is not only a scientific problem, but also a social one. Scientists involved in research on effects of pesticides, especially side-effects, have a social responsibility, and I question whether industrial scientists are living up to this responsibility.

In 1966 you published in the *New Scientist* (January 20, p. 159) an article entitled "Insecticides and Birds: Is there a Problem?" In that article, and in your present paper, you have concluded that the "problem," if any, is minimal. You have become, therefore, one of the spokesmen of the chemical industries. That is the basic source of our disagreement. Time this afternoon does not permit us to make some specific criticisms of your paper but we shall ask for an opportunity to do so during the discussion period.

NOTES ON TABLES II–XI

HHDN   Abbreviation for 1,2,3,4,10,10-
                                  hexachloro-1,4,4a,5,8,8a-hexahy-

	dro-1,4-*endo*, *exo*-5,8-dimehan-onaphthalene. Aldrin contains not less than 95% HHDN.
HEOD	Abbreviation for 1,2,3,4,10,10-hexachloro-6,7-epoxy-1,4,4a,5,6,7,8,8a-octahydro-1,4-*endo, exo*-5,8-dimethanonaphthalene. Dieldrin contains not less than 85% HEOD.
Total DDT-type compounds	Sum of concentrations of pp'-DDT, pp'-DDE and pp'-DDE (results obtained by the Schechter-Haller procedure may include other congeners of pp'-DDT).

In some surveys the same tissues were not analyzed from all birds. In these cases the type of tissue in the minority is placed in brackets.

N	Number of samples.
GM	Geometric mean.
CL	Confidence limit (P = 0.95) of geometric mean.
AM	Arithmetic mean.
NA	Results not available.
nd	None detected.
a	These concentrations for HEOD have a positive bias as some of the results are given as HEOD + pp'-DDE.
b	No residues of HHDN found.
c	No residues of heptachlor found.
d	Residues of endrin not reported, presumed to be absent.
e	Schechter - Haller colorimetric method used for determination of DDT.

f   Gas - l i q u i d chromatographic
                              method.
g   Ranges are for pp'-DDD only.
h   Residues determined by the
                              phenyl azide (colorimetric) pro-
                              cedure or by paper chroma-
                              tography.

Chapter 8

TISSUE RESIDUES OF DIELDRIN IN RELATION TO MORTALITY IN BIRDS AND MAMMALS

W. H. STICKEL, L. F. STICKEL and J. W. SPANN

ABSTRACT

An experiment was performed with *Coturnix* to learn what residue levels were indicative of death from dieldrin poisoning. Birds were fed diets containing 250, 50, 10, and 2 ppm dieldrin for periods up to 158 days. The dieldrin was 95% pure HEOD, which is 1,2,3,4, 10,10-hexachloro-6, 7-epoxy-1,4,4a,5,6,7,8,8a-octahydro-1,4-endo,exo-5,8-dimethanonaphthalene. When half of a group was dead, the other half was sacrificed for comparison of residues in dead and survivors. Dosage levels controlled time to death, but did not control residue levels in the dead. Residues in liver and carcass proved to be misleading and complicated by changes in lipid content. Brain residues correlated well with death although residues in dead and survivors overlapped. Brain residues of animals killed by dieldrin in the field and in other experiments are listed. Data agree in general for several species of birds and mammals. There is evidence, however, for species differences in average lethal brain residues. It is concluded that brain residues of 4 or 5 ppm (wet weight) or higher indicate that the animal was in the known danger zone and may have died from dieldrin. Brain residues averaged lower in wild than in experimental animals. Possible explanations include species differences, more stress and exertion in the wild, and overrepresentation in the field series of individuals that will die with low but lethal brain residues. The latter is supported by the fact that the first *Coturnix* to die in each sex and treatment group had the lowest brain residue of its group. Birds receiving 2 ppm dieldrin, and some receiving 10 ppm, were able to maintain low brain residues throughout the experiment. However, birds of the 10 ppm group could withstand little stress and mobilization of toxicant, for a few micrograms in the brain were lethal and bodies contained hundreds or thousands of micrograms.

REPORT

Introduction

DIAGNOSIS OF DEATH OF BIRDS and mammals from pesticide poisoning on the basis of residue content of victims is an important practical problem and a difficult scientific one. We have studied the problem relative to DDT (26) and have work in progress on other chemicals. Dieldrin, whose residues in nature are second in frequency only to those of the DDT group, was the second chemical studied.

An experiment was performed with Japanese quail (*Corturnix coturnix*) to determine residue levels in dead and surviving birds. Dieldrin was fed at four dietary levels, and some birds were subjected to the additional stress of shortened day length. Results are related to the field through data for two small series of animals found dead in dieldrin-treated areas. Literature records of brain residues are presented for animals for whom evidence of death from dieldrin poisoning seems conclusive.

Experiment with Coturnix

Five groups of quail totaling twenty-nine males and twenty-six females were placed on dosage on March 25, 1965. The birds were fully adult, 132 days of age. Four groups received dieldrin dissolved in edible oil and mixed into their dry diet at levels of 250 ppm, 50 ppm, 10 ppm, and 2 ppm; the fifth group received feed containing equivalent amounts of oil but no dieldrin. As soon as half of the birds of a given sex and treatment group had died, the remaining birds of that group were given clean food for two days and then sacrificed. The two-day period of clean food was designed to prevent birds already fatally poisoned from being considered survivors.

Dosage groups and times to death are shown in Table 8-I.* Birds in the 250 ppm group died soon. Four of the six males died within 9 days and the remaining two were sacrificed on day 11. Females lived longer; three died by day 14 and the remaining two were sacrificed. The difference in survival time of the sexes was even more pronounced in the 50 ppm group. Three of the six males died by day 28, but three of the five females died between days 64 and 72. By day 137, the survivors included five females

---

*Birds whose tissues were not analyzed do not appear in the table. They included six males and four females on untreated diets, none of which died; two females on 2 ppm diets that died on days 134 and 137; and five females on 2 ppm diets that were sacrificed on day 158; two of this latter group had been on short day length since day 137.

and five males on the 10 ppm diet, four males and five females on the 2 ppm diet, and six males and four females on the untreated diet. It was decided to impose a natural type of stress to see if mortality would occur in these groups. Approximately half of the birds of each sex and treatment group were moved to a room in which they received only 8 hours of light per day instead of normal August daylight. This room was kept well ventilated, but not air-conditioned, so temperature conditions were similar for both groups. In a previous study in which *Coturnix* were shifted from 14 hours to 8 hours of light per day, the birds stopped breeding, molted, and ate less; some being fed dietary dosages of DDT died, presumably as a result of drawing upon fat stores (29). In the present experiment, no bird receiving natural light died, and none of the 2 ppm group died. In the 10 ppm group that received only 8 hours of light, one male died on day 143, one male on day 146, and a female on day 153. Others lived until the experiment was terminated on day 158. Birds were frozen until analysis in early 1966. Analyses were made of the brains of all males, of the livers and remainders of males fed 250, 50, or 10 ppm of dieldrin in the diet, and of the brains of females fed these three dosages. Remainders consisted of the carcass minus brain, liver, bill, feet, wings, skin, and gastrointestinal tract.

Specimens were analyzed for dieldrin and compounds of the DDT group at the Hazleton Laboratories, Falls Church, Virginia, by methods described in the Appendix.

Characteristics of Dieldrin-killed Birds

We have pointed out (26) that birds killed by DDT typically have full flesh, no fat, and full gall bladders. Dieldrin-killed birds were less easily characterized. Some appeared to be in full flesh, others had depleted pectoral musculature. Fat was often lacking, but some remained, at least on the birds that died soonest. The gall bladder was not conspicuous in any bird examined. Feathers of the anal area tended to be caked with whitish excretions. This combination of characteristics is far from diagnostic of any one cause of death.

Weight loss was severe. Males that died on treatment averaged

77 g whereas those that survived averaged 107 g. Females that died on treatment averaged 80 g and those that survived averaged 125 g. The six untreated males averaged 116 g (108-124 g). The four untreated females averaged 125 g (109-152 g). Weight loss was also reflected in lipid content of remainders: lipid content averaged 1.2 per cent of wet weight in remainders of males that died on treatment and 6.3 per cent in those of treated survivors. The survivors tended to be heavy birds with blocky pectorals and substantial amounts of fat.

Robinson *et al.* (23) observed that loss of body weight in dieldrin-killed *Coturnix* was significantly correlated with length of time to death, being greater among birds that survived longer. This may prove to be generally true of chlorinated hydrocarbon poisoning.

Residues and Lipids in Brain

Brain residues of birds that died were significantly higher than of those that survived (Table 8-I). This was true although survivors had been on the same dosage as the dead and had only two days of clean food before being killed. Among females, brain residues in the dead averaged 21.7 ppm (wet-weight basis) as compared with 2.6 ppm in the survivors. There was no overlap between the groups. Males that died had brain levels of dieldrin that averaged 14.8 ppm and those that survived had levels averaging 4.1 ppm. The contrast was not as clear-cut as among females, however, for two surviving males had residues well into the range of residues in dead birds. Their brain residues were 11.1 and 11.9 ppm, well above the next highest value, which was 4.5 ppm. Such overlapping is not surprising, for when heavily dosed birds are sacrificed, one may expect to find some that would have died only with high brain residues, and which had not yet reached these levels. That is, the range of residues at death is wide and birds that would have died only in the upper part of the range may be found living with residues that would have been lethal to others.

The first bird to die in each sex and treatment group had the lowest brain residue for that group. This led us to inspect the data

TABLE 8-I

DATA ON BRAINS OF DIELDRIN TREATED COTURNIX

Dosage ppm	Bird No.	Sex D or S*	Day of Death	Body Wt. at Death grams	Wet wt. grams	% H₂O % wet wt.	% lipid % wet wt.	Dieldrin ppm wet wt.	Dieldrin ppm lipid wt.	Dieldrin µg
						% H$_2$O		**BRAINS**		
250	613	♂ D	5	65	0.727	73.55	10.32	7.01	67.89	5.1
250	614	♂ D	8	73	0.788	70.40	9.39	12.69	135.25	10.0
250	615	♂ D	9	88	0.751	69.20	7.06	9.99	141.84	7.5
250	616	♂ D	9	68	0.739	72.90	7.44	23.00	309.13	17.0
50	624	♂ D	9	69	0.770	73.73	7.53	6.23	82.33	4.8
50	625	♂ D	20	77	0.797	69.72	8.78	12.55	143.50	10.0
50	626	♂ D	28	91	0.798	70.17	8.52	22.56	265.25	18.0
10	635	♂ D	50	76	0.666	72.39	7.21	19.52	270.83	13.0
10	820	♂ D†	143	85	0.843	71.78	15.07‡	32.03	212.48	27.0
10	826	♂ D†	146	77	0.776	70.94	13.66‡	21.91	160.43	17.0
Means§				77	0.754	76.26	8.28	14.8¶	160.9**	
95% confidence limits					0.665 to 0.843	74.46 to 78.06	5.90 to 10.66	4.9 to 44.3	58.9 to 440.0	
250	617	♀ D	11	87	0.816	75.61	7.47	11.03	147.25	9.0
250	620	♀ D	11	67	0.776	76.15	10.82	19.33	178.37	15.0
250	621	♀ D	14	65	0.759	76.54	8.30	32.94	396.38	25.0
50	636	♀ D	64	87	0.770	75.84	7.14	19.48	273.10	15.0
50	637	♀ D	69	86	0.767	76.40	7.17	31.29	436.54	24.0
50	638	♀ D	72	89	0.763	74.96	7.20	24.90	345.83	19.0
10	821	♀ D†	153	82	0.694	77.23	7.92	21.61	272.72	15.0
Means§				80	0.764	76.10	8.00	21.7**	274.4**	
95% confidence limits					0.692 to 0.836	74.65 to 77.55	5.37 to 10.63	10.4 to 45.4	122.4 to 615.1	
250	618	♂ S	11	100	0.832	76.80	7.21	11.06	153.95	9.2
250	619	♂ S	11	102	0.758	77.44	7.38	4.49	60.97	3.4
50	627	♂ S	30	114	0.842	77.07	7.95	11.88	149.68	10.0

50	628	♂ S	30	105	0.794	77.45	7.68	3.78	49.47	3.0
50	629	♂ S	30	102	0.785	76.81	8.15	3.44	41.71	2.7
10	824	♂ S	158	118	0.763	76.80	8.91	3.93	43.77	3.0
10	825	♂ S†	158	124	0.887	76.43	8.00	1.69	21.25	1.5
10	822	♂ S†	158	91	0.716	76.53	8.51	1.54	17.62	1.1
	Means§			107	0.797	76.92	7.97	4.1	51.2	
	95% confidence limits				0.688 to 0.906	76.16 to 77.68	6.84 to 9.10	0.1 to 18.4	10.6 to 248.9	
2	817	♂ S	158	132	0.785	76.94	10.82	0.25	2.31	0.2
2	813	♂ S†	158	109	0.821	77.23	10.23	1.34	13.09	1.1
2	814	♂ S†	158	112	0.791	78.63	9.35	0.39	4.17	0.31
2	818	♂ S†	158	117	0.764	76.04	9.29	0.88	9.47	0.67
	Means§			117	0.790	77.00	9.92	0.58	5.88	
	95% confidence limits				0.742 to 0.837	74.70 to 79.30	8.45 to 11.39	0.13 to 2.67	1.22 to 13.97	
250	622	♀ S	16	125	0.786	77.35	7.76	1.65	21.26	1.3
250	623	♀ S	16	118	0.792	77.02	7.07	4.17	58.98	3.3
50	639	♀ S	74	124	0.806	76.67	8.43	5.33	62.87	4.3
50	640	♀ S	74	157	0.867	77.04	7.03	3.69	52.48	3.2
10	827	♀ S	158	134	0.710	76.19	9.15	0.87	9.50	0.62
10	828	♀ S	158	108	0.694	78.09	9.07	2.16	24.25	1.5
10	823	♀ S†	158	111	0.786	77.35	9.79	1.91	19.40	1.5
10	829	♀ S†	158	122	0.783	75.98	9.19	4.21	45.70	3.3
	Means§			125	0.778	76.96	8.44	2.6	31.1	
	95% confidence limits				0.669 to 0.887	75.60 to 78.32	6.35 to 10.52	0.1 to 8.8	8.1 to 118.8	

*Died or sacrificed.

†Received only 8 hrs of light per day after day 137.

‡As these figures are far out of line and may represent laboratory errors, they are not used in means or discussions.

§Geometric means for residues; others arithmetic.

¶Significantly different from mean of survivors at 5 per cent level. Comparisons by t-test on log-transformed data.

**Significantly different from mean of survivors at 1 per cent level.

carefully to see if brain levels at death were correlated with time on dosage before death. The data were scanty, but we found no clear indication that such trends occurred. We believe that the lower residues in the first to die simply indicate that the weakest, or the ones most vulnerable to dieldrin, went first.

Residue content also can be computed on the basis of the extractable lipid. The expression of residues as ppm on a lipid base, however, carries with it some problems that must be considered in interpretation. These are largely the result of loss of lipids as the animal loses weight. Thus the "standard" in which the units are expressed is itself changing. The problems thus produced were discussed by Langgård and Secher-Hansen (13) with particular reference to the usefulness of the term, "per gram fat-free solids," but they are general:

> The usefulness of relative quantity indications in connection with comparative investigations depends on the constancy of the reference substance. . . . The term *per unit of fat-free solids* is thus acceptable in principle only when the quantity of fat-free solids is unchanged during the experimental conditions in question or is the same in the two tissues that are being compared.

Suitably comparable methods for expressing the residue content of soils have been beset with this same basic difficulty, which has not been fully resolved for the medium.

Obviously, the demand for constancy cannot be rigorously fulfilled, for even moisture content may vary from one tissue to another, and indeed may change in certain tissues as a result of dosage with some chemicals. And of course specimens differ widely in the degree to which they have dried.

In the brain, however, comparison of dieldrin content in dead and surviving birds in terms of ppm on a lipid-weight basis gives much the same result as comparisons on a wet-weight basis. Residues in brains of males that died on dosage averaged 160.9 ppm (lipid-weight basis) as compared with 51.2 ppm in survivors. Among females, average values were 274.4 ppm for birds that died and 31.1 ppm for survivors. Residues in the dead were significantly higher than in the survivors for both males and females. Again, there was no overlap between the two groups of females.

However, residues in the two male survivors were well into the range of residues in the dead and these were the same two individuals that provided the overlap when residues were computed on the wet-weight basis.

The reasons for the general consistency in results for brain residues expressed on both the wet-weight and the lipid-weight bases lie in the fact that both weight and lipid content of brains were remarkably stable even in death.

Brains of birds of all mortality, sex, and dosage groups averaged 0.776 g, with no significant differences among the groups. The lipid percentage of the brains of birds fed 250, 50, or 10 ppm in the diet averaged 8.2 per cent for all groups combined, with no significant differences between birds that died and birds that survived. Among dead birds, there was no apparent correlation of lipid percentage with dosage level. Among survivors, however, the percentage of lipid tended to be higher in brains of birds on lower dosages. Average lipid percentages in brains of survivors were 7.4 for the 250 ppm group; 7.8 for the 50 ppm group; 8.9 for the 10 ppm group; and 9.9 for the 2 ppm group (males only in this group). Differences were significant (P <0.01) between the 10 ppm group and both the 50 ppm and 250 ppm group, which, however, did not differ significantly from each other. Lipid percentage in brains of males of the 2 ppm group was significantly different (P <0.05) from males of other dosage groups.

Thus the brain retained its weight and its lipid content even in death. As a pesticide is mobilized with stored lipids, it tends to concentrate where lipids remain. Consequently, as body weight is lost, residue levels in the brain increase until death ensues. It seems probable that in the last day or two, or in the last few hours of life, there may be an accelerated increase when birds are not eating and are undergoing intermittent convulsions.

Liver Residues and Lipids

Dieldrin residues in livers were very similar in dead and survivors (Table 8-II). They averaged slightly higher in survivors (28.8 vs. 19.7 ppm, wet weight), but the difference was not

TABLE 8-II

DATA ON LIVERS OF DIELDRIN-TREATED COTURNIX

Dosage ppm	Bird No.	Sex D or S*	Wet wt. grams	% H₂O % wet wt.	% lipid % wet wt.	Dieldrin ppm wet wt.	Dieldrin ppm lipid wt.	Dieldrin µg
						LIVERS		
250	613	♂ D	2.165	72.19	1.57	18.0	1146.49	39
250	614	♂ D	1.824	68.42	1.21	14.8	1233.33	27
250	615	♂ D	2.189	71.26	1.92	16.0	837.69	35
250	616	♂ D	2.084	71.92	1.20	25.9	2176.47	54
50	624	♂ D	2.024	69.81	1.28	19.8	1546.87	40
50	625	♂ D	2.064	68.26	1.45	29.1	2006.89	60
50	626	♂ D	2.513	67.52	2.71	51.7	1914.81	130
10	635	♂ D	2.096	71.32	1.91	5.7	300.00	12
10	820	♂ D†	1.905	68.39	1.05	24.1	2317.30	46
10	826	♂ D†	2.050	69.70	0.49	—	—	—
	Means‡		2.120	70.04	1.65	19.7	1301.0§	
	95% confidence limits		1.732 to 2.508	66.44 to 73.64	0.63 to 2.67	6.0 to 65.4	357.9 to 4727.0	
250	618	♂ S	3.287	67.99	7.70	48.7	633.28	160
250	619	♂ S	2.924	67.85	5.03	15.05	299.80	44
50	627	♂ S	2.737	68.79	3.32	36.5	1099.39	100
50	628	♂ S	4.618	56.21	21.31	140.75	660.79	650
50	629	♂ S	3.574	62.25	14.27	81.1	568.72	290
10	824	♂ S	4.811	56.45	18.33	93.5	510.09	450
10	825	♂ S	2.472	69.41	2.39	6.1	256.30	15
10	822	♂ S†	2.435	69.36	0.82	2.7	329.26	6.6
	Means‡		3.357	64.79	9.14	28.8	491.0	
	95% confidence limits		1.511 to 5.200	53.39 to 76.19	<0.01 to 24.75	1.7 to 474.2	185.9 to 1297.0	

*Died or sacrificed.
†Received only 8 hours of light per day after day 137.
‡Geometric means for residues only.
§Significantly different from mean of survivors at 1 per cent level.
Comparisons by t-test on log-transformed data.

significant and the distributions overlapped through most of the range of both groups.

Livers of the dead had much less substance and much less lipid than those of the survivors. Livers of dead birds averaged 2.1 g, those of survivors 3.4 g. The only overlap between the groups was produced by one dead bird whose liver weighed essentially the same as the lightest livers of survivors.

Lipid content of livers of dead birds averaged 1.6 per cent, ranging from 0.5 to 2.7 per cent. Lipid content of survivors was extremely diverse. Lipid percentages were 0.8, 2.4 and 3.3 in three birds, not greatly different from amounts in livers of dead birds, but percentages for the other four ranged from 5.0 to 21.3, three exceeding 14 per cent.

We assume that the low lipid content of the livers of dead birds represents a depletion that preceded death. The diversity of the lipid content of livers of survivors is more difficult to understand. It is probable that livers of some survivors became enlarged and fatty, perhaps adaptively (4, 7, 11, 28).

Thus, both liver weight and liver lipid content differed between birds that died from dieldrin poisoning and those that survived. Hence, neither wet weight nor lipid weight fulfills Langgård's criterion of constancy in the reference standard.

When residues of dieldrin in the liver were expressed on a lipid basis, residues in the dead averaged considerably higher (1301 ppm) than in the living (491 ppm). Differences between means were highly significant. However, residues in two males that died nearly spanned the range of values of survivors.

The differences between dieldrin concentration in livers expressed on a wet-weight basis and concentrations expressed on a lipid-weight basis appear to be largely the result of the increased concentration of dieldrin in a greatly reduced amount of lipid in the dead. Livers of dead birds actually contained fewer micrograms of dieldrin (12-130 μg, average 49μg) than did those of many survivors, three of which contained 290, 450, and 650 μg. the survivors' great diversity in micrograms of dieldrin in livers (7-650 μg) was similar to their diversity in percentage lipid.

In fact, there was a striking correlation between the lipid content of the liver and the micrograms of dieldrin in survivors (r = 0.98, 6 degrees of freedom) . Data for dead birds are inadequate for evaluation of this trend, but there are no contrary indications.

Residues and Lipids in Remainders

Dieldrin residues in remainders averaged 17.2 ppm (wet weight) in dead birds and 26.7 ppm in survivors (Table 8-III) . The two groups overlapped broadly and there was no statistically significant difference between the means.

When residues in remainders were expressed on a lipid basis, those in the dead averaged 1640 ppm, conspicuously higher than in survivors (459 ppm) . Only one survivor contained lipid residues as high as those in the dead bird that had the lowest residues of its group.

Total micrograms of dieldrin in remainders ranged from 300 to 1900 μg (average 921 μg) in dead birds and from 510 to 2800 μg (average 1646 μg) in survivors; the difference between the means was not statistically significant.

Remainders of dead birds weighed less and had lower lipid content than those of survivors. Those of dead birds averaged 46 g; those of survivors, 54 g. Lipid content averaged 1.2 per cent in the dead, 6.3 percent in survivors. There was no overlap in percentages of lipid in remainders of dead and surviving birds.

Residue Levels, Dosage Levels, and Survival

One interesting finding was that dosage levels from 250 ppm to 10 ppm had little if any effect on residue levels in the dead. The effect of dosage level was to govern length of time to death but not to vary residues in the dead. This held true for residues in brains, livers, and remainders. Apparently death occurs when residues reach the level lethal for the individual. This may occur quickly with high doses or slowly with low doses. It may occur when stored residues are mobilized at times of starvation, illness, or exertion. It means that the same figures define the danger zone of residues regardless of severity of exposure or length of time required to reach a lethal level. The practical importance of this is great, for it would be infeasible to define danger levels if diag-

TABLE 8-III
DATA ON CARCASS REMAINDERS OF DIELDRIN-TREATED COTURNIX*

Dosage ppm	Bird No.	Sex D or S†	Wet wt. grams	% H₂O % wet wt.	REMAINDERS % lipid % wet wt.	Dieldrin ppm wet wt.	Dieldrin ppm lipid wt.	Dieldrin µg
250	613	♂ D	35.3	73.55	0.76	15.9	2092.10	560
250	614	♂ D	40.6	70.40	1.28	17.0	1328.12	690
250	615	♂ D	54.5	69.20	2.24	23.85	1066.96	1300
250	616	♂ D	37.3	72.90	0.56	8.04	1428.57	300
50	624	♂ D	39.7	73.73	0.56	7.80	1392.85	310
50	625	♂ D	55.1	69.72	1.64	25.40	1548.78	1400
50	626	♂ D	56.0	70.17	1.68	33.92	2017.85	1900
10	635	♂ D	45.2	72.39	0.52	9.95	1913.46	450
10	820	♂ D‡	51.8	71.78	1.12	21.23	1892.85	1100
10	826	♂ D‡	39.5	70.94	1.44	30.4	2111.11	1200
		Means§	45.5	71.51	1.16	17.2	1640.0¶	
		95% confidence limits	28.4 to 62.6	67.86 to 75.16	<0.01 to 2.46	5.8 to 50.4	1039.0 to 2614.0	
250	618	♂ S	53.5	70.28	2.80	22.42	800.00	1200
250	619	♂ S	43.2	69.52	6.60	34.72	525.75	1500
50	627	♂ S	54.8	71.29	3.72	51.10	1373.65	2800
50	628	♂ S	54.1	69.24	5.00	9.43	188.00	510
50	629	♂ S	53.9	65.98	7.80	50.09	642.30	2700
10	824	♂ S	58.8	68.74	6.68	35.71	534.43	2100
10	825	♂ S	69.2	63.69	10.36	21.67	209.45	1500
10	822	♂ S‡	47.7	67.04	7.20	18.03	250.41	860
		Means§	54.4	68.22	6.27	26.7	458.6	
		95% confidence limits	39.1 to 69.7	63.23 to 73.21	1.47 to 11.07	8.5 to 83.9	113.7 to 1875.0	

*See Table 8-I for day of death and body weight.
†Died or sacrificed.
‡Received only 8 hours light per day after day 137.
§Geometric means for residues only.
¶Significantly different from mean of survivors at 1 per cent level.
Comparisons by t-test on log-transformed data.

nostic residues were closely correlated with length or severity of exposure.

Another interesting finding was that residue levels in brains of some birds remained low despite long periods of dosage. If kept under stable conditions, perhaps these birds never would have reached lethal levels. None of the birds receiving 2 ppm of dieldrin for 158 days had brain residues over 1.34 ppm and one had only 0.25 ppm. Only brains were analyzed for the dosage group, so we do not know how high body residues were.

However, the low brain residues in the 2 ppm group correlate well with the fact that none of these birds died when they were changed to 8 hours of light. The single male of this group that died during the experiment was a very thin, much pecked bird that probably died from natural causes, for its brain level was only 1.1.

Brain residues remained low even in some birds of the 10 ppm group. Although three males and one female on this dosage died with typically high brain residues, three males and four females survived. Their brain residues varied from a low of 0.87 to a high of 4.2 ppm. As these birds were on dosage for the full 158 days and some survived the stress of shortened day length, it is likely that some of them could have continued indefinitely on 10 ppm. On the other hand, survivors of the 10 ppm group probably were in a poor position to withstand mobilization of stored residues. Brains of the three surviving males contained only 1.1, 1.5, and 3.0 μg of dieldrin. Their remainders, however, contained 860, 1500, and 2100 μg of dieldrin. About 1 per cent of the stored dieldrin moving into the brain would have resulted in death. Brains of all birds that died on dosage contained dieldrin residues of 4.8 to 27.0 μg, averaging 14.8. Surviving birds, by contrast, contained from 0.2 to 10.0 μg with a mean of 2.7. As survivors of the 10 ppm group had only 1.1 to 3.0 μg, they seem to have maintained low brain residues for the entire 158-day period despite large amounts of dieldrin in their bodies. Others on the same dosage died with relatively high brain levels. Apparently 10 ppm is a marginally tolerated dosage, one that might well be employed in searching for long-term effects on behavior or physiology in *Coturnix*.

The principal finding in the *Coturnix* study was the confirmation of the brain as a useful tissue in appraising probability of death from dieldrin poisoning. Comparisons (wet-weight basis) showed the liver and carcass remainders to be of no value in this appraisal, for residues in dead and survivors did not differ significantly. Even in the brain, however, the separation was not perfect; a few survivors had residues in the range of the dead birds. Hence it is necessary to consider the range of brain residues in the dead to represent a danger zone. Residues in this zone must be considered to be above safe levels. As dieldrin levels in brains do not always distinguish between dead and living, one seldom can say with certainty that a given individual died from dieldrin. As the work of Robinson *et al.* (23) has shown, pigeons and *Coturnix* can live with surprisingly high brain levels of dieldrin.

When dieldrin residues were expressed on a lipid basis, brain residues still provided a useful criterion; liver residues appeared to be far better on a lipid basis than on a wet basis. The lipid base made a nearly complete separation between carcass residues of dead and survivors. Although use of a lipid base does not fulfill the requirement of a stable reference standard, it could have practical value in diagnosis—if consistent.

Residues in wild birds that died in dieldrin-treated areas proved important in this connection. Their brain residues proved reasonably consistent with results of the *Coturnix* study, but their liver and carcass residues expressed on a lipid base were far different from those of *Coturnix*. They were, in fact, quite similar to those of *Coturnix* survivors. These data will be given in detail in the next section.

Residues and Lipids in Animals Dying in Dieldrin-treated Areas

Mortality of wild animals is high in areas treated with dieldrin at two or more pounds per acre. Many die in the same time period, and in these circumstances there is little likelihood of confusing dieldrin-induced mortality with mortality from other causes. Wild animals live under quite different conditions from caged ones, and many different species may be exposed. Hence, the residues in *Coturnix* dying from dieldrin dosage in captivity were compared with those in field-collected dead animals to check the applicability of the conclusions. The samples of wild species came from two areas. One of these was an area in Tennessee heavily treated with granular dieldrin to eliminate an infestation of the white-fringed beetle. According to Mr. Summer Dow, United States Bureau of Sport Fisheries and Wildlife, Atlanta, Georgia, the weather was dry at the time of application and no mortality of wildlife occurred for some time, but when rain came and released dieldrin from the granules, the birds and the mammals began to die. Some of these were sent to Patuxent by Mr. Dow and later were sent to the Stoner Laboratories, Campbell, California, for analysis.

The other was a large area around Erie, Pennsylvania, treated early in 1966 with granular dieldrin at the rate of three pounds per acre for elimination of the European chafer beetle. Numbers of birds were found dead in the treated area. Some of these were sent to Patuxent through the cooperation of Mr. James Stull of Waterford, Pennsylvania, and Mr. C. R. Studholme, United States Bureau of Sport Fisheries and Wildlife, University Park, Pennsylvania. Eleven birds of three species for which we had no dieldrin brain analyses were selected, and brains were sent to the Wisconsin Alumni Research Foundation, Madison, for analysis. Data for the field series (Tables 8-IV and 8-V) will be discussed together. All brain residues for these series were within the 95 per cent confidence limits calculated for *Coturnix*. All but two were within the actual range of brain residues of dieldrin-killed *Coturnix*. A cotton rat with 5.62 ppm was barely below the 6.2 ppm of a *Coturnix*. A robin with 5 ppm in the brain was seen dying in convulsions, which are usual in dieldrin poisoning.

Residues for the field series fell generally in the lower portion of the *Coturnix* range and had lower means. The differences among them were not statistically significant (analysis of variance), however, unless female *Coturnix* were included, or unless all wild species were combined for the comparison. Residues in brains of female *Coturnix* that died were significantly higher than in males

(P <0.05 on a lipid basis; P <0.20 on a wet-weight basis). The array of means was as follows:

Coturnix females	21.7 ppm (wet)	274 ppm (lipid)
Coturnix males	14.8	161
Cottontail rabbit	13.8	—
Starling	12.5	205
Robin	9.6	152
Meadowlark	9.3	137
Cotton rat	7.9	94
Woodcock	7.5	126

Since residue readings for liver and carcass were available only for meadowlarks, cotton rats, and rabbits, brain residues for these three also were compared separately; in this comparison, dieldrin residues in brains of rabbits proved to be significantly higher (P <0.05) than those of cotton rats; dieldrin residues in brains of *Coturnix* males were significantly higher (P <0.05) than those of meadowlarks or cotton rats.

Thus, although it appears that dieldrin poisoning can be judged from brain residues, the variation among species, and particularly the tendency toward lower residues in several wild species than in captive *Corturnix,* suggest that the degree of danger represented by the lower levels is greater than would have been anticipated from the experiment. Several reasons for lower levels in field specimens can be hypothesized:

1. Species differences.
2. Greater stress and activity in the wild.
3. Tendency for individuals that will die with low brain residues to be over-represented among field dead. This will be discussed later.
4. Postmortem breakdown of dieldrin.

This seems unlikely, for dieldrin is stable and specimens were in good condition, not decayed.

Dieldrin residues (wet weight) in livers of *Coturnix,* cotton rats, meadowlarks and rabbits did not differ significantly. On a lipid basis, however, residues in *Coturnix* livers were significantly higher than those in cotton rats or meadowlarks (data not available for rabbits). In fact, lipid-base residues in livers of wild species were similar to those in livers of *Coturnix* that survived.

TABLE 8-IV

RESIDUES IN ANIMALS FOUND DEAD IN A DIELDRIN-TREATED AREA IN TENNESSEE

BRAINS

Species	Sex	Per cent lipid % wet wt.	Dieldrin ppm wet	Dieldrin ppm lipid	DDT ppm wet	DDE ppm wet	Lindane ppm wet	H.E.* ppm wet
Meadowlark (*Sturnella magna*)	♂	5.8	8.75	150.9	0.37	0.49	0.57	nd†
	♂	7.9	8.85	112.0	nd	2.84	0.86	nd
	♂	6.5	8.65	133.1	nd	0.95	0.24	nd
	♂	7.6	8.40	110.5	nd	1.40	1.43	nd
	♂	6.3	12.10	192.1	nd	0.86	2.36	nd
		6.8‡	9.26§	136.7§				
Cotton rat (*Sigmodon hispidus*)	♂	8.3	11.10	133.7	1.31	.0.67	0.25	0.50
	♂	9.3	8.08	86.9	0.44	0.17	0.19	0.17
	♂	8.1	5.62	69.4	0.83	0.26	0.29	0.34
	♂	7.4	9.91	133.9	0.62	0.36	0.21	0.25
	♀	9.3	6.20	66.7	0.48	0.17	0.30	0.21
		8.5‡	7.92§	93.6§				
Cottontail rabbit (*Sylvilagus floridanus*)	♂	—	17.80	—	0.20	0.05	0.02	0.11
	♂	—	9.25	—	0.09	0.03	nd	0.06
	♀	—	8.45	—	0.20	0.05	0.02	0.15
	♀	—	18.70	—	0.03	0.03	0.01	0.03
	♀	—	19.10	—	0.09	0.02	0.02	0.05
			13.78§					

LIVERS

Species	Sex	Per cent lipid % wet wt.	Dieldrin ppm wet	Dieldrin ppm lipid	DDT ppm wet	DDE ppm wet	Lindane ppm wet	H.E.* ppm wet
Meadowlark (*Sturnella magna*)	♂	2.0	7.9	394.5	nd†	1.12	nd	nd
	♂	2.5	14.5	580.0	1.39	4.53	0.43	nd
	♂	2.6	15.9	611.5	nd	2.00	nd	nd
	♂	2.9	13.5	465.5	nd	1.30	nd	nd
	♂	2.3	15.6	678.3	0.53	0.70	nd	0.22
		2.5‡	13.1§	535.8§				
Cotton rat (*Sigmodon hispidus*)	♂	3.7	26.6	718.9	5.50	nd	0.67	nd
	♂	2.7	18.7	692.6	0.07	0.03	0.03	0.06
	♂	4.5	15.1	335.6	0.26	nd	1.16	0.20

			CARCASS REMAINDERS				
♂	3.7	23.5	635.1	0.21	0.14	0.06	0.11
♀	3.3	36.8	1115.1	nd	nd	0.35	nd
	3.6‡	23.0§	652.6§				
Cottontail rabbit (Sylvilagus floridanus)							
♂	—	103.0	—	0.05	0.02	0.01	0.01
♂	—	71.3	—	0.12	0.06	nd	0.05
♂	—	28.2	—	0.04	0.02	0.01	0.01
♀	—	32.6	—	0.07	0.03	0.02	0.04
♀	—	39.4	—	0.06	0.03	nd	0.03
	—	48.4§	—				
Meadowlark (Sturnella magna)							
♂	3.4	2.78	81.8	nd†	0.14	nd	nd
♂	2.9	3.33	114.8	nd	1.22	nd	nd
♂	2.0	5.54	277.0	nd	0.78	nd	nd
♂	2.6	5.22	200.8	nd	0.72	0.04	nd
♂	2.5	4.18	167.2	nd	0.10	nd	nd
	2.7‡	4.07§	154.3§				
Cotton rat (Sigmodon hispidus)							
♂	2.4	4.52	188.3	nd	nd	nd	nd
♂	3.1	9.13	294.5	nd	nd	nd	nd
♂	2.5	2.38	95.2	tr¶	nd	nd	nd
♂	2.9	3.27	112.7	0.12	0.14	0.07	0.15
♀	4.1	5.62	137.1	tr	tr	nd	nd
	3.0‡	4.48§	152.2§				
Cottontail rabbit (Sylvilagus floridanus)							
♂	—	14.90	—	0.05	0.03	0.02	0.02
♂	—	11.30	—	0.05	0.03	0.02	0.03
♀	—	4.14	—	0.04	0.02	0.01	0.02
♀	—	4.15	—	0.03	0.02	0.01	0.02
♀	—	4.22	—	0.04	0.01	0.01	0.03
	—	6.57§					

*Heptachlor expoxide.
†nd = none detected.
‡Arithmetic mean.
§Geometric mean.
¶tr = trace.

Wild species died with more liver lipids (2.5%, meadowlarks; 3.6, cotton rats) than did *Coturnix* (1.6%). It seemed possible from *Coturnix* data that lipid-based liver residues might be of diagnostic value, but this did not hold true across species lines.

Dieldrin residues in remainders averaged 17.2 ppm in *Coturnix,* but only 4.1 in meadowlarks, 4.5 in cottonrats, and 6.6 in rabbits. Residues in remainders of the wild species were much like those of surviving *Coturnix.* Differences among the three wild species were not statistically significant. The same result prevailed for residues expressed on a lipid basis. Cotton rats and meadowlarks died with a higher lipid content (3.0 and 2.7% respectively) than did *Coturnix* (1.2%). Thus, the relatively clear-cut differences in lipid-based residues in remainders of dead and surviving *Coturnix* are not useful in translation to species in the field.

Literature Records

Robinson, Brown, Richardson, and Roberts (21, 23) conducted important experiments with *Coturnix* and pigeons and

TABLE 8-V
RESIDUES IN BRAINS OF BIRDS FOUND DEAD IN A DIELDRIN-TREATED AREA IN PENNSYLVANIA

Species	Per cent lipid % wet wt.	Dieldrin ppm wet	Dieldrin ppm lipid	DDE ppm wet	DDD ppm wet
Robin	6.7	13.0	194.0	1.60	nd*
(*Turdus*	6.7	17.0	253.7	3.40	nd
migratorius)	6.2	9.8	158.1	1.50	nd
	5.9	5.0	84.7	0.50	0.22
	6.5	10.0	153.8	1.50	0.13
	6.6	7.3	110.6	0.77	0.15
	6.0	9.8	163.3	3.10	0.19
	6.4‡	9.6§	151.5§		
Starling	6.3	12.0	190.5	1.20	nd
(*Sturnus vulgaris*)	5.9	13.0	220.3	1.50	tr†
	6.1‡	12.5§	204.8§		
Woodcock	6.2	6.3	101.6	0.84	nd
(*Philohela minor*)	5.8	9.0	155.2	0.62	nd
	6.0‡	7.5§	125.6§		

*nd = none detected.
†tr = trace.
‡Arithmetic mean.
§Geometric mean.

reached many of the conclusions that we have, including those concerning diagnostic values of brain residues. They found no significant differences in residues between birds dying on different dosages. This was true whether single doses or chronic doses were used. They also noted that there was no correlation between brain residues and time to death. They found no significant difference between sexes in brain residues. Liver residues in dead and survivors were more distinct in their study than in ours, but were significantly correlated with time of death, which reduces their value in diagnosis of death among animals whose level and duration of exposure are unknown.

Mean brain levels for *Coturnix* were 17.4 ppm, somewhat higher than in ours which were 14.8 for males and 21.7 for females. Surviving *Coturnix* proved to have brain residues of from 3.1 to 15 ppm with a geometric mean of 6.9, also higher than those in our study. These surprisingly high residues occurred in the living because the birds were on dosages of 20 to 40 ppm for 50 to 72 days and were sacrificed while on treatment.

Domestic pigeons were similarly studied and the dead were found to have a mean brain residue of 20.0 ppm. Pigeons that survived a single dose of 100 mg/kg and had ten days for cleanup varied in brain residues from 1.6 to 7.5 ppm, averaging 3.8. Four pigeons that survived 180 days on a diet of 50 ppm contained brain residues ranging only from 2.15 to 8.5 ppm, averaging 5.8.

In another study (24) Robinson and his co-workers fed pigeons 50 ppm dieldrin for six months and then sacrificed them at intervals while on a normal diet. They estimated that the brains of these pigeons contained an average of 7.7 ppm dieldrin at the end of dosage. The rate of decline on cleanup indicated a biological half-life of dieldrin in the brain of 57.1 days.

The pigeon data in these two papers are of much interest in showing that although lethal brain residues for pigeons are not above those known for other experimental animals, pigeons are able to withstand for long periods doses that would kill other birds, including *Coturnix,* much sooner.

The pigeon should not be considered representative of birds generally in its ability to tolerate pesticides. We believe that few species of birds could tolerate 50 ppm of dieldrin for six months. Chickens — also commonly used in pesticide work — can be misleading in the same way. Gannon (6) has shown that chickens are far more resistant to heptachlor, heptachlor epoxide, dieldrin, and DDT than are *Coturnix* or English sparrows.

Koeman and van Genderen (12) present significant brain residues for birds found dead or dying. We list in Table 8-VI the four of these that are most convincing of death from dieldrin and agree with the authors that records of convulsions significantly support the residue evidence. The authors mention terns that were seen falling from the air, crying, trembling, and unable to fly. Where brain residues are available for these birds, some dieldrin appears, but the authors suspect that the endrin reported was causative, and we concur. Endrin was not recorded in the brains for which we list the dieldrin levels.

The perceptive paper by Harrison *et al.* (8) points out that brain levels were diagnostic in dogs; that there was no correlation between clinical effects and dieldrin levels in liver; and that quickly killed dogs had relatively low levels in fat. Dogs unquestionably killed by dieldrin or aldrin had brain residues of from 2.8 to 9.4 ppm. These figures are probably too low, for, as the authors point out, they were not corrected for recovery, which ran about 60 to 70 per cent.

Conclusion

Despite the admitted dangers of comparing residues determined by different laboratories using different methods, it is clear that brain residues are far superior to those of other tissues in determining the likelihood that animals were killed or seriously endangered by dieldrin or aldrin poisoning. It is also apparent that there is a large measurement of agreement between species, between birds and mammals, and between laboratories in respect to brain residues. Indeed, most of the brain residues that are demonstrably associated with death from dieldrin fall within limits established for *Coturnix*. In contrast, residues in other tissues often are misleading and their use should be avoided where possible.

TABLE 8-VI
SELECTED PUBLISHED RECORDS OF DIELDRIN BRAIN RESIDUES

Species	Wild or Experimental	ppm, wet weight	Remarks	Source
Dog (*Canis familiaris*)	Experimental	2.8 to 9.4, av. 6.2	Survivors had 0.2 to 3.8, av. 1.3	(8)
Dog	Experimental	10, 12 Includes aldrin		(10)
Heifer (*Bos taurus*)	Experimental	12		(10)
Cow (*Bos taurus*)	Experimental	16		(10)
Steer (*Bos taurus*)	Experimental	18		(10)
Ewe (*Ovis aries*)	Experimental	30		(10)
Cottontail rabbit (*Sylvilagus floridanus*)	Nearly wild	26, 31, 31	Large open pens, treated 2 lbs dieldrin per acre Survivors had less than 1 ppm	(1)
Japanese quail (*Coturnix coturnix*)	Experimental	95% in 8.7-34.6 range; geometric mean 17.4	Survivors had 3.1 to 15.0; geometric mean 6.9	(23)
Domestic pigeon (*Columba livia*)	Experimental	95% in 13.5-32.5 range; geometric mean 20.0	Survivors had 1.6 to 8.5; geometric mean 3.6	(23)
Pheasant (*Phasianus colchicus*)	Experimental	10, 18		(15)
Shoveler duck (*Spatula clypeata*)	Wild	10, 20	Dead at lake receiving pesticide from factory	(25)
Green-winged teal (*Anas carolinensis*)	Wild	9	Dead at lake receiving pesticide from factory	(25)
Lesser scaup duck (*Aythya affinis*)	Wild	7.7, 10.6, 13.3, 16.0	Dead at lake receiving pesticide from factory	(25)
Kestrel, British (*Falco tinnunculus*)	Wild	9.8, 12.0	Found dead	(30)
Peregrine falcon (*Falco peregrinus*)	Captive	7.8	Convulsions	(12)
Buzzard (*Buteo buteo*)	Wild	4.4	Heavy convulsions	(12)
Sparrow hawk (*Accipiter nisus*)	Wild	6.7	Found dead	(12)
Barn owl (*Tyto alba*)	Wild	8.1	Convulsions	(12)

Although specimens may lie in the field so long that brains are in no shape for analysis, in our experience with both small birds and eagles, brains of many animals found dead in the field can be removed in adequate condition for analysis, even after long periods in the freezer. One precaution is necessary: the brain of a frozen specimen should be removed before the body is thawed. Long thawing of previously frozen brains promotes liquefaction.

If we assume that the residues reported by Harrison *et al.* (8) for dog brains are a little too low, and if we accept the convulsing Dutch *Buteo* with 4.4 ppm, we may conclude that the lowest demonstrably lethal brain residues for dieldrin are around 4 or 5 ppm. Lower levels may be valid, but they have not been satisfactorily demonstrated. To judge from unpublished data on experimental American kestrels at Patuxent, and from hawk records in Table 8-VI, members of the hawk group are not exceptions to what has been said.

It is important to note that brain residues recorded for wild animals usually do not come up to means established for *Coturnix* or pigeons. This may represent species differences, or it may reflect the greater stresses and exertions to which wild animals are subject. Another factor may be indicated by the fact that the first bird to die in each of our *Coturnix* sex and treatment groups had the lowest brain residues of its group. This suggests that the individuals that will die with the lowest brain residues tend to be the ones that die first. Such individuals may be over-represented among animals found dead in the field unless mortality is sweeping. Complete kills of certain bird populations under heavy treatment are well documented, but dangerous applications generally cause only partial mortality in the wild. Examples of partial mortality after heavy dieldrin applications came from Norfolk, Virginia, and Battle Creek, Michigan, both of which were treated with two pounds per acre of granular dieldrin for control of white-fringed beetle (Norfolk) or Japanese beetle. State game biologists who were present stated that although it was easy to pick up dead birds, the extent of the loss would not have been guessed from the numbers that remained (18). Under such conditions, one might expect an over-representation of low but lethal brain residues and under-representation of truly high residues.

Even in our *Coturnix* experiment, there may have been a tendency for the lower brain residues to be over-represented, for as soon as half of a group died the rest were given clean food and sacrificed after two days. By this system, the birds that would have died with the highest brain residues presumably were among the survivors. This is not certain, however, for resistance to poisoning depends on the liver, lipid levels, and other defense mechanisms (which help keep brain levels low) at least as much as it depends on the resistance of the brain itself to the toxicant.

Many workers have shown that dieldrin is broken down and excreted from the body and have given considerable information on rates and metabolites (14, 19, 20, 22, 24). In addition, there are at least three studies showing the surprising fact that dieldrin residues declined sharply after long periods on dosage and while dosage continued. This was found for rats, dogs, and chickens (3, 17, 22). As long time periods occurred before residues fell, especially in dogs and chickens, it seems likely that a process other than the simple induction of liver enzymes was involved. Certainly our usual concept of a given intake resulting in a given equilibrium level applies poorly to this chemical. Long-term projections are therefore risky.

On the other hand, as Cummings *et al.* demonstrated (5), dieldrin is one of the most accumulative pesticides over periods of a few weeks and its rate of loss is one of the slowest. Heath and Vandekar (9) have shown that repeated doses may have lethal effects by causing mobilization of residues from earlier doses. Sublethal effects, such as suppression of spontaneous motor activity (31), presumably could be brought on in the same way.

Clearly, dieldrin is a truly dangerous chemical and residues of, let us say, 1 ppm in the brain cannot be regarded lightly, for when brain residues are that high, body residues are many times as great. A *Coturnix* that had 1.5 μg (1.7 ppm) in the brain had 1500 μg (21.7 ppm) in the carcass remainder. It must be remembered that a few micrograms in the brain represent a fatal amount for a small bird, and that the body of a well-contaminated bird contains hundreds or thousands of micrograms, many of which would tend to enter the central nervous system at any time of fat depletion. Seasonal fat cycles are important, for birds tend to lose weight in spring or summer and pesticides in their tissues then become concentrated. This may occur at a time when stresses of maintaining territories or rearing young are pressing. It is well known that a fat animal or an animal that is gaining weight can stand more pesticide than an animal that is thin or is losing weight. One flight — or one fight — by a bird that was none too fat to begin with could result in lethal mobilization. That this is more than a theoretical consideration is shown by records (12, 27) of pesticide-laden birds falling from the sky.

APPENDIX

Hazelton Laboratories: *Sample size:* livers and brain in entirety; carcasses 25 g subsamples of homogenate. *Cleanup:* acetonitrile partitioning (acetonitrile-petroleum ether) and passage through florisil column with two elutions, the first 6% ethyl ether in petroleum ether, the second 15% ethyl ether in petroleum ether.

Analysis: Glowall Corporation "Chromalab" gas chromatograph equipped with a 22.5 μCi radium 225 electron capture detector; column ID glass, 6 feet, 3 mm, packed with 10% DC 200 (12,500 centistokes) on 80 to 90 mesh Anakrom ABS; column temperature 210°C; airflow 120 ml per minute with nitrogen.

Stoner Laboratories: *Sample size:* livers and brains in entirety; carcasses 20 g subsamples of homogenate. *Drying and Extraction:* sodium sulfate, isopropyl alcohol, and petroleum ether in a blender. *Cleanup:* acetonitrile partitioning (16) and passage through florisil column with two elutions, the first 5% ethyl ether in petroleum ether, the second 15% ethyl ether in petroleum ether.

Analysis: Dohrman microcoulometric gas chromatograph; column 4 to 5 foot, $\frac{1}{4}$-inch, packed with Epon 1001 2.5% on acid washed chromosorb; temperature 210 to 215°C; airflow 150 ml per min with nitrogen.

Wisconsin Alumni Research Foundation: *Sample size:* livers and brains in entirety; carcasses, 10 g subsamples of homogenate. *Dry-*

ing and Extraction: air oven at 40 to 45°C followed by blending with sodium sulfate; 8 hours in Soxhlet extractor with ethyl ether-hexane (1:3). *Cleanup:* two elutions through a florisil column, the first 5% ethyl ether in hexane, the second 15% ethyl ether in hexane.

Analysis: Barber-Coleman Model 5000 gas chromatograph with ^{90}Sr detector; column, glass, 4 feet, $\frac{1}{4}$-inch, packed with 5% DC 200 (12,500 centistokes) on Cromport XXX 60 to 70 mesh; column temperature 210°C; airflow 120 ml per min with nitrogen.

REFERENCES

1. ALLEN, S.H.: Effects of Dieldrin on Reproduction of Confined Cotton-tails. M.A. thesis, Univ. Mo., 1968, p. 82.

2. BLACKMORE, D.K.: *J Comp Path Ther, 73*:391, 1963.

3. BROWN, V.K.; RICHARDSON, A.; ROBINSON, J., and STEVENSON, D.E.: *Food Cosmet Toxicol, 3*:675, 1965.

4. CONNEY, A.H.; WELCH, R.M.; KUNTZMAN, R., and BURNS, J.J.: *Clin. Pharmacol Ther, 8* (1) (pt. 1) :2, 1967.

5. CUMMINGS, J.G.; EIDELMAN, M.; TURNER, V.; REED, D.; ZEE, K.T., and COOK, R.E.: *J Assoc Offic Agr Chemists, 50* (2) :418, 1967.

6. GANNON, N.: *Proc N Cent Branch ESA, 15*:39, 1960.

7. VAN GENDEREN, H.: *Mededel Landbouwhogeschool Opzoekingssta Staat Gent, 33* (3) :1321, 1965.

8. HARRISON, D.L.; MASKELL, P.E.G., and MONEY, D.F.L.: *New Zeal Vet J, 11*:23, 1963.

9. HEATH, D.F., and VANDEKAR, M.: *Brit J Industr Med, 21* (4) :269, 1964.

10. KITSELMAN, C.H.; DAHM, P.A., and BORGMANN, A.R.: *Amer J Vet Res, 11*:378, 1950.

11. KLION, F.: *Fed Proc, 23* (2) (pt. 1) :2707, 1964.

12. KOEMAN, J.H., and VAN GENDEREN, H.: *Mededel Landbouwhogeschool Opzoekingssta Staat Gent, 33* (3) :1879, 1965.

13. LANGGÅRD, H., and SECHER-HANSEN, E.: *Acta Pharmacol Toxic, 25*:267, 1967.

14. LUDWIG, G.; WEIS, J., and KORTE, F.: *Life Sci, 3*:123, 1964.

15. McEWEN, L.C., et al.: *US Fish Wildlife Serv Circ, 167*:68, 1963.

16. MILLS, P.A.: *J Assoc Offic Agr Chem, 42* (4) :734, 1959.

17. MITCHELL, L.E.: Pesticides: properties and prognosis. *In Organic Pesti-cides in The Environment.* Advan. Chem. Ser. 60, 1966, chapt. 1.

18. Oral communications from Robert B. McCartney (Va.) and C. T. Black (Mich.).

19. QUAIFE, M.L.; WINBUSH, J.S., and FITZHUGH, O.G.: *Food Cosmet Toxic, 5*:39, 1967.

20. RICHARDSON, L.A.; LANE, J.R.; GARDNER, W.S.; PEELER, J.T., and CAMP-BELL, J.E.: *Bull Environ Contam Toxic, 2* (4) :207, 1967.
21. ROBINSON, J.: *Chem Industr,* 1967 p. 1974.
22. ROBINSON, J.: Papers presented by Shell Res. Ltd., VI Intern. Plant Protec. Congr., 1967, p. 88.
23. ROBINSON, J.; BROWN, V.K.H.; RICHARDSON, A., and ROBERTS, M.: *Life Sci, 6* (11) :1207, 1967.
24. ROBINSON, J.; RICHARDSON, A., and BROWN, V.K.H.: *Nature, 213* (5077) : 734, 1967.
25. SHELDON, M.G., et al.: *US Fish Wildlife Serv Circ, 167:*73, 1963.
26. STICKEL, L.F.; STICKEL, W.H., and CHRISTENSEN, R.: *Science, 151* (3717) : 1549, 1966.
27. STICKEL, W.H.: *Mass Audubon, 51* (3) :112, 1967.
28. TINSLEY, I.J.: *J Agr Food Chem, 14:*563, 1966.
29. *US Bur Sport Fisheries Wildlife Resource Publ, 23:*52, 1966.
30. WALKER, C.H.; HAMILTON, G.A., and HARRISON, R.B.: *J Sci Food Agr, 18:*123, 1967.
31. WEISS, L.R., and BRODIE, R.: *Toxic Appl Pharmacol, 11* (2) :7, 1968.

DISCUSSION

HERMAN: Is it true that the actual number of micrograms of dieldrin in the brain is not a better measure than ppm wet weight in brains of birds which have survived and which have died?

STICKEL: That is correct. There were survivors with as much as 9 or 10 μg of dieldrin in the brain, and some died with less than that amount.

RISEBROUGH: Was the sex of the bird a significant factor?

STICKEL: It seemed to be, although we do not wish to make this generalization from our small treatment groups. We found that the mean content of dieldrin in the male brain was 14.8 ppm, wet weight, while that for the female was 21.7 ppm. The 95 per cent confidence limits were as low as 4.9 ppm for the males and 10.4 ppm for the females.

RISEBROUGH: Does not then the sex differences appear to be significant?

STICKEL: The difference between sexes was significant for certain treatment groups, but not overall. On a dose of 50 ppm, the males died long before the females. A more extensive study probably would show that males are more susceptibe to poisoning with dieldrin. This is usually true of poisoning by pesticides.

RISEBROUGH: I would suggest a link with the hormone balance of each sex.

RUDD: It seems to me that there can be some doubt about the applicability of methodology. Is it possible to extrapolate information from a captive species whose conditions do not represent the wild?

STICKEL: It is necessary to keep comparing data from experiments with data from field series. In the present study, the range of values established for dieldrin in brains of laboratory animals was applicable to findings from the field. When we found this was true for a variety of species, both birds and mammals, we felt it was fair to extrapolate to other species. There is no great difference between birds and mammals as far as brain levels of dieldrin are concerned. This also appeared to be true in our study of DDT.

RUDD: Is there not an unstated but implicit purpose in using a threshold residue level in wild species? It is essentially a managerial one. Would you conduct your research in a different way if this implicit goal of regulating were absent?

STICKEL: I would not say that the goal was managerial. One of the main reasons we determine lethal levels is to enable us to answer specific, practical questions on the death of animals in the wild. It is to the advantage of applicators and chemical companies as well as conservationists to be able to determine whether or not animals were killed or endangered by a given chemical. Another reason for determining lethal levels is to give us perspective for judging low brain levels.

WURSTER: With regard to the different susceptibility of males and females to pesticides, there is evidence that both DDT and dieldrin interfere with the function of calcium on the nerve axon [Gordon, H.T., and Welsh, J.H.: *J Cell Comp Physiol, 31:* 395, 1948]. Both are nerve poisons. Higher calcium concentrations in the blood give a lowered susceptibility to the poison. Females secrete estrogen which increases blood calcium; therefore, nerves of females are bathed in a higher calcium concentration than are those of males. We should, therefore, expect the females to be more resistant to DDT or dieldrin than the males, and to succumb at a higher concentration.

WELCH: Is it not possible for a small amount of insecticide to create a more subtle biochemical alteration leading to a change in a normal physiological process?

STICKEL: As far as I know, this is not characteristic of chlorinated hydrocarbons. When a chlorinated hydrocarbon, which affects the nervous system, is used in sufficient quantity to kill, a corresponding lethal level is found in the nervous system. However, such data are not available for all chemicals of the group.

WELCH: DDT is metabolized to DDE which is then stored in fat. What per cent of the total DDT ingested is converted to DDE? It might be only a small percentage, therefore the DDE levels in fat may be a poor index of the total amount of DDT consumed by the bird. The DDT intake may be large enough to change a physiological process in a way which will kill the animal with a delay of weeks or months. Is this a plausible concept? When the bird dies only the DDE residue would now be used to account for the death.

SPENCER: Many of the metabolites in such cases were ingested as metabolites to begin with and were not a result of metabolism within the bird.

WELCH: Do we really know that?

STICKEL: Yes, this is well established. Another partial answer to the question is that for DDE to cause death or injury it must be present in large amounts in the brain, in levels much higher than would be needed for a combination of DDT and DDD. I know of no case where a subtle lingering effect caused death at a later time when brain residues were low.

WELCH: When fields are sprayed with DDT, is it not the case that birds consume a large dose of DDT?

STICKEL: Birds may consume both DDT and DDE. Insects covert DDT to DDE rapidly and in large proportions; animals in turn, eat these DDE-laden insects and consequently may have extremely high DDE residues.

WELCH: The percentage of each DDT metabolites forward from DDT is not really known?

STICKEL: The percentage usually is not known, because DDT compounds are received through the food chain and the food chain causes metabolic shifts. To know, one would have to analyze specified foods at specified times.

RUDD: Are you trying to stress, Dr. Welch, that DDE is supposedly the less toxic fraction and yet may be responsible for death?

WELCH: I think it is very important to establish whether animals are consuming DDT or DDE.

RUDD: This will vary with location and food item.

STICKEL: I know of no instance of delayed death accompanied by low brain residue levels in the chlorinated hydrocarbon group. Lethal brain levels of endrin are very low, but delayed effects are unlikely for endrin is degraded and excreted rapidly.

BERG: Is there any correlation between the concentration of an organochlorine compound, for example, dieldrin, in the brain of a female quail and in the yolk of the eggs from the same bird?

STICKEL: I wouldn't want to try to answer this positively, but I would guess there is not, because the chemical that goes to the yolk represents current intake in large part. It represents in part stored residues from the fat of the animal. If the animal is functioning normally, its brain level will be held very low in comparison with the yolk.

BERG: Then the yolk is not as protected as the brain. Quite the opposite actually, it serves as a route of excretion of lipids and of lipid-soluble poisons for the egg-laying bird.

RADOMSKI: Has it been firmly established that birds can metabolize DDT to DDE?

STICKEL: Yes; this occurred when birds were fed p,p'-DDT; some DDE appeared almost immediately. DDE is continually generated in the body from stored residues of DDT, even when the bird is on clean food.

WELCH: In your studies, what per cent of the total DDT fed to birds is converted to DDE?

STICKEL: This would depend on what time after the initial

exposure a sample was taken. The change proceeds continually but slowly over a period of months.

ROBINSON: We observed a strong correlation among residues found in blood, brain, liver and eggs in chickens fed 20 ppm, 10 ppm, and also lower doses of dieldrin for 9 months.

STICKEL: Weren't the brain levels very low compared with the levels in the other organs during the time the animal was still healthy and alive?

ROBINSON: Yes, but there is still a step-like progression in concentration of residue correlated with the concentration in the feed.

Chapter 9

THE SIGNIFICANCE OF DDT RESIDUES IN ESTUARINE FAUNA

P. A. BUTLER

ABSTRACT

A nationwide program was initiated in 1965 to monitor residues of ten synthetic, chlorinated hydrocarbon pesticides in estuarine populations of fish and shellfish. About 160 stations have been established where samples are collected at thirty-day intervals for analysis by gas chromatography with electron capture. The summarized data show that estuarine pollution levels reflect the intensity of agriculture in the associated river basin. Most of the positive analyses show residue levels in the range of 10 to 200 μg/kg of DDT, DDE, or DDD; dieldrin and endrin residues are typical of a few estuaries. Because of occasional residues in the range of 10 to 20 μg/g in fish and oysters, experiments were undertaken to determine effects of a DDT-contaminated diet on fish and crustaceans. Dietary levels of 2 to 5 μg/g of p,p'-DDT caused 35 to 100 percent mortality within two to ten weeks in laboratory populations of shrimp, crabs, and fish.

Animals killed by the diet usually contained significantly lower body residues of DDT than randomly selected living animals on the same diet. There was, however, essentially no correlation between the amount of DDT residue and the size of the animal or the length of time it fed on the contaminated food. The experimental and monitoring data indicate that existing widespread pesticide pollution is causing significant decreases in productivity of estuarine populations of fish and shellfish. Resistant surviving animals are instrumental in concentrating and transmitting lethal amounts of pesticide residues in the food web.

REPORT

Introduction

CONCERN ABOUT POLLUTION by the persistent, or so-called hard pesticides has increased greatly and with good reason in the past few years. Substantial evidence is accumulating that these chemicals, particularly DDT, dieldrin, and endrin, are being recycled and concentrated in the food web to the extent that mortality in nontarget animal populations is increasing significantly at sites and times far removed from those of the initial application of the pesticide.

Reproductive failure of the lake trout in New York (1) and high mortalities in hatchery-reared coho salmon have been clearly associated with DDT residues in the eggs. A similar cause has been postulated for declining reproductive success in the Bermuda petrel (8) and other carnivorous birds. The accumulation of heavy burdens of DDT in marsh sediments (7), its persistence for more than fifteen years in soils (5), its aerial dispersal in particulate form (6), and its worldwide usage help explain its nearly universal occurrence in the world biota.

The apparently low mammalian toxicity of DDT and its recognized value in the control of disease vectors as well as insect pests of agricultural importance have supported the thesis that we must continue its use despite the occasional damage it inflicts on beneficial species.

The direct application of DDT and other pesticides to the salt-marsh and estuarine environment has concerned fishery biologists because of the known sensitivity of crustaceans to insecticides. Shrimp make up the single most valuable marine harvest in the United States, and several species of crabs and lobsters have regional importance.

In our early investigations at the Gulf Breeze Biological Laboratory, we demonstrated that each of ten synthetic pesticides evaluated interfered significantly with oyster growth after 24 hours exposure to toxicant levels of 0.1 $\mu g/g$ (4). More sophisticated investigations (3) revealed that oysters were sensitive to as little as 0.0001 $\mu g/g$ of DDT in their environment. At this and even greater dilutions, the oyster stores DDT and may concentrate it as much as $70,000 \times$ the environmental level. These body residues are flushed away at uniform rates when the oyster is placed in unpolluted water.

The usefulness of oysters as a yardstick to bioassay pesticide pollution in the estuary prompted us to undertake a nationwide monitoring program to determine existing levels of pesticide pollution and to identify areas in which marine productivity might be affected already to a significant extent.

This report summarizes seasonal and regional trends in the pesticide residues of estuarine mollusks. Data from controlled laboratory experiments are presented to help in the interpretation of the importance of these residues.

Estuarine Monitor Program

The first samples of oysters were collected in July 1965. The program now has about 160 permanent stations in estuaries with shellfish populations, in fifteen coastal states (2). Populations of any one of 9 species of clams, mussels, or oysters — depending on the estuary — are monitored at each station at thirty-day intervals. About three thousand samples have been collected in duplicate and more than four thousand analyses performed. Gas-liquid chromatography with electron capture is employed with two columns of different polarity so that the following compounds are clearly resolved and can be quantified: aldrin, BHC, dieldrin, DDT, DDD, DDE, endrin, heptachlor, heptachlor epoxide, lindane, and methoxychlor. In samples of special interest, we search occasionally for other pesticides. Identified compounds are measured at levels above 9 μg/kg; when necessary, identifications are confirmed with thin-layer chromatography.

In the present report, all references to DDT include its metabolites DDD and DDE, which are uniformly present when the parent compound is identified. We interpret the presence of a high percentage of DDT as indicating direct exposure to pollution; when the metabolites are present alone or at disproportionately high levels we suspect that the residues have been transmitted through the food web. Despite the wide array of persistent pesticides used in the United States, only DDT, dieldrin, and endrin are commonly present in our monitor samples. Occasionally the residues can be shown to have been derived from a single source. In one analysis, unusually large residues of a pesticide that was not used locally could have been derived only from the effluent of a manufacturing plant in the drainage basin. The presence of Mirex as a residue can be associated directly with fire ant control programs.

The monitor program has demonstrated some clearly defined regional differences in the levels of pesticide pollution in United States estuaries. Table 9-I summarizes data collected during calendar year 1967 in five states where the mollusks have characteristic levels of pesticide residues.

Although the bioassay animals differ — the soft clam is used in Maine and either the eastern or Pacific oyster in the other tabulated states — they are similar in their ability to detect and store the persistent pesticides. Reported residue levels should not be interpreted as precise indices of pesticide pollution levels but rather as indicating relative levels in different areas or at different times. We suspect that most estuarine pesticide pollution is derived from agricultural practices and expect accordingly that estuaries in nonagricultural areas will tend to be less contaminated than those draining large agricultural regions. Samples from Maine and Washington are less contaminated than samples from other states in the monitor program. Conversely, samples collected in California consistently contain relatively high residues as a result of the intensive farming in that state (Table 9-I).

TABLE 9-I
PESTICIDE RESIDUES IN ESTUARINE MOLLUSKS COLLECTED
AT MONITORING STATIONS IN FIVE STATES IN 1967

State	Number of Stations	Number of Samples*	Percentage of Positive Samples	Maximum Level Detected ($\mu g/kg$)		
				DDT†	Dieldrin	Endrin
Washington	18	211	12	98	0	0
Maine	7	78	13	31	0	0
Virginia	10	123	86	256	40	0
Texas	12	129	90	335	85	19
California	12	140	91	1600	39	19

*Collected at 30-day intervals.
†DDT includes the metabolites DDE and DDD.

As an example of the existing extent of pesticide pollution in estuaries, 134 (82%) of 190 mollusk samples collected in the nationwide monitor program in April 1967 contained residues of two or more of the chlorinated hydrocarbon pesticides.

Some understanding of relative changes in the levels of pesticidal pollution from year to year in a given estuary may be gained by arbitrarily averaging the monthly DDT residues in consecutive periods and comparing them on a seasonal or annual basis. Our records of DDT residues are complete for the period July through February for 3 consecutive years at one station in a drainage basin where there is considerable truck farming. The average amount of DDT residue (μg/g) each year was as follows: 1965 to 1966 — 95, 1966 to 1967 — 55, and 1967 to 1968 — 185. The reason for these differences is not now known, but by the end of the projected four years of this monitor program, we expect to be able to correlate observed residues with agricultural practices, meteorological conditions, and basin geography.

That seasonal differences in pesticide residue are susceptible to some degree of interpretation now is shown by Figure 9-1, in which the monthly DDT residue data have been plotted for three representative drainage areas. For each area, the data for each month in 2 consecutive years have been averaged and a bimonthly moving average has been graphed. River basin "A" on the mid-Atlantic coast, is characterized by moderate agricultural development and has no large industries or cities on the watershed. The moderate increase in residues after the spring plantings is predictable. River basin "B" is adjacent to "A"; agricultural development is similar, but in addition, there are several large cities and some light industry. The residues show the same pattern as in "A" but are consistently higher — a possible indication that domestic sewage is an important source of pesticide pollution in the estuary. River basin "C" is on the Texas coast, in an area characterized by intensive truck farming and (because of the warm climate) double cropping. These circumstances are reflected clearly in the pattern and levels of DDT residues in the local population of oysters.

Nonperiodic and random monitoring of estuarine populations have produced data that are difficult to interpret. It is apparent, however, that animals accommodate physiologically to pesticide residues acquired gradually which would be lethal had they been acquired in a short time. We have made numerous analyses of apparently healthy fish that contained residues ten to fifteen times

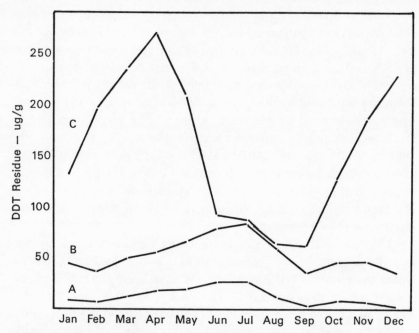

FIGURE 9-1. Seasonal changes in DDT residues in the eastern oyster, *Crassostrea virginica,* in three representative estuaries. "A" represents an estuary draining a river basin with a moderate level of agricultural activity; "B" is an adjacent estuary which receives, in addition, the effluent from several cities and some light industrial plants; "C" is typical of an estuary draining a river basin where there is intensive truck farming and at least two harvests annually.

greater than residues observed in fish killed as a result of gross pesticide pollution. Random samples from populations of shad, menhaden, anchovy, mullet, and oysters have contained DDT residues ranging up to 16 $\mu g/g$. We are concerned about the effect these residues have on predator populations, and whether we can estimate the losses in productivity resulting from the transfer of such pesticide residues into higher trophic levels.

Laboratory Investigations

The Gulf Breeze Laboratory has accumulated a large amount of data in the past ten years that show the relative toxicity of most of the commonly used pesticides when estuarine test animals are

exposed to them for periods of one to four days. Tests are conducted in flow-through systems in which the seawater supply is
continuously replenished and a stock solution of the toxicant is
dripped into the system at a uniform rate. Such test procedures
show that 48-hour TL_{50} values for DDT, dieldrin, and endrin for
various species of crustaceans and fish are usually 1 $\mu g/kg$ or less
within the normal range of environmental salinity and temperature.

Since it appeared that estuarine animals would be more likely
to acquire pesticide residues from their food rather than directly
from the water, a testing "protocol" was established to determine
the effects of a DDT-contaminated diet on contained populations
of fish and crustaceans. This arrangement made it possible to observe small changes in mortality.

The food was prepared by exposing the oysters to a stock solution of p,p'-DDT in a flow-through system until they acquired a
DDT residue of about 27 $\mu g/g$. The oyster meats were then
homogenized and diluted with homogenized, uncontaminated
oysters to obtain the approximate desired level of contamination
in the diet. The food mixes were analyzed by gas chromatography
to determine the actual level of DDT, and then frozen in 1-cm
cubes and stored until needed.

Test populations were maintained in 20-gallon aquaria with a
supply of flowing seawater adequate to ensure their well-being.
The salinity was about 26 pp thous. and the temperature ranged
from 26 to 31°C. The animals quickly learned to accept the frozen
food. They were fed as much as they would consume completely
during a short period while under observation. The amount of
food consumed by individuals varied, but the average consumption of food per gram of animal per day was calculated.

Treatment of dead animals for analysis varied in different experiments. In some tests, dead animals were analyzed individually,
and in others, all animals that died over a period of days were
analyzed as a group. In all experiments, dead animals were first
frozen and then the whole body was homogenized for analysis. In
some tests, apparently healthy animals were removed from the
experimental aquaria and quick-frozen for comparison with

TABLE 9-II

SUMMARY OF TEST CONDITIONS IN FEEDING EXPERIMENTS 1-6

Experiment Number	Animal*	Average Length (mm)	Average Weight (g)	Number of Animals		DDT in Food (μg/g)		Food Eaten per Gram of Animal per Day (g)
				Test	Control	Test	Control	
1	Shrimp	73	3.2	60	60	2.00	trace	1.2
2	Blue crab	50†	12.4	50	50	2.00	0.033	0.3
3	Pinfish	44	3.0	44	44	27.12	0.080	0.3
4	Pinfish	44	3.0	49	49	4.07	0.080	0.3
5	Pinfish	55	4.8	50	50	4.70	0.023	0.2
6	Croaker	99	15.3	53	0	2.57	—	—

*Brown shrimp, *Penaeus*; Blue crab, *Calinectes sapidus*; Pinfish, *Lagodon rhomboides*; Atlantic croaker, *Micropogon undulatus*.
†For blue crabs average width (rather than length) is the measurement given.

animals that had died during the same experiment. Table 9-II summarizes pertinent data for the different tests.

The effects of feeding the DDT-contaminated diet to shrimp, crabs, and fish are summarized in Table 9-III. In all experiments, mortality increased abruptly in the experimental group as compared to the control animals. The mortality in control groups of shrimp and crabs was the result of cannibalism. (In later experiments, cannibalism was avoided by holding crustaceans in individual compartments.)

In the Atlantic croaker experiment, the fish were fed as much as they would readily consume twice daily but the average amount eaten per fish was not calculated. The cumulative mortality reached 50 per cent on the forty-third day, and 100 per cent on the sixty-seventh day, when the last three fish died. No control group was maintained in this experiment. Dead specimens were frozen and analyzed individually for DDT residues. The average residue was calculated for all fish that died in the fifth and tenth weeks.

The experimental evidence is clear that residues in animals dying because of a DDT-contaminated diet may be substantially lower than DDT residues in natural populations of fish and shellfish. The experimental conditions were artificial and biased by the fact that the entire diet of the test animals was contaminated. Undoubtedly, the added stress of test conditions may have contributed to the increased mortality. Although test animals fed readily, behavioral changes were obvious. The crustaceans, especially, were unnaturally quiet when undisturbed. The shrimp were so inactive that silt collected on them. On the other hand, they were much more sensitive to mechanical and visual stimuli than animals in control groups. Such behavioral changes may contribute substantially to changes in the pesticide-induced death rate in the natural environment.

The pattern of mortality demonstrated in the several experiments is of considerable importance. In the work with crustaceans, increased mortality in both experimental and control groups due to cannibalism masks the gradual mortality increase due to DDT. The tests with pinfish show that mortality increase resulting from a massive dietary intake of DDT (experiment 3) parallel instances

TABLE 9-III
EFFECTS OF A DDT-CONTAMINATED DIET ON
CRUSTACEANS AND FISH*

Experimental Number, Animal, and Elapsed Time (Days)	Experimental Animals			Control Animal	
	DDT Residue (µg/g)† Live	Dead	Percentage Cumulative Mortality	DDT Residue (µg/g)	Percentage Cumulative Mortality
No. 1, Shrimp					
0	0.009(24)	—	—	—	—
5	—	0.490(4)	6.7	trace(3)	5
7	0.341(10)	—	—	—	—
10	—	0.410(16)	33.3	trace(3)	10
14	0.410(10)	—	—	—	—
15	—	0.262(14)	57.0	trace(2)	13
No. 2, Blue crab					
0	0.031(10)	—	—	—	—
7	—	1.300(8)	24.0	0.071(8)	22
14	—	0.930(8)	54.0	0.059(8)	38
No. 3, Pinfish					
0	0.012(10)	—	—	—	—
1	—	—	13.5	—	—
2	—	14.560‡			
		7.890(30)	63.5	0.134(10)	0
No. 4, Pinfish					
0	0.134(10)	—	—	—	—
7	—	1.970(1)	2.0	—	—
14	—	3.300(4)	10.0	—	—
21	—	5.620(12)	35.0	0.153(10)	0
No. 5, Pinfish					
0	0.066(10)	—	—	—	—
5	—	0.640 (5)	10.0	—	—
7	2.700(10)	—	—	—	—
10	—	0.550 (17)	44.0	—	—
14	4.230(10)	—	—	—	—
15	—	0.780(3)	50.0	0.067(10)	4.0
No. 6, Croaker					
0	0.127(10)	—	—		§
14	—	—	2.0		
21	—	—	6.0		
28	—	—	19.0		
35	—	0.641(9)	36.0		
42	—	—	49.0		
49	—	—	57.0		
56	—	—	64.0		
63	—	—	80.0		
67	—	1.023(11)	100.0		

*See Table 9-II for experimental conditions.
†Average residue per fish, whole body, wet weight; includes DDD and DDE; number of fish analyzed shown in parentheses.
‡Residue in gastrointestinal tract and remainder of body respectively.
§No control fish maintained.

of gross pollution in nature that result in mass mortalities of fish within a one- or two-day period.

In pinfish tests in which the dietary intake of DDT was below 5 μg/g (experiments 4 and 5) mortality increases were slight in the first week but increased markedly as feeding continued. The day-to-day increases in death rate, however, remained so low that they would be almost impossible to detect in natural populations of motile aquatic animals.

In experiment 5, residue analyses were made on fish that succumbed, and also, at the end of the first and second weeks, on randomly selected experimental fish. This sacrifice of live experimental fish before the experiment was terminated may have decreased the calculated total mortality but did not otherwise affect the experiment. DDT residues in these apparently healthy fish were about five times as great as in fish killed by the diet. Obviously, the level of DDT residues in fish does not, in itself, define the past exposure of the fish to pesticide pollution or indicate the degree of its endangerment.

The residue analyses of the Atlantic croaker show that, although there is some tendency for total DDT residues to increase with length of exposure to the contaminated diet (Fig. 9-2) the relationship is not predictable. It is also apparent that there is little correlation between the size of the fish and the amount of residue (Fig. 9-3), although it might have been predicted that larger fish would eat more and therefore contain larger amounts of DDT.

Conclusion

The early death of sensitive individuals exposed to pesticides is predictable. The survival of resistant individuals, however, implies not only the selection of resistant stocks but also the accumulation of greater body residues of persistent pesticides that will be passed on to predators in the food web. The restriction of breeding pairs of animals to resistant stocks may cause the segregation of other characteristics in the gene pool that are of unpredictable survival value.

I assume on the basis of these experiments that high residues

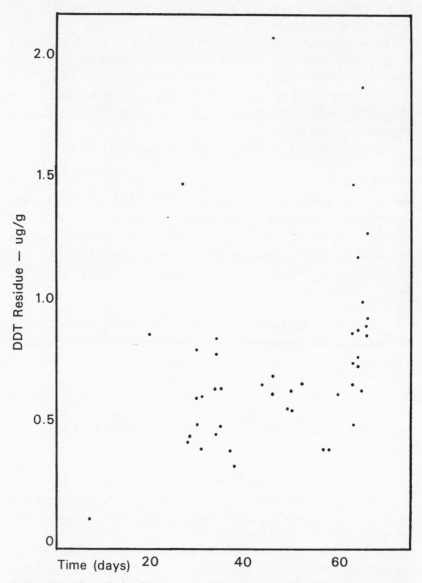

FIGURE 9-2. The relationship between DDT body residues in Atlantic croaker and the length of time they were fed a DDT-contaminated diet. Each dot represents an individual fish.

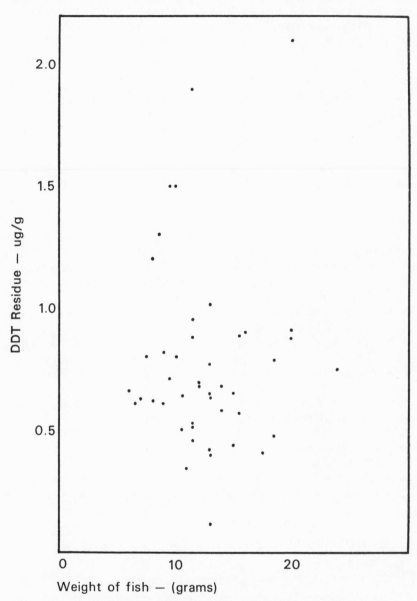

FIGURE 9-3. The relationship between DDT body residues and the weight of Atlantic croaker fed on a DDT-contaminated diet. Each dot represents the weight of one fish.

of persistent pesticides in randomly sampled natural populations represent resistant individuals. It is more than probable that in pesticide polluted estuaries, significant but undetected increases occur in the mortality of sensitive life stages of fish and shellfish populations. Stocks of some animals can compensate for such losses, but pollution by the persistent pesticides may be the decisive factor in eliminating already endangered species.

ACKNOWLEDGMENT

I thank Erwin H. Schroeder and Alfred J. Wilson, Jr., for conducting the feeding experiments and the chemical analyses.

REFERENCES

1. BURDICK, G.E. *et al.*: The accumulation of DDT in lake trout and the effect on reproduction. *Trans Amer Fish Soc, 93* (2) :127, 1964.
2. BUTLER, P.A.: Pesticide residues in estuarine mollusks. *National Symposium on Estuarine Pollution, Stanford University, August 1967.* Department of Civil Engineering, Stanford University, pp. 107-121.
3. BUTLER, P.A.: Pesticides in the environment and their effects on wildlife. *J Appl Ecol, 3* (suppl.) :253, 1966.
4. BUTLER, P.A., *et al.*: Effect of pesticides on oysters. *Proc Nat Shellfish Assoc, 51:*23, 1960.
5. NASH, R.G., and WOOLSON, E.A.: Persistence of chlorinated hydrocarbon insecticides in soils. *Science, 157* (3791) :924, 1967.
6. RISEBROUGH, R.W., *et al.*: Pesticides: Transatlantic movements in the northeast trades. *Science, 159* (3820) :1233, 1968.
7. WOODWELL, G.M., *et al.*: DDT residues in an east coast estuary: a case of biological concentration of a persistent insecticide. *Science, 156* (3776) : 821, 1967.
8. WURSTER, C.F., JR., and WINGATE, D.B.: DDT residues and declining reproduction in the Bermuda petrel. *Science, 159* (3818) :979, 1968.

DISCUSSION

RUDD: Would you please define your use of the word resistance?

BUTLER: I refer to individual resistance, not species or population resistance. When we monitor the environment and find healthy-looking animals with high residues, these are the resistant specimens. The sensitive ones have probably been killed off.

RUDD: The alternative is species resistance. An influx of low-level residues over a period of years into the estuaries may well produce a resistant population which still shows a spectrum of responses but at higher levels of the insecticide.

BUTLER: We have traced one species of fish through six generations and found that resistance varies from generation to generation, and depends on the time span between the exposure of mothers and the birth of the young. The resistance of the first offspring generation, is dependent on the stage of development of the maternal gonads at time of exposure to the pesticide.

In an area where residues are consistently high (for example in the lower part of the Laguna Madre, Texas) there has been no observed breeding in the sea trout for the last 2 years. An analysis of the gonads of the females showed levels as high as 8 ppm DDT.

HERMAN: What do you find in the San Francisco Bay? Do you have a sample station there?

BUTLER: We are sampling oysters primarily in California. Mr. John Modin of the State Fish and Game Department, in cooperation with us is also sampling the Asiatic clam in the San Francisco Bay area.

HUNT: We are trying to assess the hazard of trout food that was contaminated with pesticides, so we contaminated some trout food ourselves. The lethal dose on oral administration of DDT approached that of mammals (500 to 700 ppm). On the other hand, with dermal exposure, the toxic level is less than 1 ppm. Do you have data on contact toxicity of DDT in oil?

BUTLER: None. We have not been able to buy any commercial fish food in the United States that is not contaminated with DDT. We are getting one brand now from West Germany that is quite clean.

FASSETT: Do you have a control experiment in which you feed uncontaminated oyster meat to the fish?

BUTLER: Except for the last experiment which was performed for 100 per cent mortality, all other experiments had controls with the same type of fish, same water supply, and an almost uncontaminated food supply. All oysters in our area have a background level of contamination with DDT residues. The Mississippi River drains half of the United States and pesticide pollution of the Gulf of Mexico fauna is to be expected.

WELCH: Was the 8 ppm DDT found in the gonads of sea trout measured as DDE or DDT?

BUTLER: It was primarily DDE with some DDD and DDT; dieldrin and endrin were also present.

WELCH: Did you analyze other parts of the fish?

BUTLER: The whole body, in all the samples I mentioned, but in this particular case, only gonads from the female fish were collected.

STICKEL: When whole bodies are analyzed, one often finds that residues in the living are greater than in the dead.

BUTLER: In our case, the difference was a factor of 4.

BERG: This would apply to total body burden, but not to pesticide concentration per kilo, as I understand.

What is the relation betwen the exposure of the gravid female and the fate of the offspring? Are the offspring from an exposed ovary more or less likely to succumb?

BUTLER: If the ovary is in the formative phase when the exposure takes place, the first generation is likely to be more sensitive.

BERG: There is then a cumulative process of poisoning here and not a selective process favoring resistance. I suppose there is no gene for resistance to be favored. This would account for the downward fluctuations in resistance from generation to generation.

BUTLER: In one experiment a population of fish was exposed to enough DDT to kill 98 per cent of them. The remaining 2 per cent were bred, then their offspring were tested. We have continued this for six generations, and sensitivity or resistance depended only on the time of the year that we exposed the fish.

Chapter 10

MERCURY CONTAMINATION OF THE ENVIRONMENT IN SWEDEN

A. G. Johnels and T. Westermark

ABSTRACT

Mercury is present everywhere in nature although it is primarily found in minute amounts. However, mercury and mercury compounds have a wide use in agriculture and industry, and an increase in natural levels of mercury content has been recorded.

By means of activation analysis, the mercury content in feathers of eleven terrestrial bird species has been studied. Of these the goshawk (*Accipter gentilis*) is represented in this paper. The material has been derived from museum collections and from living or freshly killed birds; it covers a period of more than 100 years. These studies reveal nearly constant mercury levels from the middle of the previous century until 1940. After that, a well documented increase in mercury concentration occurs, amounting to at least ten to twenty times the previous level. In 1940 alkylmercury compounds began to be used as seed dressings. The synchronous appearance of increased mercury accumulations in birds indicates that the seed dressing agents are a main source of contamination in the terrestrial environment.

Activation analysis of the mercury content in axial musculature of pike (*Esox lucius*) has been used for estimating the level of mercury in the aquatic environment. The mercury concentration factor in pike is 3000 or more. The relation between the mercury content and the weight or age of fish specimens is described. A graphical method has been used to establish the figure for the mercury content in pike specimens of 1 kg. Those figures have been used for geographical comparison. The effect of water contamination from industrial waste is demonstrated in some cases. There are indications that mercury also appears as an airborne pollution, mainly affecting the water environment. In contrast with the goshawk, where a sudden rise of mercury concentration in feathers occurred together with the onset of the use of methylmercury compounds in agriculture, two fish-eating bird species show a gradual increase in mercury content since the previous century. This indicates that the mercury contamination of water follows the general increase in industrial activity. A number of sources of mercury contamination are discussed, and it is concluded that in many areas human activities have raised the mercury level of the environment far above natural levels.

221

REPORT

Introduction

MERCURY IS A COMPONENT of the earth's crust. Owing to evaporation, it is distributed by aearial circulation. Thus it is present everywhere in nature, although it is primarily found in minute concentrations (17). Mercury and mercury compounds have a wide use in agriculture and industry, and unintended and intended losses to the environment occur. The natural levels consequently are obscured by additions from human activities.

About 1955, several ornithologists in various parts of Sweden observed a decrease in the populations of some seedeating birds. The increase in the number of birds of prey found ill or dead also attracted attention. In 1958, Borg (3) drew attention to the fact that birds found dead and sent for examination to the State Veterinary Medical Institute had remarkably high residues of mercury in their livers and kidneys (4-200 mg/kg). Studies on shot and trapped birds also revealed remarkable figures (1-53 mg/kg) (5). The conclusion was that the large numbers of dead specimens of seedeating birds (pheasants, partridges, pigeons, finches) and birds of prey (eagles, buzzards, hawks, falcons and owls) in particular, and of corvine birds indicated a very general poisoning in birds. The source of the mercury was attributed to the use of mercury compounds as agent in seed-dressing in agriculture (5, 6). The decline in the population of several bird species in Sweden was attributed to the use of pesticides in Sweden (13).

The results of Borg *et al.* (4-6) are partly summarized in Figure 10-1. The curves demonstrate that the incidents of specimens having more than 2, 5, and 10 mg/kg of mercury in the liver will increase during May to June and October to November. The peaks of the curves are caused by the use of seed dressed in mercury compounds during certain periods in spring and early autumn.

In 1964 the present authors studied the mercury content of axial muscle of pike from a few Swedish lakes by means of activation analyses (7, 16, 20). Although it was known that fish contain comparatively large amounts of mercury in their tissues as compared with most other living organisms (14, 15, 16), the first levels recorded (300 to 800 nanograms per gram (ng/g) caught our interest (11, 19).

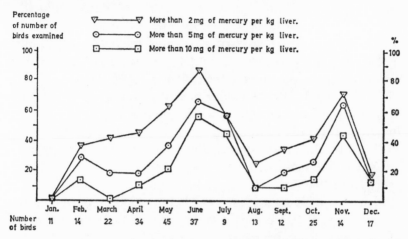

FIGURE 10-1. Seasonal variation of mercury levels in seedeating birds picked up dead in the field during 1964. After Borg, Wanntorp, Erne and Hanke, 1968.

The present authors with co-workers demonstrated in 1965 that the mercury levels in birds of prey from central Sweden from 1860 to 1870 was less than one tenth of the levels in 1964 and 1965 (1). Feathers are formed during a relatively short period in juvenile life and subsequently replaced according to patterns that differ with the species. They register the level of mercury for the formative periods. The ratio of the mercury content of breast muscle to that of feathers is about 1 to 7. The feather content thus roughly corresponds to the dry-weight content of muscle. There are variations in mercury content within a feather but they are not greater than a factor of 1 to 2.

Terrestrial Environment

The study of birds' feathers for estimating levels of mercury content enables the historical development of mercury contamination in the environment to be traced (1). For that purpose museum specimens of several bird species from the previous century to the present have been studied: pheasant (*Phasianus colchicus*), partridge *(Predix perdix)*, willow grouse *(Lagopus lagopus)*, corn bunting *(Emberiza calandra)*, peregrine *(Falco peregrinus)*, eagle owl *(Bubo bubo)*, white-tailed eagle *(Haliaetus albicilla)*, long-eared owl *(Asio otus)*, tawny owl *(Strix aluco)*, buzzard *(Buteo buteo)* and hen harrier *(Circus cyaneus)*. In all these species, with the exception of the *Lagopus lagopus,* an increase in mercury content was observed after 1940 or 1950 (1).

Female goshawks *(Accipiter gentilis)* shot in the months of May and June at their nests provide comparatively good material for study (Fig. 10-2). Although evidence is lacking from 1940 to 1945, we think it is safe to conclude that the rise of the mercury content is due mainly to the use of methylmercury as a dressing agent from 1940 to 1966. There is no indication in the diagrams that the use of inorganic mercury before 1930 or phenylmercury from 1930 to 1940 has caused an increase in the mercury content of these terrestrial bird species.

That there shows no increase in the mercury content in the willow grouse is in accordance with the fact that this species lives in the mountainous border between Norway and Sweden where no influence from the use of seed-dressing mercury can be expected.

In the spring of 1966 regulations prohibited the use of alkyl-mercury compounds as seed-dressing agents. Instead the compound alkoxyalkylmercury was used and at the same time the quantity of dressed seed used was reduced only by 15 to 20 per cent of the original extent. In many predatory bird species studied by us in 1966 and 1967, the mercury levels dropped abruptly to a very low level compared with those from 1940 to 1965. Our observations are corroborated by studies on pigeons *(Columba palumbus)* by Borg *et al.* (4) ; in this species a general decrease in the mercury content of the liver was noted in 1966. The seed-dressing agents seem to be responsible for the greater part of the increase in the terrestial bird fauna. The reason that the level in our material from 1966 and 1967 is about 50 per cent above the typical of the previous century may be explained by a simultaneous influence of other sources which are discussed below.

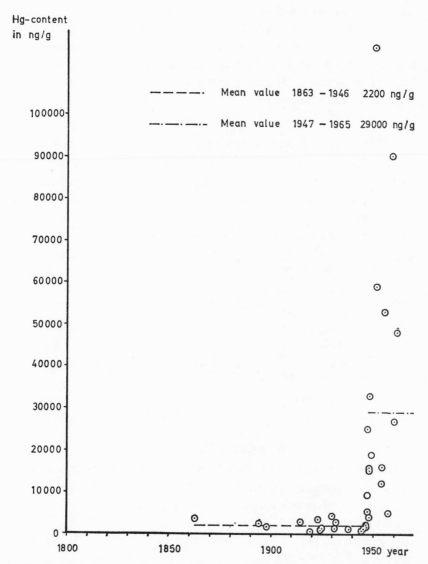

FIGURE 10-2. Mercury content of feathers from goshawk females shot at the nest in May and June during the period 1863 to 1965.

As shown in Table 10-I there are also examples from the Swedish terrestrial fauna where a low level of mercury prevails. These bird and mammalian species are all true herbivores; their mercury level is noticeably lower than that of the seedeaters and predatory birds. It should be added that examples of high levels have been found also among small rodents and predatory mammals such as the fox.

TABLE 10-1
MERCURY CONTENT OF MATERIAL FROM 1964 AND 1965
IN MIDDLE SWEDEN

Species	Locality	Mercury Content ng/g
Cow, muscle	Närke	8
Horse, muscle	Närke	7
Moose (Alces alces), muscle	Rockhammar, Västmanland	10
Roe deer (Capreolus capreolus), muscle	Rockhammar, Västmanland	55
Roe deer (Capreolus capreolus), muscle	Sorunda, Södermanland	6
Roe deer (Capreolus capreolus), muscle	Tallnäs, Södermanland	7
Brown hare (Lepus evrupaeus), muscle		11
Yellownecked field-mons (Sylvaemus flavicollis), muscle	Kvismaren, Närke	8
Capercaillie (Tetrao urogallus), feather	Kilsbergen, Närke	60
Capercaillie (Tetrao urogallus), feather	Kilsbergen, Närke	<40
Capercaillie (Tetrao urogallus), feather	Sikfors, Västmanland	75

The Aquatic Environment

In terrestrial animals the increase in mercury level is caused by an accumulation from the food. It has been demonstrated experimentally that fish accumulate mercury directly from the surrounding water (2, 8), although food also may play a role in this connection (8, 14). The concentration factor of aquatic mercury in pike is 3000 or more.

The sources of contamination of water are many. In agricultural use alkyl- or alkoxyalkylmercury is distributed in the fields, some of which will find its way to water. From paper and pulp industry, phenylmercury is lost to the environment. In several other industrial processes there are losses of metallic or inorganic mercury.

In 1966 Westermark *et al.* (1) demonstrated that part of the mercury content in bird's feathers is metallo-organic. In 1967 Norén and Westöö (12, 21) demonstrated that almost all mercury present in fish muscle appears in the form of methylmercury. This seems to be the most noxious of all mercury substances lost to the environment (1).

Aquatic Organisms

In our work (11) the northern pike (*Esox lucius*) was selected as a test organism. Its advantages are as follow:

1. A stationary habit which provides good geographic information.
2. A lifespan of several years which serves to integrate temporal variations in the occurrence of accumulative substances in the environment.
3. A wide species' distribution which permits comparative studies over extensive geographic areas.
4. No populations with stunted growth.

Even if the growth rate of the individuals is somewhat reduced in the northern parts of Scandinavia, the growth is still rapid and comparable all over the country.

Of all the fish tissue studied, muscular tissue normally has the highest mercury content (Table 10-II). In laboratory experiments which are short-term compared with the cases from nature and in

TABLE 10-II
HG CONTENT IN ORGANS OF PIKE SPECIMEN
(WEIGHT 3120 g, TIDÖ- LINDÖ, LAKE MÄLAREN, 1/7 1964)

Organ	Hg ng/g
Heart muscle	1000
Axial muscle	850
Liver	780
Kidney	640
Ovary	560
Intestine	610
Epidermal finrays	390
Gill	300
Brain	290
Spleen	280
Scales	104

cases where high contamination occurs, the proportions may be different. In the present study of fish, the figures refer to the mercury content of the axial musculature.

There are variations in the mercury content in pike from one and the same locality. Some regularity in this variation is apparent when individual values of mercury content are plotted against the weight, length, or age of the fish specimens. The relation of mercury content to age is demonstrated in Figure 10-3. However, scale reading is a somewhat unreliable method for determining age in pike; therefore, the mercury content in the present paper is related to weight (Fig. 10-4). Within the weight limits studied there seems to be a linear relation between age or weight on one

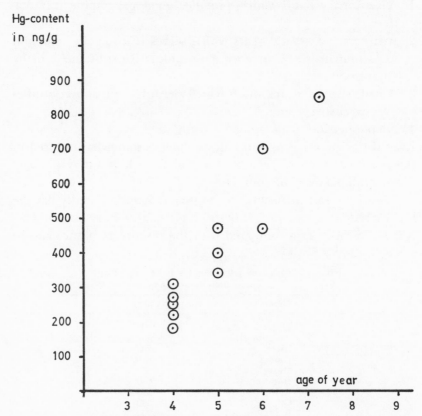

FIGURE 10-3. Mercury content of axial muscle of pike in relation to age of the specimens. Tidö-Lindö, Lake Mälaren, 1964.

hand and mercury content in axial musculature or other organs on the other. For low levels of mercury in the sample (below 200 ng/g), little variation occurs among the individuals; a very moderate increase, or none at all, occurs with increasing weight of the fish specimens. As the mean level of mercury content rises, the mercury level in relation to weight increases noticeably (Fig. 10-4). Extremely high levels of mercury, caused by a manifest contamination by waste-water outlets, appear to bear no relation to age or weight. Evidently the degree of exposure to contamination is more an influential factor than increasing age.

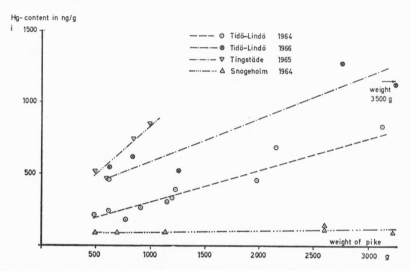

FIGURE 10-4. Relation between mercury content of axial muscle and total weight in pikes. Tidö-Lindö, Lake Mälaren, Lake Tingstäde, Gotland and Lake Snogeholm, Skåne.

Since about 1946 phenylmercury acetate has been added to mechanical pulp in Sweden to improve its storage properties (18). Part of the substance is lost to the waste water. The effect of this contamination from a pulp factory in central Sweden to a small stream is demonstrated in Table 10-III. The mercury content in pike caught above and below the dam increased by a factor of 5 to 10.

TABLE 10-III
HG LEVELS IN AXIAL MUSCULATURE OF PIKE
(PULP FACTORY, CENTRAL SWEDEN)

Above Dam:		*Below Dam:*	
Weight, g	Hg ng/g	Weight, g	Hg ng/g
1964			
395	160	150	2300
625	260	475	3000
725	330	600	1600
7000	830	1050	1500
1966			
245	340	1000	2400
350	280	1075	2700
475	390	1400	3100
575	620		

In the same year similar compounds were used to prevent formation of slime in the machinery of paper mills (17). Part of the substance was lost to the waste water. Tables 10-IV and 10-V demonstrate the effect of this loss in a stream in southern Sweden. In this case not only pike but also a number of invertebrates and a few plants were studied. It is evident from these figures that not only fish but also other organisms are strongly influenced by mercury in water.

TABLE 10-IV
EFFECT OF PAPER MILL ON THE Hg CONTENT IN AQUATIC ORGANISMS
(PAPER MILL LOCATED AT A STREAM IN SOUTHERN SWEDEN, APRIL 1965)

Material	*Locality*	*Hg ng/g*
Tricoptera	15 km above mill	52
Tricoptera	14 km above mill	54
Tricoptera	6 km below mill	10700
Tricoptera	1 km below mill	17000
Tricoptera	5 km below mill	5600
Plecoptera sl. Isoperla	15 km above mill	72
Plecoptera sl. Isoperla	17 km below mill	2400
Asellus aquaticus	1 km above mill	65
Asellus aquaticus	15 km above mill	59
Asellus aquaticus	20 km below mill	1900
Sialis	20 km above mill	52
Sialis	1 km above mill	49
Sialis	1 km below mill	5500
Sialis	6 km below mill	4800
Fontinalis	15 km above mill	76
Fontinalis	below mill	3700
Water lily, piece of caulis	1 km above mill	16
Water lily, piece of rhizome	1 km below mill	520

TABLE 10-V
EFFECT OF PAPER MILL ON THE Hg CONTENT IN AQUATIC ORGANISMS
(THE SAME PAPER MILL AND LOCALITY AS IN TABLE 10-IV—
MATERIAL FROM 1965 AND 1966)

Material 1965			Locality	Date	Hg ng/g
Hirudinea:					
Helobdella sp.			above dam	24/8	25
Helobdella sp.			below dam	23/8	3100
Helobdella sp.			below dam	19/8	2600
Helobdella sp.			below dam	20/8	2350
Helobdella sp.			below dam	18/8	4400
Insecta, Tricoptera:					
Hydropsyke pellucidula			below dam	18/8	1220
Hydropsyke pellucidula			below dam	20/8	1400
Hydropsyke pellucidula			below dam	23/8	1700
Pike	♂	490 g	8 km above dam	30/9	1200
Pike	♂	575 g	4 km above dam	27/9	1100
Pike	♀	2040 g	4 km above dam	29/9	1680
Pike	♂	580 g	4 km above dam	29/9	1220
Pike	♂	300 g	8 km below dam	30/9	5700
Pike	♂	725 g	8 km below dam	30/9	5650
Pike	♂	910 g	8 km below dam	29/9	3400
Pike	♀	1575 g	8 km below dam	26/9	8000
Atlantic Salmon		4770 g	30 km below dam	autumn	220
Atlantic Salmon		2000 g	30 km below dam	autumn	151
1966					
Pike	♀	1200 g	20 km below dam	Sept	9800
Pike	♀	170 g	20 km below dam	31/5	6300
Pike	juv	120 g	20 km below dam	Sept	5600

The use of mercury ceased at this paper mill in June 1965. As shown in Table 10-V the high levels of mercury remained in the autumn of that year. The age of the insect larvae in this case was only about two months, while in the previous table the larvae were about eight months old. The figures are for that reason not strictly comparable. However, the figures for pike from 1966 are as high as those from 1965. There seems to be no sign of a decrease in the contamination. In paper fibers collected outside the mouth of the river after a flood in the spring of 1966, the mercury content was found to be high (2,400 ng/g in a wet sample). The high content of mercury in fiber deposits in the river may serve to explain the persistence of the contamination. The fiber or mud deposits in the river may gradually contaminate the water.

For a geographic comparison the figures of the mercury levels must be standardized in relation to weight; the figures in the present paper are for pike weighing 1 kg. These figures have been made by simple graphical interpolation (see Fig. 10-3). The figures for the various localities studied by us (Table 10-VI) are summarized in Figure 10-5. In the mountainous area along the border between Sweden and Norway the few localities studied by us give very low figures (Fig. 10-5, I). In a number of lakes situated in typical farmland where mercury compounds have been used as seed-dressing a slight increase may be traced (Fig. 10-5, II). Higher levels are found in many lakes which receive no direct waste-water contamination. These lakes are situated, however, in highly industrialized areas and the most plausible explanation of the high mercury levels seems to be an aerial fallout of mercury (Fig. 10-5, III). Finally (Fig. 10-5, IV), many localities known to be contaminated by waste water have been studied. The mercury levels in the axial musculature of pike are the highest.

TABLE 10-VI

CALCULATED VALUES FOR Hg CONTENT IN AXIAL MUSCULATURE IN STANDARD PIKE SPECIMEN OF 1 Kg FOR 49 LOCALITIES
(Number of specimens for calculation in column n. For categories see text.)

Nr	Locality	Year	Hg ng/g	n	Category
	Lapland				
1	Ladtjojaure	1965			
2	Hornavan	1964	75	4	I
	Hornavan	1966	140	11	I
	Norway		170	5	
3	Lianvatnet, Trondheim	1964			
	Jämtland		60	5	I
4	Storsjöns utl.	1966			
	Dalarna		250	7	III
5	Aspan	1965			
	Uppland		250	5	III
6	Gårdskär, coast	1965			
7	Gräddö	1964	700	4	III
8	Stockholm, archipelago	1965–66	480	6	III
9	Tämnaren	1966	2300	7	IV
10	Lårstaviken, Mälaren	1966	400	4	III
11	Ullnasjön	1966	620	3	III
12	Lovö, Mälaren	1966	200	3	II
	Västmanland		500	3	III
13	Sagån, Orresta	1965			
14	Tidö-Lindö, Mälaren	1964	900	5	III
	Tidö-Lindö, Mälaren	1966	300	11	III
15	Galten, Mälaren	1964	600	5	
16	Aspen	1964	500	7	IV
	Närke		750	4	III
17	Hemfjärden, Hjälmaren	1965			
18	S. Hjälmaren	1965	370	5	III

TABLE 10-VI (Continued)

Nr	Locality Lapland	Year	Hg ng/g	n	Category
	Värmland		250	6	II
19	Gapern	1965			
20	Lusten	1965	1100	3	III
21	Bergsjön	1966	480	3	IV
22	Glafsfporden	1966	1100	4	IV
23	Rottnan	1966	950	5	IV
24	Mellanfryken	1966	480	2	III
25	Kattfjorden, Vänern	1964	2500	4	IV
	Västergötland		1000	3	IV
26	S. Dalbosjön, Vänern	1964	600	5	IV
27	Ymsen	1965	110	4	II
	Södermanland				
28	Yngaren	1966	600	3	III
	Östergötland				
29	Bråviken	1966	2300	5	IV
30	Tåkern	1964	300	5	II
	Bohuslän				
31	Bullaren	1965	580	5	III
	Gotland				
32	Tingstäde träsk	1965	850	4	III
	Småland				
33	Bolmen	1965	140	3	II
34	Örken	1966	700	3	IV
35	Varetorpssjön	1966	1600	3	IV
36	Helgasjön	1966	1400	4	IV
37	Skatelövsfjorden, Åsnen	1966	450	4	IV
38	S. Åsnen	1966	200	4	II
39	Granö kraftverksdamm	1965	1400	4	III
	Blekinge				
40	Pukaviksbukten coast	1966	1000	4	IV
41	Vambåsa coast	1966	480	3	III
	Skåne				
42	Ulkesjön	1965	1000	3	III
43	Sövdesjön	1964	130	5	II
44	Snogeholmssjön	1964	100	5	II
45	Krageholmssjön	1965	300	3	II
	Finland				
46	Kiskon, Kirkkojärvi, Åboland	1965	300	3	
47	Österö, Vasa	1965	500	5	
	Denmark				
48	Tömmerby fjord	1964	350	5	
	Switzerland				
49	Greifensee	1965	100	4	

The influence of various kinds of bedrock containing minute amounts of mercury has not been sufficiently studied. For the moment there is no information to prove that such an influence may cause high levels. This particular problem is, however, under study.

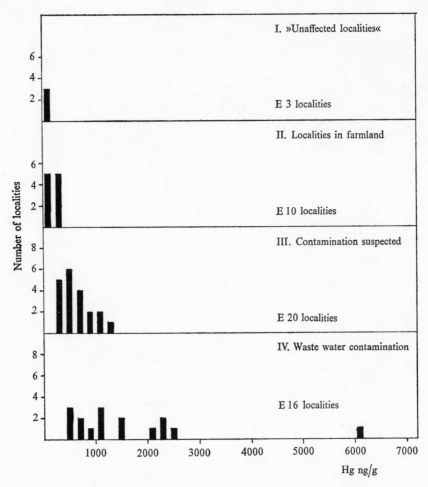

FIGURE 10-5. Mercury level in axial musculature of pike from 48 localities in Sweden and 1 in Norway. Material from Table 10-III, 10-V and 10-VI.

Fish-eating Organisms

The osprey (*Pandion haliaëtus*) has a diet which consists almost exclusively of fish of about 0.3 kg. The great crested grebe (*Podiceps cristatus*) has a diet which is about two-thirds small fish specimens and one-third insect larvae, etc. Feathers of museum specimens of these two species from the previous century to the present have been examined. The result of these studies is represented in Figure 10-6. For comparison, the figures for goshawk are represented in Figure 10-7. Even if the material is sparse, it seems safe to conclude that the increase of mercury level in the two fish-eating birds follows another course than that demonstrated by the goshawk and other terrestrial birds. Already about 1900, a rise in the mercury content was observed and has gradually increased to the present time. Also dead specimens of the white-tailed eagle (*Haliaetus albicilla*) demonstrate very high levels of mercury content. It should be noted that in these cases the amounts of chlorinated hydrocarbons are also high (DDT and related substances and polychlorinated biphenyls, PCB [10]). In this species there are indications that the normal propagation of the species fails in regions where an influence from contamination is present. Mercury compounds and chlorinated hydrocarbons are known to cause such an effect (5, 6, 9).

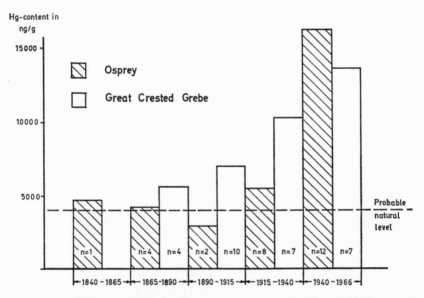

FIGURE 10-6. Mercury levels in feathers of osprey (*Pandion haliaëtus*) and great crested grebe (*Podiceps cristatus*) in periods from 1840 to 1966.

The osprey is a migratory bird and will return to Sweden in spring to propagate. The feathers of the young thus register the general mercury level in the surroundings of the nest (Fig. 10-8). In autumn the young migrate to the Mediterranean area and then to South Africa. The young do not return to Sweden until their second year. The adults return every year to the same locality for nesting. The adults replace their feathers according to a pattern which is now under study. In this case, the feathers should register the general level of mercury contamination of the place where the

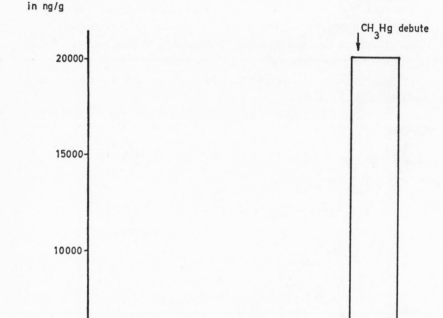

Figure 10-7. Mercury levels in feathers of goshawk (*Accipter gentilis*) in periods from 1815 to 1965. See also Figure 10-2.

birds fed during the time when the feathers were forming and growing. The results seem to be in accordance with the moulting pattern. The feathers formed in Sweden have the same mercury content as those of the nestling.

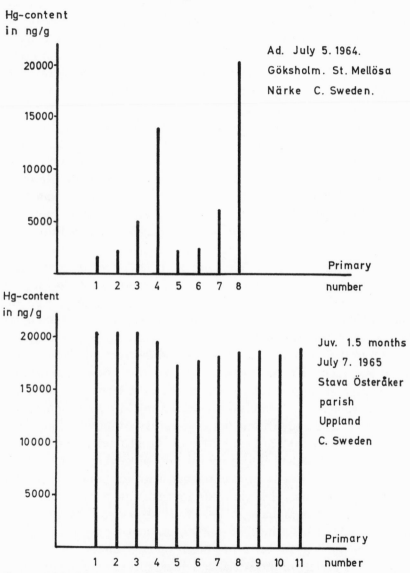

FIGURE 10-8. Mercury content of primaries of wing of osprey (*Pandion haliaëtus*). Top of adult, bottom of juvenile.

Sources of Mercury Contamination

In the chlorine-alkali industry, mercury is used as an electrode and losses to air and water occur. This industry was introduced in Sweden about 1900. The burning or heating of various kinds of raw material in industrial processes (oil, coal, sulfide and other ores, clay, etc.) will release mercury. Gas and cokeworks were started in the last half of the 1800's in several places in Sweden where treatment of ore has an ancient history. In dentistry, the use of mercury amalgam started before the turn of the century, and there were various other uses of mercury compounds in medicine, laboratories, leather industry and in industries manufacturing chemical products. The products of chemical industries may also contain mercury (e.g., alkali, sulfuric acid) which may be released in other places when the products are used. Mercury contamination from industry takes place via the waste water or gas emissions. In agriculture, a part of the distribution of mercury is intentional but some further addition may be produced through the distribution of chemical fertilizers (e.g., superphosphate). In paper and pulp industry, phenylmercury acetate has been used since 1946 as a fungicide. Part of the substance is lost to the environment. There are other common uses of mercury, e.g., in thermometers, other instruments and electrical articles such as lamps, batteries and switches. Substantial quantities of mercury must find their way to rubbish heaps and destruction devices.

It is evident that the aquatic environment is more susceptible to mercury contamination in general than the terrestrial environment; a study of the feathers of the fish-eating bird specimens gives evidence of and shows influence of mercury contamination. The curves in Figure 10-6 are indicative of the general industrial development in Sweden. The investigation of mercury contamination in the Swedish environment began with the assumption that the intentional use of mercurials in agriculture was the main source. The progress made in 1964 to 1965 proved that there is a more complex background of sources, particularly for the aquatic environment. The problem has taken on new dimensions and is related to a variety of industrial activities. It is not confined to the areas hitherto studied.

ACKNOWLEDGMENT

The cooperation of W. Berg, C. Edelstam, S. Jensen, M. Olsson, V. Olsson, J. Sondell and others, is gratefully acknowledged. This work was supported by the Royal Commission of Natural Resources, the Technical Research Council, the Agricultural Research Council, the Knut and Alice Wallenberg's Foundation, and the Ekhaga Foundation.

REFERENCES

1. BERG, W.; JOHNELS, A.G.; SJÖSTRAND, B., and WESTERMARK, T.: *Oikos, 17*:71, 1966.
2. BOËTIUS, J.: *Medd. Danmarks Fiskeri-og Havundersögelser, 3* (4) :93, 1960.
3. BORG, K.: *Proc. VIII Nord. Veterinärmötet,* Helsingfors, 1958, p. 394 (in Swedish).
4. BORG, K.; WANNTORP, H.; HANKO, E., and ERNE, K.: *Svensk Jakt, 8*:364, 1967, (in Swedish).
5. BORG, K.; WANNTORP, H.; ERNE, K., and HANKO, E.: *J Appl Ecol, 3* (suppl.) :171, 1966.
6. BORG, K.; WANNTORP, H.; ERNE, K., and HANKO, E.: *Viltrevy, 5* (5) :1, 1968.
7. CHRISTELL, R.; ERWALL, L.G.; LJUNGGREN, K.; SJÖSTRAND, B., and WESTERMARK, T.: *Proc. Int. Conf. Modern Trends in Activation Analysis.* College Station, Texas, 1965, vol. 1, p. 380.
8. HANNERZ, L.: *Rep. Inst. Freshw. Res. Drottningholm, 48*:120, 1968.
9. JEFFERIES, D.J.: *Ibid, 109*:266, 1967.
10. JENSEN, S.: *New Sci, 32*:612, 1966.
11. JOHNELS, A.G.; WESTERMARK, T.; BERG, W.; PERSSON, P.I., and SJÖSTRAND, B.: *Oikos, 18*:323, 1967.
12. NOREN, K., and WESTÖÖ, G.: *Vår Föda, 2*:13, 1967.
13. OTTERLIND, G., and LENNERSTEDT, I.: *Vår Fågelvärld, 23*:363, 1964 (with an English summary).
14. RAEDER, M.G., and SNEKVIK, E.: *Kongl Vidensk Selsk Forhandl, 13* (42) : 169, 1941.
15. RANKAMA, K., and SAHAMA, TH. G.: *Geochemistry.* Chicago, U. of Chicago, 1950.
16. BJÖSTRAND, B.: *Anal Chem, 36*:814, 1964.
17. STOCK, A., and CUCEL, F.: *Naturwissenschaften, 22/24*:390, 1934.
18. VALLIN, S.: *Svensk Papperstidning, 19*:1, 1947, (with an English summary).
19. WESTERMARK, T., et al.: *Kvicksilverfrågan i Sverige.* Stockholm, 1965 (in Swedish).
20. WESTERMARK, T., and SJÖSTRAND, B.: *J Appl Rad Isotopes, 9*:1, 1960.
21. WESTÖÖ, G.: *Acta Chem Scand, 20*:2131, 1966.

DISCUSSION

SUZUKI: Have you any reports of abnormal behavior of fish in your country? Fishermen in Minamata Bay observed fish that were swimming in a vertical position and fish that were splashing on the surface on their sides. This was never seen before. These were not dying spasms.

JOHNELS: We have not observed anything like that. Hannerz experimented with fish which carried the levels of mercurials that we observed in nature. The fish appeared normal, but when he tried to move them from one tank to another, they suddenly died. His conclusion was that it was the stress effect of the transfer which caused them to die.

JERNELÖV: In a way I think we could use the relation between the liver and the body muscles to say in which direction the situation is changing. If residue values in the liver are higher than in the muscle, we can be quite sure that fish are exposed to pollution and we will see a rapid increase in the amount of mercury in the fish. If, on the other hand, higher amounts are in the muscles, then the increase in fish will not be so rapid.

JOHNELS: This is correct. In situations of acute pollution the amount of mercury in the liver is much higher than that of the musculature. The distribution of mercury in various tissues, which I showed, is for lakes where the fishes have less than, say, 1,000 ng of mercury per g muscle.

BERG: You showed that fish do not accumulate mercury when they are exposed to measurable but low concentrations, but they accumulate mercury progressively when the concentrations are a little higher. Does this not illustrate the difference between the natural composition of the environment, which is wholesome since we evolved to live in it, and the polluted environment, which is unphysiological? We hear that mercury is everywhere, so a little bit cannot hurt you. Would you take the observation that a fish is accumulating mercury as evidence of too much pollution?

JOHNELS: Yes, an increase in concentration in fish in relation to age could be judged as a sign that there are raised levels of mercury in the water. In the case of low or moderate pollution of mercury this relation is very distinct. But when approaching high levels of pollution, it is possible to find one fish with 8,000 ng/g in muscle and another with 2,000 ng/g in one locality and there is no relation to age or weight. The waste is always discharged some place in high concentration, and you see the effects of local exposures to the pollutant before it is uniformly mixed in the water body.

RUDD: I am intrigued by this idea of tracing the life history of an individual by mercury levels in its primary feathers. Can we use other persistent chemicals for a measure of this sort?

JOHNELS: Yes, you can use the same method for DDT, the PCB's and related substances. In fact feathers of birds were used for the identification of PCB. Jensen found this substance and was able to prove that it was not DDT or its metabolites, since the feathers dated before 1940. This is an example of an important task of museums: to collect and store materials that can be studied chemically in the future, without knowing what chemical analysis will be desired. Perhaps freeze-drying is a partial solution to this problem.

RUDD: Is this method valid for studying the life history of a single individual in relation to, say, DDT?

JOHNELS: Yes, I think that it is quite possible. For example, sea eagles shed feathers periodically and you can collect a sequence of feathers from one bird at, for instance, the breeding site. It is also possible to collect feathers from nestlings without killing the specimens.

RADOMSKI: I should think this would be very difficult to do with pesticides for two reasons: first, the pesticide levels in the feather would be extremely low, and second, the possibilities of external contamination would be very high.

JOHNELS: It is not so difficult. It has been done with PCB.

RUDD: This report shows the value of museum collections, which is not generally understood. The carefully stored old specimens were of great value in studies of effects of pesticides, both in this study of feathers and in the studies of egg shells.

SESSION III

MECHANISMS OF ACTION: DOSES TO INDIVIDUALS AND THEIR RELATION TO ACUTE TOXIC DOSE

Chapter 11

NEUROLOGICAL SYMPTOMS FROM CONCENTRATION OF MERCURY IN THE BRAIN

T. SUZUKI

ABSTRACT

A hypothesis on the critical concentration of mercury in the brain, which is necessary to produce neurological symptoms, was tested. Mice which had been given a single, or repeated, peroral dose of different amounts of methylmercury chloride showed neurological symptoms for a peak concentration of mercury of more than 20 $\mu g/g$ of wet tissue. Death followed the occurrence of neurological symptoms for peak concentrations of more than 30 $\mu g/g$.

To determine the biological half-life of mercury in methylmercury, a solution of methylmercury acetate, labeled with radioactive mercury (^{203}Hg), was subcutaneously injected into mice seven times. The peak concentrations of mercury in the human brain poisoned by methyl or ethylmercury were calculated on the assumption that the decrease in mercury content in the brain and blood was similar to the decrease observed in experiments on mice. The calculated peak concentrations in humans were more than 50 $\mu g/g$ for the cases that died from methylmercury poisoning and 15 $\mu g/g$ in the cases which survived ethylmercury poisoning.

REPORT

Introduction

ALTHOUGH THE TOXICOLOGICAL effects of mercury depend on the chemical properties of the compounds used (3, 8, 10, 13, 14, 23, 29, 37), the problem can be simplified by correlating the manifestation of toxic effects from mercury with the concentration of mercury in each organ and tissue. The difference in the toxicological effects of various mercury compounds is presumably dependent on the bodily distribution and metabolism of these compounds.

Therefore, a study of the toxic action of mercury on the brain should include an analysis of the mercury content in the brain.

245

This analysis should determine firstly, whether the concentration of mercury is critical enough for neurological symptoms to manifest themselves; and secondly, whether any process of cumulative, unrecognizable changes in the brain tissue, followed by a manifestation of neurologicial symptoms after a longer lapse of time, is detected.

In this study, methylmercury compounds were used for analysis, because they (as well as ethylmercury and n-propylmercury compounds) are retained in the brain in higher concentration than other mercury compounds (2, 6, 7, 18, 19, 21, 25, 26, 27, 28, 30, 31, 35) , and because they induce a severe neurological disturbance which has not been observed in human or animal poisoning from other mercury compounds (1, 11, 12, 14, 15, 17, 20, 23, 24, 32, 36, 38) .

Poisoning of Mice by Methylmercury Chloride

Methylmercury chloride suspended in olive oil was administered perorally, by a gastric tube, in varying doses and frequencies. Table 11-I shows diagrammatically the results of the experiment. Neurological symptoms appeared in two groups given one dose of 46 and 32 mg Hg/kg, respectively. In groups which received a dose larger than 46 mg Hg/kg, death occurred before any neurological symptom could be detected. Mice receiving a dose of less than 32 mg Hg/kg showed no neurological symptoms up to the thirtieth day after dosage was administered. Seven daily doses of 14.7 and 10 mg Hg/kg administered to groups A and B,

TABLE 11-I
RESPONSE OF MICE WITH REGARD TO DOSE AND FREQUENCY OF
METHYL MERCURY CHLORIDE

Frequency	100	68	46	Dose 32	22	mg Hg/kg 14.7	10	6.8	4.6
1	I	I	II	IV	V	V			
7						II	III	V	V

The response of each group is patternized as follows;
  I: Death without neurological symptom (NS) in all the mice
 II: Death with NS in all the mice
III: Survival with NS in all the mice up to the 30th day
 IV: Survival with NS in some mice
  V: Survival without NS in all the mice

respectively, induced neurological symptoms; all mice from group A died during the 5 days after the last dose administered, while only one out of ten mice died from the latter dosage.

TABLE 11-II
CRITERIA FOR DIAGNOSIS OF NEUROLOGICAL SYMPTOM IN MICE

		0	1	2
A	Postural change on being laid on its back	Impossible to lay on its back	Possible, but turn back instantly	Continue to be lying
B	Staying on a slender rod	Stable	Unstable, but not fall	Fall
C	Maintaining the head on horizontal position	Frequent		Drop-head
D	Paralytic posture of hind-legs	Negative	Absence of extension	Not move, semi-flexed
E	Abnormal movement of hind-legs on being hung	Negative		Positive, crossing of hind-legs

The criteria used for diagnosing neurological symptoms is shown in Table 11-II and the results observed are recorded in Table 11-III. The dominant neurological symptoms are related to cerebellar, or vestibulo-cerebellar, regulation of movement and

TABLE 11-III
RESULTS OF ETHOLOGICAL OBSERVATION ON MICE ADMINISTERED
WITH METHYL MERCURY CHLORIDE

		Single Administration					Repeated Administrations (7 times)			
Dose, mg Hg/kg		46	32		22	14	14	10		6.8	4.6
Time after adm.		2-3 d	4	30	30	30	2-3	5	30	30	30
No of animals		5	3	5	5	5	10	10	9	10	10
A	0		3	5	5	5		9	8	10	10
	1							1	1		
	2	5					10				
B	0		3	5	5	5		5		10	10
	1							4	3		
	2	5					10	1	1		
C	0		2	xx	5	5		xx	3	10	10
	2	5	1				10		6		
D	0		3	5	5	5		3	3	10	10
	1							6	6		
	2	5					10	1			
E	0	✱	3	3	5	5	✱	5	4	10	10
	2			2				5	5		

✱; no movement of hind-legs
xx; no record

posture. Early symptoms in mice hung by their tail are an abnormal movement and posture of hind legs. In an advanced stage, mice drop their heads on being hung, and are unable to return to normal posture when laid on their backs. In this stage "crossing of hind legs," (first described by Takeuchi *et al.*) (33), and an unskillful ability to remain on a slender rod were observed. Symptoms fluctuated with the passage of time. Figure 11-1 shows an example of this fluctuation in a mouse that received seven daily doses of 10 mg Hg/kg.

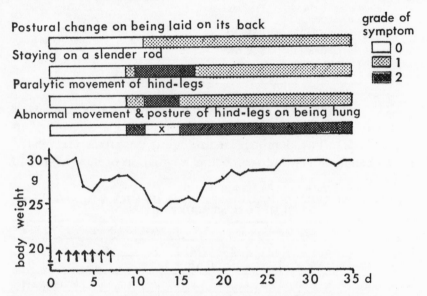

FIGURE 11-1. Occurrence and fluctuation of neurological symptoms in a mouse that received seven daily doses of 10 mg Hg/kg. The arrow in the figure indicates peroral administration of methylmercury chloride. Table 11-II shows the grade of symptom in detail. The mark (×) in the figure means no movement of hind legs.

Figure 11-2 shows the concentration of mercury in the brain as a function of the time after administration. For those groups in which all the mice died, the concentration of mercury was estimated from an autopsy on one specimen. It should be noted that a mercury concentration of more than 20 μg/g of wet tissue was found only in those groups which had shown some neurological

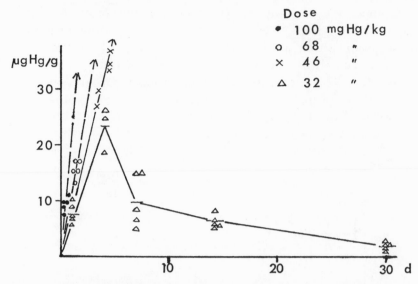

FIGURE 11-2. Mercury concentrations in the brain after a single peroral administration of methylmercury chloride. Broken line with arrow is an assumed trend after the death of all mice.

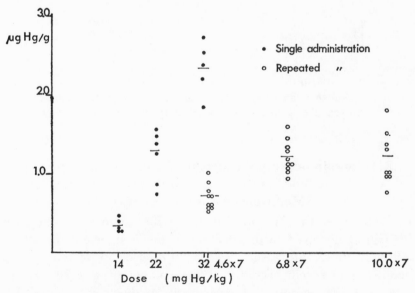

FIGURE 11-3. Mercury concentrations in the brain for a survival of 30 days after the last administration of methylmercury chloride.

symptoms. We can only estimate the peak concentration of mercury for the groups given 22 and 14 mg Hg/kg, from an autopsy of a specimen that survived 30 days (Fig. 11-3). However, if the mercury content in the brain decreases in the same manner as it does when 32 mg Hg/kg are administered, the peak concentrations would probably not exceed 20 μg/g in these groups. Peak concentrations of mercury of more than 20 μg/g occurred in those groups given repeated dosages (Fig. 11-4).

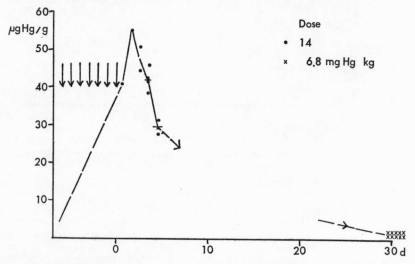

FIGURE 11-4. Mercury concentrations in the brain after repeated administrations of methylmercury chloride. The arrow shows an administration and the broken line is an assumed trend of change of mercury contents in the brain.

Relationship Between the Mercury Content in the Brain and That in the Blood in Repeated Subcutaneous Injections of Methylmercury Acetate in Mice

Methylmercury acetate labeled with radioactive mercury (^{203}Hg) is dissolved in distilled water and is injected subcutaneously 10 times with a successive daily dose of 50 or 100 μg per mouse. The specific activity of the compound was 20 μC/mg Hg/ml at the beginning of experiment. As shown in Figures 11-5 and 11-6, the concentrations of mercury in various organs and

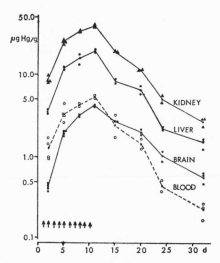

FIGURE 11-5. Change of mercury concentrations in kidney, liver, brain and blood after repeated subcutaneous injections of methylmercury acetate labeled with radioactive mercury. A daily dose was 50 μg per mouse. Arrow indicates injection.

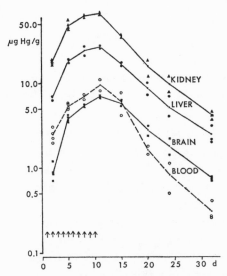

FIGURE 11-6. Change of mercury concentrations in the kidney, liver brain tate labeled with radioactive mercury. A daily dose was 100 μg per mouse. and blood after repeated subcutaneous injections of methylmercury ace-Arrow indicates an injection.

tissues had risen in the course of administration and did not saturate, in other words, did not reach the constant value. The concentration of mercury in the blood was constantly higher than that in the brain in the course of administration and for a short period after it, but the decrease of mercury concentration in the blood was faster than that in the brain. The biological half-life of mercury in the brain was calculated to be 7.2 and 6.2 days respectively, and that in the blood 4.2 and 4.1 days respectively for 50 and 100 μg per mouse. Table 11-IV shows the values for the half-life of mercury in several organs and tissues.

TABLE 11-IV

THE TIME OF HALF-LIFE OF MERCURY IN VARIOUS ORGANS AND
TISSUES AFTER REPEATED INJECTIONS OF
METHYL MERCURY ACETATE

Organs and Tissues	100 µg, 10 times	50 µg, 10 times
Brain	6.18 d	7.18 d
Cerebrum	6.17	7.05
Brain stem	6.16	6.94
Cerebellum	6.21	7.49
Liver	6.23	5.79
Kidney	4.85	5.52
Adrenal	5.83	6.62
Adipose tissues	5.39	5.54
Blood	4.12	4.24

The Mercury Concentration in the Human Brain and its Relation to the Time From Onset of Neurological Symptoms to Death

There have been several reports describing the mercury content in the brain of patients who showed characteristic neurological symptoms and died from poisoning, supposedly from an alkylmercury compound (1, 12, 34, 39). Figure 11-7 shows the relationship of the mercury concentration in the brain at the onset of neurological symptoms to that at the time of death. Assuming that the decrease of the mercury content in the human brain is similar to that in the mouse, namely, a half-life of 6.5 days, the mercury concentration at the onset of the symptom could be calculated for those cases who died not too long afterwards, by comparison with our data on mice, which, of course, is confined to about 30 days after the last administration of mercury. Thus, the lowest extrapolated value for the onset is about 50 μg/g. On the other hand, the graphically demonstrated relation shows another correlation between the mercury concentration from onset to that at time of death. The mercury in the human brain could have two phases in its decrease: (a) a phase of rapid decrease shortly after the exposure has ceased, similar to that observed in the experiment on mice; and (b) a late phase of slow decrease, as is seen in Figure 11-7.

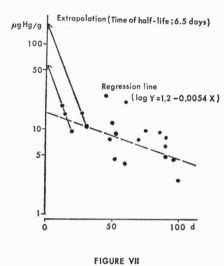

FIGURE VII

FIGURE 11-7. Mercury concentrations in the human brain as determined from autopsy materials and the time from onset of neurological signs due to alkylmercury poisoning to death.

Mercury Concentration in the Blood of Patients Who Survived Ethylmercury Poisoning

We experienced two actual and two suspected cases of ethylmercury poisoning in 1961 (16), although the symptoms were limited to the peripheral neural, even in the real cases. At the first examination, 15 days after the end of exposure to mercury, the mercury concentrations in the blood of the two patients were 1.5 and 1.72 μg/g in the actual cases and 0.65 and 1.00 μg/g in the suspected cases. If the half-life of mercury in the blood is 4 days and the ratio of the mercury concentration in the blood to that in the brain is 1:0.8, according to results from the experiment with mice, the concentrations of mercury in the brain of the two patients after exposure would be 13.6 and 15.6 μg/g in the actual cases and 5.9 and 9.0 μg/g in the suspected cases.

Conclusion

The experimental results on mice given peroral doses of methylmercury chloride suggest the probable existence of a threshold or a critical concentration of mercury in the brain of about 20 μg/g of wet tissue when neurological symptoms manifest themselves. The relationship between the mercury concentration and its effect has been reported by Clarkson *et al.* (4, 5) for mercurial diuretics. In that case, diuresis occurred when the amount of mercuric ion released from organic mercurial exceeded 35 μg/g in the kidney.

In our experiment on mice, death following neurological symptoms was associated with a concentration of mercury of more than 30 μg/g in the brain, and survival after neurological symptoms with a concentration of 20 μg/g. For human cases of methyl- or ethylmercury poisoning, the calculated peak concentrations of mercury in the brain, based on the assumption that the mercury in the brain and blood decreases, were more than 50 μg/g in patients who died and about 15 μg/g in patients who survived.

Of course, this calculation is only based on a trial for detecting or testing the existence of a threshold, or a critical concentration. But, the hypothesis of critical concentration does not seem to be refuted; in fact, it seems to be supported as far as the present experimental results and analyses for human cases are concerned.

In this report, we have treated the brain as a whole, but the neurological symptoms observed in mice were concerned primarily with the cerebellum. Though the mercury of methylmercury distributes evenly in the human brain, there are slight differences in its concentration in different parts of the brain, i.e. the mercury content is approximately the same in the cerebrum and the cerebellum, but lower in the brain stem and the spinal cord. The mercury content decreases in various parts of the brain of mice injected with methylmercury acetate in the following order: (a) in the cerebrum, (b) in the cerebellum and (c) in the brain stem. The effects of mercury concentration on different parts of the brain also vary. Further studies are needed to know if the cerebellum is the most susceptible to mercury.

ACKNOWLEDGMENT

The experimental studies on mice in this report were carried out in collaboration with Miss T. Miyama (Department of Public Health, Faculty of Medicine, University of Tokyo) and Miss Y. Ohta (Tokyo Atomic Industrial Research Laboratory Ltd.). The author wishes to thank Professor H. Katsunuma (Chief of the author's department) and Associate Professor A. Koizumi (Department of Public Health, Faculty of Medicine, University of Tokyo) for reviewing his paper.

REFERENCES

1. AHLMARK, A.: *Brit J Industr Med, 5*:117, 1948.
2. BERLIN, M., and ULLBERG, S.: *Arch Environ Health, 6*:610, 1963.
3. BIDSTRUP, P.L.: *Toxicity Of Mercury And Its Compounds.* Amsterdam, Elsevier, 1964.
4. CLARKSON, T.W., and GREENWOOD, M.: *Brit J Pharmacol, 26*:50, 1966.
5. CLARKSON, T.W.; ROTHSTEIN, A., and SUTHERLAND, R.: *Brit J Pharmacol, 24*:1, 1965.
6. FRIBERG, L.: *Arch Industr Health, 20*:42, 1959.
7. FRIBERG, L.; ODEBLAD, E., and FORSSMAN, S.: *Arch Industr Health, 16*:163, 1957.
8. FRIEDMAN, H.L.: *Ann NY Acad Sci, 65*:461, 1957.
9. GAGE, J.C., and SWAN, A.A.B.: *Biochem Pharmacol, 8*:77, 1961.
10. HAGEN, U.: *Naunyn Schmiedberg Arch Exp Path Pharm, 224*:193, 1955.
11. HEPP, P.: *Naunyn Schmiedberg Arch Exp Path Pharm, 23*:91, 1887.
12. HÖÖK, O., *et al.*: *Acta Med Scand, 150*:131, 1954.
13. HUGHES, N.L.: *Ann NY Acad Sci, 65*:454, 1957.
14. HUNTER, D.; BOMFORD, R.R., and RUSSELL, D.S.: *Quart J Med, 9*:193, 1940.
15. IRUKAYAMA, K., *et al.*: *Jap J Med Progress, 50*:491, 1963.
16. KATSUNUMA, H.; SUZUKI, T.; NISHII, S., and KASHIMA, T.: *Rep Institute Sci Labor,* 1963, No. 61, p. 33.
17. LUNDGREN, K-D., and SWENSSON, A.: *J Industr Hyg, 31*:190, 1949.
18. MILLER, V.L., *et al.*: *Toxic Appl Pharmacol, 3*:459, 1961.
19. MILLER, V.L.; KLAVANO, P.A.; JERSTAD, A.C., and CSONKA, E.: *Toxic Appl Pharmacol, 2*:344, 1960.

20. Okinaka, S., *et al.*: *Neurology, 14:*69, 1964.
21. Rothstein, A., and Hayes, A.D.: *J Pharmacol Exp. Ther, 130:*166, 1960.
22. Saito, M., *et al.*: *Jap J Exp Med, 31:*277, 1961.
23. Sebe, E., and Itsunu, Y.: *Jap J Med Progress, 49:*607, 1962.
24. Sera, K., *et al.*: *Kumamoto Med J, 14:*65, 1961.
25. Suzuki, T., *et al.*: *Jap J Exp Med, 33:*277, 1963.
26. Suzuki, T., *et al.*: *Jap J Exp Med, 34:*211, 1964.
27. Suzuki, T.; Miyama, T., and Katsumuma, H.: *Jap J Med Progress, 48:*
    716, 1961.
28. Swensson, A., *et al.*: *Arch Industr Health, 20:*467, 1959.
29. Swensson, A.: *Acta Med Scand, 143:*365, 1952.
30. Swensson, A.; Lundgren, K.D., and Lindström, O.: *Arch Industr Health,*
    *20:*432, 1959.
31. Swensson, A., and Ulfvarson, U.: *Int Arch Gewerbepathol Gewerbehyg,*
    *24:*12, 1967.
32. Takeuchi, T., *et al.*: *Kumamoto Igaku Zasshi, 40:*1003, 1966.
33. Takeuchi, T., *et al.*: *Kumamoto Igaku Zasshi, 34* (suppl. 3) :531, 1960.
34. Takeuchi, T.: *Pathology of Minamata-disease.* Kotsuna, M. (Ed.) :
    Minamata-byo. Kumamoto, 1966, chapt. 4.
35. Ulfvarson, U.: *Int Arch Gewerbepathol Gewerbehyg, 19:*412, 1962.
36. Ushigusa, S.: *Kumamoto Igaku Zasshi, 39:*33, 1965.
37. Weiner, I.M.; Levy, R.I., and Mudge, G.H.: *J Pharmacol Exp Ther,*
    *138:*96, 1962.
38. Yoshino, Y.: *Clin Neurol, 5:*130, 1965.
39. Yoshino, Y.: *Sogo Rinsho, 14:*580, 1965.

DISCUSSION

CLARKSON: Are neurological symptoms delayed in mice, or do they appear very quickly after you inject the mercury? There is a delay period of a week or two in man.

SUZUKI: The length of delay of the symptoms depends on the dose and the frequency of the administration. With higher doses, the symptoms appear 2 days after the last injection, which is a total time of ten days from the first administration. With single administrations there is also a delay.

WELCH: Did you see signs that could be attributed to changes in sensory function, or were all your findings related to disturbed motor function of the nervous system?

SUZUKI: We considered the possibility that the loss of ability to lift the head was due to a lesion in the vestibular systems of the brain, but pathological studies located the lesions in the cerebral hemispheres.

GAGE: Did you investigate other sensory systems with such stimuli as food?

SUZUKI: No, but we observed other lesions to the nervous system. For example, peripheral neuritis is very common in moderate cases of methylmercury poisoning.

BERGLUND: Would not you consider the first symptom as the important one? In man the first symptom was not an overt symptom like ataxia, but numbness of hands and feet, which could be very difficult to detect in experimental animals. In your data on mice, as in our tests of methylmercury poisoning in rats, the first symptom was loss of weight, compared to control animals. This happened in the first week.

SUZUKI: In our experiments, the loss of body weight has been very well correlated with the appearance of some other neurological findings.

BERGLUND: But the loss of weight was apparent before the other symptoms.

SUZUKI: They were correlated; using the word "first" in a wider sense, there might be some changes in the enzymatic activities or some biochemical changes. We need more sensitive indicators of methylmercury poisoning.

FASSETT: Was the exposure of people to methylmercury due to the use of an ointment? Were some of them occupational exposure?

SUZUKI: Yes, two or three patients suffered from methylmercury poisoning because it was in a fungicidal dressing, but there were no occupational exposures.

Chapter 12

RISK OF METHYLMERCURY CUMULATION IN MAN AND MAMMALS AND THE RELATION BETWEEN BODY BURDEN OF METHYLMERCURY AND TOXIC EFFECTS

F. Berglund and M. Berlin

ABSTRACT

The distribution and elimination of methylmercury compounds in mammals are reviewed. Some recent, yet unpublished, data are included. The relationship among body burden, concentration in the critical organ—the brain—and neurological signs are discussed. Injuries to the central nervous system, with neurological signs of varying type depending on the species studies, seem to appear at a mercury concentration in the brain of the order of 8 $\mu g/g$ or more. The fetal brain seems to be more sensitive to methylmercury than the adult brain. The body burden of methylmercury varies in different species when the toxic concentration is reached in the brain. It is concluded that, due to the slow elimination of methylmercury—its biological half-life varying between twenty to seventy days in different species—there is a considerable risk for cumulation of methylmercury in the tissue of mammals at prolonged exposure to methylmercury compounds. Accumulation and excretion can be described by an exponential equation as long as serious intoxication has not occurred. Consequently, there is a linear correlation between intake of methylmercury and concentration of mercury in different organs of the body at equilibrium. Little is known about the dose at which the first detectable signs or changes appear. Damage to the germ cells with genetic consequences cannot be excluded at low levels of body burden. The mercury concentration in the erythrocytes seems to be a good diagnostic index for the body burden of methylmercury while the mercury concentration in hair is a record of the body burden of methylmercury at the time of development of the pile.

REPORT

IN 1965, A CONSIDERABLE concern was expressed by Swedish hygienists and toxicologists when (a) it became clear to them that due to pollution of our lakes and rivers the mercury content in our freshwater fish had risen about two hundred times and when (b) through the work of Westöö (34), the mercury in fish meat was identified as methylmercury. It was recognized that a possibility existed for a toxicological outbreak in Sweden of the same type which occurred in Japan at Minamata Bay (28) and Niigata (15). A search through the literature for data on methylmercury toxicology was made in order to estimate a threshold value for body burden of methylmercury in man and a threshold value for mercury concentration in fish meat for consumption.

In 1963, Berlin (3) evaluated the scientific possibilities for estimating a hygienic threshold value for body burden of methylmercury. He concluded that the available data at that time were insufficient for any estimation based on scientific data. New efforts were then made to evaluate those data which appeared in the literature since 1963, but available data were still insufficient. A number of projects were started to improve the basis for evaluations. Many of these projects are still in progress and with others the material is under analysis. In the following, an attempt has been made to summarize the data presently available and the conclusions that can be based on them.

Absorption of Methylmercury Through the Skin, the Lung and the Digestive Tract

Methylmercury salts, if applied on the skin dissolved in an ointment e.g., may give rise to clinical toxicity in man (21, 25, 32). From the practical point of view the most important routes of absorption are the lung and the digestive tract. For example, in connection with the occupational handling of seed-dressings there is a considerable risk of absorption of methylmercury in the respiratory tract. Several cases of human intoxication have been described (2, 17, 18). Methylmercury salts are volatile at room temperature. Data on the degree of retention or on the diffusion coefficient in the alveoli for methylmercury salts are lacking. Also, there is little available data on the intestinal absorption of methylmercury in the food. Data from the rat and man indicate that more than 90 per cent of the methylmercury in the food is absorbed (1, 10).

Distribution of Methylmercury in the Body and Tissue Tolerance to Methylmercury

The distribution of methylmercury at different dose levels has been studied in a number of mammalian species, including man. The most extensive data are obtained from rats, which are quantitatively summarized in Table 12-I; the mean values are expressed in μg/g and calculated from data in the references. One column gives the mercury concentration in the brain. In most cases a modification of combustion and dithizon binding of mercury with spectrophotometric analysis of the dithizon mercury complex has been used for analysis. In some cases a tracer technique with ^{203}Hg and scintillation counting has been used. In the human cases it cannot be excluded that exposure to other mercury compounds has occurred which will interfere with the liver and kidney values.

Autoradiographic studies on the methylmercury distribution have been made on the mouse (5). After intravenous injection, the methylmercury is distributed in the blood, and during the first 24 hours the methylmercury diffuses into the tissues, resulting in an even intracellular distribution in the whole body. The concentration differences seen between different organs and cell types are small compared to those observed after administration of the inorganic mercury salts (3) or inhalation of mercury vapor (7). Mercury also accumulates in hair to a considerable degree (14, 15). The distribution and diffusion into the tissues can be accelerated by the simultaneous administration of dimercaprol (BAL) which shortens the distribution time to about 1 hour (6). The placenta constitutes, in contrast to what is seen with other heavy methyl salts, no barrier to methylmercury (5). In the brain however, peak concentrations do not appear until several days after its appearance in other body organs. The blood-brain barrier is thus a relative hindrance to methylmercury penetration, which can be accelerated by the simultaneous administration of dimercapol (6). In all species studied, 90 per cent of the methylmercury in the blood is bound to the red cells. The fraction in plasma varies among the species; in man 10 per cent (30); in the rat 4.5 per cent (14, 33); in the rabbit 10 per cent (4); and in the squirrel monkey 9 per cent (8).

TABLE 12-I

METHYLMERCURY DISTRIBUTION IN MAMMALS

Species	Brain μg/g	Liver μg/g	Liver/ Brain	Kidney μg/g	Kidney/ Brain	Blood μg/g	Blood/ Brain	Number of Animals	Reference
Mouse	28*	72	2.6	64	2.3			8	Saito et al. 1961 (23)
	0.49	1.7	3.4	4.2	8.5	0.4	0.82	5	Suzuki et al. 1963 (24)
Rat	2.7	14	5.2	55	21	44	16	15	Berlin et al. 1965 (6)
	4	16	4	51	13	80	20	6	Friberg 1959 (13)
	1.7	7.3	4.3	24	14	21	12	6	Gage 1964 (14)
	0.19	0.92	4.8	4.6	24	3	16	5	Ulfvarson 1962 (33)
Rabbit	1.5	2.9	2	2.9	2			4	Ulfvarson 1962 (33)
Cat	9	52	5.8	15	1.7			3	Swensson 1952 (26)
Dog	33*					12	0.36	3	Takeuchi 1961 (28)
Squirrel								3	Yoshino et al. 1966 (35)
Monkey	13*	10	0.77	6	0.46	3.4	0.26	2	Berlin, Nordberg 1968 (8)
Man	5*	20	4	30	6			1	Ahlmark 1948 (2)
	5*	14	2.8	3	0.6	4	0.8	1	Lundgren, Swensson 1948 (18)
	12*	39	3.3	27	2.3			1	Höök et al. 1954 (17)
	11*	40	3.6	66	6			10	Takeuchi 1961 (28)
	25*	21	1	51	2			1	Tsuda et al. 1963 (32)

*Toxic signs.

In rabbit plasma more than 99 per cent of the methylmercury is bound to plasma protein (4). The relative hindrance of the blood brain barrier to methylmercury has been confirmed through quantitative measurements on the mouse by Suzuki (24). Cases of clinical intoxication have been described in which the neurological signs culminated up to several weeks after the cease of exposure to methylmercury (18). Similar observations on experimental intoxication in squirrel monkeys have been made (8).

There are considerable differences in distribution patterns of methylmercury among different mammalian species (Table 12-I). The rat differs markedly from other species with its low content of mercury in the brain compared to other organs. There are also variations in the distribution pattern among the other species. The amount of methylmercury in the different organs when related to the brain in Takeuchi's clinical material (28), is considerable. The mercury concentration in the kidney and liver is generally higher than in the brain in his material. In the squirrel monkey the concentration of mercury in the brain is higher than in any other organ. Ekman *et al.* (10) measured by scintillation counting the amount of methylmercury in the head of three volunteers thirty days after they had orally received a small tracer dose of methylmercury[203]. Although they found 15 to 20 per cent of the total body burden in the head, analysis of the hair from these cases at this time showed negligible amounts of mercury.

Table 12-II summarizes the mercury concentrations which have been reported from the brain in animals and man with definite neurological signs. Although there are large differences in the distribution patterns between different species, the concentration of methylmercury in the critical organ, the brain, is of the same order of magnitude when neurological signs appear in all studied species. Sensitivity to methylmercury is more pronounced in the fetal stage of brain development than in any other stage. Fetal injury has been reported by Engelson and Herner (12) and from Minamata by Matsumoto *et al.* (20). Experimental data elucidating the fetal tolerance to methylmercury are still lacking.

The distribution of methylmercury in the brain has been studied in the dog (35), the squirrel monkey (8), and man (22).

TABLE 12-II
METHYLMERCURY CONCENTRATION IN THE BRAIN AT
NEUROLOGICAL SIGNS

Species	Range μg/g	Number of Individuals Based Upon	Reference
Mouse	10–61	10	Saito *et al.* (23)
Rat	49	10	Takeshita *et al.* (27)
Cat	8–18	7	Takeuchi *et al* (28)
Dog	8–50	5	Yoshino *et al.* (34)
Squirrel Monkey	12–14	2	Berlin *et al.* (8)
Man	3– 9	2	Lundgren, Swensson (18)
	12	1	Höök *et al.* (17)
	9–24	3	Takeuchi *et al.* (28)
	15–66	2	Okinaka *et al.* (35)
	22–48	1	Tsuda *et al.* (32)

Available data indicate that there are concentration differences of a factor of 2 or more between different parts and structures of the brain. Yoshino *et al.* (35) report that in a dog with neurological symptoms, the calcarine cortex has a higher concentration of methylmercury than other parts of the brain. As in man and primates, the dog shows signs of visual field constriction and impaired visual acuity with pathological changes of the calcarine cortex in methylmercury intoxication. In autoradiographic studies on the squirrel monkey (8), accumulation of mercury has been observed in certain cortical layers in the cerebellar and calcarine cortex. Swensson and Lundgren (18) report a relatively higher concentration of mercury in the basal ganglia of one of their clinical cases. Our present knowledge therefore, is not contradictory to the hypothesis that there is a correlation between the degree of mercury accumulation and pathological changes.

Elimination of Methylmercury

All studied mammals show a very slow excretion of methylmercury. No data indicate that the carbon-mercury bond can be broken in the body. The biological half-life for methylmercury varies between 20 and seventy days in the studied species. There are elimination data from the rat (1, 33), the squirrel monkey (8), and man (9, 11). The excretion of methylmercury occurs mainly by the feces, with only a small part excreted by the urine.

Studies in the rabbit (4) and man (19) indicate a correlation between the blood concentration and the urinary excretion. Autoradiographic observations in the mouse (4) indicate that methylmercury is excreted in the bile, but also via the intestinal mucous membranes. The relative importance of different parts of the intestinal tract for excretion of methylmercury cannot be evaluated on the basis of present data.

The very slow excretion of methylmercury and the even distribution seen in autoradiography with the absence of distinct barriers between the different compartments of the body, except for the central nervous system, indicate that the turnover among the general body compartments is not slower than the excretion rate. There is no reason to believe that complicated metabolic processes are extensively involved in the turnover of methylmercury among different types of tissues, except for the central nervous system. If the transfer of methylmercury between different compartments of the body occurs freely and with a rate which is rapid in comparison with the excretion rate, it may be assumed that there is an equilibrium among the compartments of the body with respect to mercury concentration under nontoxic conditions. It may also be assumed that the excretion is related to the total body burden of methylmercury, i.e. that a defined fraction of the total body burden of mercury is excreted per unit of time. If that is the case, the course of the accumulation can be expressed by an exponential equation as follows:

$$A = \frac{\alpha}{\beta} \; (1 - e^{-\beta t}) \tag{1}$$

in which A is the body burden of methylmercury, a is the daily intake of methylmercury and β the fraction of body burden which is excreted.

The decrease of body burden, when intake has ceased, can be expressed according to the formula

$$A = A_o \cdot e^{-\beta t} \tag{2}$$

in which A_0 is the body burden at the cessation of intake.

Ulfvarson tested this hypothesis in a rat experiment (33). He injected twelve rats every other day subcutaneously with methylmercury salt and found a good adjustment of his accumulation curve to Equation 1 with an excretion constant of 0.03. However his study lasted not more than eighteen days, which was thus too short a duration of time to allow the methylmercury to reach equilibrium.

If Equation 1 correctly describes the methylmercury accumulation curve, the body burden would increase towards a limit value when excretion is equal to intake. Ninety-seven per cent of this level should have been reached after one hundred days. If there is a constant relation between excretion and body burden of methylmercury, until toxic disturbances appear there will be (a) a linear correlation at equilibrium between daily intake of methylmercury and body burden and (b) a linear correlation between intake and concentration of methylmercury in each organ or tissue. It follows, then, that the mercury concentration in the red cells, which are easily obtainable for diagnostic purposes, should be a good index of the total body burden of methylmercury at all times when equilibrium in the body is postulated, and that the mercury concentration in hair gives a good record of the body burden of methylmercury at the time of the development of the particular part of the pile. The low methylmercury concentration in urine and feces makes these parameters less suitable for diagnostic purposes or as practical indices for body burden of methylmercury.

To further test these hypotheses and, if possible, to titrate a zero effect dose in the rat, the uptake, accumulation curve and excretion curve, were studied in 250 rats given specific concentration of methylmercury in their food for nine months. To make it possible to determine the body burden continuously during the accumulation and to determine adequately the mercury concentration in the organs after the animals had been sacrificed, the methylmercury in the food was labeled with radioactive mercury. Three dose levels and one control group were studied, with the highest dose being 5 mg methylmercury/kg food, the middle dose — 1 mg methylmercury/kg food, and the lowest dose —0.2 mg of methylmercury/kg food. To exclude effects due to irradiation one group was given the middle dose of unlabeled methylmercury. At all three dose levels, the equilibrium level was reached after about six months, corresponding to an excretion constant of 0.02 or a daily excretion of 2 per cent of the body burden. The accumulation curve seems to be adjusted to Equation 1 and a linear correlation was observed between dose and body burden of methylmercury when the results in the three groups were compared. The material is now under data processing, and the detailed values will be available in June 1968. During the study, no change of body weight occurred and no pathological-anatomical changes and/or effect on conditioned operant behavior could be observed at any dose level. The highest concentration of methylmercury in the brain of the animals which were given the highest dose of methylmercury was found to be about 8 $\mu g/g$ brain tissue.

Of special interest is whether the central nervous system, which has a delayed uptake of methylmercury in comparison with the rest of the body organs, may also show a delayed elimination. Results from the above mentioned experiments, in which the mercury concentration in the brain was determined at different times, do not support the idea that there is any appreciable delay in its elimination. Ekman *et al.* (10) observed no delay as compared to the decrease of the total body burden in the elimination of methylmercury from the brains of three volunteers given tracer doses of labeled methylmercury. In the rat experiment of Ahlborg *et al.* (1), no increase of mercury concentration in the brain could be shown between the seventh and eighth month of constant intake of methylmercury, which indicates that the brain at this point had practically reached equilibrium with the rest of the body with respect to methylmercury concentration. The results of Ekman *et al.* (11) from human volunteers indicate that the excretion of methylmercury in man is considerably slower than in the rat and is about 1 per cent per day with a half-life of seventy days. A similar value of the biological half-life was observed by Birke *et al.* (9) from cases in which individuals had consumed fish in Sweden containing methylmercury. This means that 97 per cent of the equilibrium level will be reached according to Equation 1 after about one year.

Toxic Effects of Methylmercury Other Than Neurological Disturbances or Injuries of the Central Nervous System

Disturbances other than the neurological signs described by Hunter *et al.* (16) and Takeuchi (28, 19), with pathological-anatomical changes in the brain or physiological or biochemical changes caused by methylmercury intake have not been described. There is no evidence for a carcinogenic effect of methylmercury although such an effect cannot be excluded. Methylmercury has been shown by Ramel (22) to have a considerable potency as a mitosis-disturbing agent, causing polyploidy and chromosome disjunction. He made his observations on root cells of *Allium cepa* and on larvae of *Drosophila melanogaster*. The methylmercury concentration in the substrate which produced an effect was as low as 50 ng/g in some experiments. This compares with the methylmercury level in the red blood cells of a Swedish control population (31) which was of the order of 10 ng/g. Tejning's (30) epidemiological study on Swedish fish-eaters from a region with freshwater fish containing methylmercury showed a 5.5 fold increase of methylmercury in the blood cells after a calculated intake of 20 mg/year of methylmercury. No chromosome abberations have been yet reported from human cases. In the material from the rat investigation of Ahlborg *et al.* (1) no chromosome abberations have been found in any blood cells. These studies are, however, not yet finished and the analysis of blood cells from people exposed to methylmercury is still in progress.

Conclusion

Available evidence indicates that there is a considerable risk for cumulation of methylmercury in the tissue of mammals at prolonged exposure to methylmercury compounds. This risk is mainly due to their very slow elimination and a biological half-life which varies between twenty and seventy days in different species. Injuries to the central nervous system, with neurological symptoms depending on the species studied, seems to appear at a mercury concentration in the brain of the order of 8 μg/g or more. The fetal brain seems to be more sensitive to methylmercury than the adult brain. The body burden of methylmercury varies in different species when the toxic concentration is reached in the brain. The accumulation and excretion can be described by an exponential equation as long as serious intoxication has not occurred. Consequently, there is a linear correlation between intake of methylmercury and concentration of mercury in different organs of the body. Little is known about the dose level at which the first detectable signs or changes appear. Damage to the germ cells with genetic consequences cannot be excluded at low levels of body burden. The mercury concentration in the red corpuscles seems to be a good diagnostic index of body burden of methylmercury and the mercury concentration in hair a record on the body burden of methylmercury at the time of development of the pile.

REFERENCES

1. AHLBORG, U.; BERGLUND, G.; BERLIN, GRANT C.; V. HAARTMAN, U., and NORDBERG, G.: To be published.
2. AHLMARK, A.: *Brit J Industr Med, 5:*117, 1948.
3. BERLIN, M.: *Acta Med Scand* (suppl.) 1961, p. 396.
4. BERLIN, M.: *Arch Environ Health, 6:*626, 1963.
5. BERLIN, M., and ULLBERG, S.: *Arch Environ Health, 6:*610, 1963.
6. BERLIN, M.: JERKSELL, L.-G., and NORDBERG, G.: *Acta Pharmacol Toxic, 23:*312, 1965.
7. BERLIN, M.: JERKSELL, L.-G., and VON UBISCH, H.: *Arch Environ Health, 12:*33, 1966.
8. BERLIN, M., and NORDBERG, G.: To be published.
9. BIRKE, G.; JOHNELS, A.G.; PLANTIN, L.O.; SJÖSTRAND, B., and WESTERMARK, T.: *Läkartidningen, 64* (37) :3628, 1967.
10. EKMAN, L.; GREITZ, U.; MAGI, A.; SNIHS, J.-O., and ABERG, B.: *Nord Med, 79:*456, 1968.
11. EKMAN, L.; GREITZ, U.; PERSSON, G., and ÅBERG, B.: *Nord Med, 79:*450, 1968.
12. ENGLESON, G., and HERNER, T.: *Acta Paediat, 41:*289, 1952.
13. FRIBERG, L.: *Arch Industr Health, 20:*42, 1959.
14. GAGE, J.C.: *Brit J Industr Med, 21:*197, 1964.
15. HOSHINO, O.; TANZAWA, K.; TERAO, T.; UKITA, T., and OHUCHI, A.: *J Hyg Chem, 12:*94, 1966.
16. HUNTER, D.; BOMFORD, R.R., and RUSSELL, D.S.: *Quart Med, 33:*193, 1946.

17. Höök, O.; Lundgren, K.-D., and Swensson, Å.: *Acta Med Scand, 150*:131, 1954.
18. Lundgren, K.-D, and Swensson, Å.: *Nord Hyg Tidskr, 29*:1, 1948.
19. Lundgren, K.-D.; Swensson, A., and Ulfvarson, U.: *Scand J Clin Lab Invest, 20*:164, 1967.
20. Matsumoto, H.; Koya, G., and Takeuchi, T.: *J Neuropath Exp Neurol, 24*:563, 1965.
21. Okinaka, S.; Yoshino, M.; Mozai, T.; Mizuno, Y.; Terao, T.; Watanabe, H.; Ogihara, K.; Hirai, S.; Yoshino, Y.; Inose, T.; Anzai, S., and Tsuda, M.: *Neurology, 14*:69, 1964.
22. Ramel, C.: *Hereditas, 57*:445, 1967.
23. Saito, M.; Osono, T.; Watanabe, J.; Yamamoto, T.; Takeuchi, M.; Ohyagi, Y., and Katsunuma, H.: *J Exp Med, 31*:277, 1961.
24. Suzuki, T.; Miyama, T., and Katsunuma, H.: *Jap J Exp Med, 33*:312, 1963.
25. Suzuki, T.; Miayama, T., and Katsunuma, H.: *Jap J Med Progress, 48*:716, 1961.
26. Swensson, Å.: *Acta Med Scand, 143*:365, 1952.
27. Takeshita, A., and Uchida, M.: *Kumamoto Med J, 16*:178, 1963.
28. Takeuchi, T.; Proc. of VIIth Int. Congr. of Neurology in Rome, 1961, p. 1.
29. Takeuchi, T.; Kambara, T.; Morikawa, N.; Matsumoto, H.; Shiraiishi, Y., and Ito, H.: *Acta Path Jap, 9*:769, 1959.
30. Tejning, S.: Report No. 67 08 31 from the Department of Occupational Medicine in Lund, 1967 (in Swedish).
31. Tejning, S.: Report No. 67 02 06 from the Department of Occupational Medicine in Lund, 1967 (in Swedish).
32. Tsuda, M.; Anzai, S., and Sakai, M.: *Yokohama Med Bull, 14*:287, 1963.
33. Ulfvarson, U.: *Int Arch Gewerbepathol Gewerbehyg, 19*:412, 1962.
34. Westöö, G: *Acta Chem Scand, 20*:2131, 1966.
35. Yoshino, Y.; Mozai, T., and Nakao, K.: *J Neurochem, 13*:397, 1966.

DISCUSSION

WELCH: Have you been able to identify the chemical nature of the mercury in the brain of any of the species studied?

BERLIN: Yes, to some extent. It is in the form of methylmercury. The carbon-mercury bond is unbroken. The evidence comes from double labeling experiments with carbon 14 and mercury 203; double labeling gives identical radiograms for the distribution in the brain [Ullberg, *Personal communication*]. Analysis of rat brains has been made with the method of Westöö, showing that there is methylmercury in the brain. I am convinced that the carbonmercury bond is stable in mammalian tissue.

WELCH: If the rat could make a metabolite of mercury which was different from that formed by other species, it might explain why mercury traverses the blood-brain barrier faster in this species.

BERLIN: Another factor which may be of importance is that the fraction of methylmercury in the plasma of the rat is lower than in other species by a factor of 3. Also, there is a large difference between man and rat in elimination rates of methylmercury from the body. Methylmercury is excreted mainly by the feces, only a few per cent through urine. A considerable amount is excreted in the bile. However, how much is reabsorbed is unknown. Autoradiograms of mice intestines show an accumulation in the mucosal epithelium. Some methylmercury is certainly reabsorbed with shed epithelial cells in proportion to their turnover rate. Some of this amount could, of course, also be reabsorbed in the intestines; we know from experimental observations in rat and man that methylmercury is efficiently absorbed in the intestines. There is a big difference between species in turnover rates of epithelial cells. The turnover is five times faster in the rat as compared to man. This seems to correlate with the difference between the two species in rate of elimination of mercury.

NORSETH: What was the highest amount of total mercury you found in the rat?

BERLIN: The highest figure for the brain was 8 μg.

BERG: With the procedure of tracing mercury compounds through the food chain and through the body you arrived at a target organ. You also seem to have demonstrated a threshold level for damage. Dr. Stickel traced organochlorine compounds to the same target organ and also found a no-effect level and what could be a threshold. Is it inherent in the toxicological approach, that when you study a target organ you find a threshold? Or is it just the nature of the compounds, that there is some resemblance in the mechanism of action of these two toxic groups of compounds?

BERLIN: Well, I think our methods to detect effects in the central nervous systems are rather crude. We do not have a selection of graded effects to study. We looked for behavioral changes by studying avoidance response and conditioned reflexes in rats, and found no effects of poisoning until there was an irreversible deterioration of nerve cells.

FASSETT: It seems to me there is a fundamental difference between the ingestion of methylmercury and the ingestion of organohalogens. The methylmercury presumably had reacted with the SH group in the food proteins, and the animal is actually eating a protein to which methylmercury is bound very strongly. Is it still bound to an amino acid when it is absorbed?

BERLIN: Methylmercury has been introduced in combination with many organic and inorganic ions and this brought no difference in distribution. There was one exception only, when the complex used was exceptionally stable in the body. I think that methylmercury moves freely in the body from one binding substance to the next because the stability of the binding is approximately the same.

NORSETH: I observed a difference in distribution between injections of pure inorganic mercury and those of inorganic mercury contaminated with protein. This was four hours after intravenous injection.

BERLIN: This is a shorter observation time than I discussed. Such protein contamination may have been in a particulate form. In that case it was probably taken up by the reticuloendothelium.

NORSETH: Is the mercury in feces in the form of inorganic mercury or methylmercury?

BERLIN: This is not known yet.

JERNELÖV: So the properties of the methylmercury can be changed according to what its mixed with.

BERLIN: These were *in vitro* tests with high concentrations of methylmercury and a controlled medium. When methylmercury is injected into the body I think that it reacts with such a variety of SH-groups in the tissue that the injected form in which it is matters little, unless it is not of very high stability, as for example, the methylmercury-BAL complex.

DEICHMANN: Have you tried to alter the rate of excretion of the mercury compound?

BERLIN: Yes, we have injected mice with both the methyl-mercury compound and BAL (British anti-lewisite), but this had no influence on the amount of methylmercury retained in the body. It did have some influence on the distribution of mercury shortly after injection. The BAL and the methylmercury compound were injected daily, and the body burden was determined by wholebody counting of labeled mercury. BAL, however, forms a very stable complex with the methylmercury, even compared with possible complexes in the body. Thus BAL administered together with methylmercury compounds will temporarily cause a different methylmercury distribution [ref. Berlin *et al.*, 1965]. I think that methylmercury moves freely in the body from one binding site to another as there doesn't seem to be any large difference in stability between different possible complexes in the body.

DEICHMANN: Was there an effect on the kidney, or any other effect on excretion?

BERLIN: One observed effect is that BAL forced mercury into the brain quicker. There was some indication of increased secretion in urine, but this was probably matched by a lower excretion in the feces. When we measured the whole body count in the animals given only methylmercury and compared that with animals given both BAL and methylmercury, we could not find a statistical significant difference. But this was in mice, which have a rapid excretion compared to man. In man the effect on excretion may be quite different. BAL may be useful there, but you have to consider that it will enhance the penetration of methyl-mercury to the brain. In acute situations this may not be good.

BERGLUND: There are two papers which indicate a difference in the effect of different methylmercury compounds, depending on the anion part; but we hesitate to accept them at full value because of the high doses that were used. Yamashita [*J Jap Soc Intern Med, 53:*529, 1964] tested CH_3-HgS-CH_3 and found significant differences in toxicity compared to other methylmercury compounds, but he also used high doses. A pathologist [Morikawa, N.: *Kumamato Med J, 14:*71, 1961] found different locations of brain lesions with methylmercury bound to sulfhydryl compounds. We gave labeled methylmercury hydroxide intravenously to rats and we collected the bile for about a twelve-hour period through a bowel fistula. We measured both the labeled mercury and the methylcercury compound, and there was complete agreement. So that, at least in the bile, all the mercury that was excreted was in the form of methylmercury. But we don't know if that methylmercury is absorbed again in the gut. Perhaps the fecal mercury derives from lower down in the intestine. We have not investigated whether methylmercury is excreted in the feces. In the bile, all of the mercury excreted is in the form of methylmercury [Berglund, F., Nordberg, G., and Westöö, G., unpublished].

Chapter 13

ISOTOPE EXCHANGE METHODS IN STUDIES OF THE BIOTRANSFORMATION OF ORGANOMERCURIAL COMPOUNDS IN EXPERIMENTAL ANIMALS*

T. W. CLARKSON

ABSTRACT

Very little work has been done on the biotransformation of heavy metals. This paper describes studies on the biotransformation of a variety of organomercurial compounds used commercially as pesticides and clinically as diuretics. The biotransformation reaction selected for study was the splitting of the carbon-mercury chemical bond resulting in the release of inorganic mercury (Hg^{++}) in animal tissues.

The major problem is to distinguish between the organic and inorganic compounds of mercury. Two methods, based on an isotope exchange reaction, will be described.

Preliminary results indicate that different compounds release Hg^{++} at widely different rates, paramercuribenzoate being most rapidly and methylmercury most slowly metabolized of the seven compounds tested. Species difference in rates of metabolism were also observed. These differences were most marked with the compound paramercuribenzoate. Some of the compounds tested produced severe kidney damage when given at LD 50 doses. The nephrotoxic action of these compounds appears to be related to their rate of biotransformation.

REPORT

Introduction

The Biotrasformation of Organomercurial Compounds in Experimental Animals

THE METABOLISM OF FOREIGN organic compounds has been studied intensively in the past fifteen or so years (Brodie, 1964). Some compounds are metabolized to biologically inert products which are rapidly excreted while others give rise to high-

Note: This paper is based on work performed under contract with the United States Atomic Energy Commission at the University of Rochester Atomic Energy Project and is Report No. UR–49– 958.

ly toxic metabolites. Studies on the different rates of biotransformation in different animal species, on the toxicities of the
metabolites, on the effects of one drug on the rate of metabolism
of another, have given a new insight into a variety of important
phenomena in toxicology and pharmacology; for example species,
strain, and sex differences in toxicity, side effects of drugs, tolerance, potentiation and antagonism between two or more compounds, the marked sensitivity of the newborn to certain chemicals. This paper describes new methods involved in studies on
the biotransformation of a variety of organomercurial compounds.
Some preliminary results are also reported.

The biotransformation reaction selected for study was the
splitting of the carbon-mercury chemical bond resulting in the
release of inorganic mercury (Hg^{++}) in animal tissues. The major
problem is to distinguish between the organic and inorganic compounds of mercury. Two methods, based on an isotope exchange
reaction, will be described.

Preliminary results indicate that different compounds release
Hg^{++} at different rates, paramercuribenzoate being most rapidly
and methylmercury most slowly metabolized of the seven compounds tested. Species differences in rates of metabolism were also
observed. These differences were most marked with the compound
p-mercuribenzoate. Some of the compounds tested produced severe
kidney damage when given at LD50 doses. The results suggest that
the nephrotoxic action of a compound is related to its rate of
biotransformation.

Methods

Analysis Using Organomercurials Labeled with the ^{203}Hg Isotope

The principle of the method is illustrated in Figure 13-1.
When a solution of mercuric (Hg^{++}) salt labeled with the ^{203}Hg
isotope, is exposed to an atmosphere containing mercury vapor, a
rapid isotopic exchange takes place resulting in the movement of
the isotope into the vapor phase. On the other hand, ^{203}Hg present
in solution as an organomercurial compound, covalently bonded
to a carbon atom, does not undergo isotopic exchange with the
vapor or does so very slowly. Thus the amount of inorganic mer-

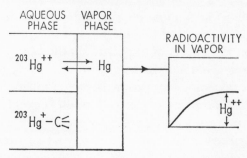

FIGURE 13-1. The measurement of inorganic mercury (Hg^{++}) in the presence of organomercurial compounds (Hg-C) by isotope exchange.

cury present in mixtures with organomercurials is equal to the quantity of mercury exchangeable into the vapor phase. The Conway microdiffusion unit (Conway, 1947) provides a simple means of following this reaction. The solution containing radioactive mercury is placed in the outside chamber and a 0.1 ml of metallic mercury is placed in the center chamber. To facilitate removal, the metallic mercury is contained in a small glass cup, Figure 13-2. When the lid is placed over the unit a small fraction of the metallic mercury volatizes to saturate the gas phase with mercury vapor and isotope exchange with the mercuric mercury commences. At isotopic equilibrium, the exchangeable ^{203}Hg isotope will be at the same specific activity in the metallic mercury and in the mercuric salt present in the solution. In practice all the radioactivity present in the outside chamber as $^{203}Hg^{++}$ is transferred to the droplet of metallic mercury in the central chamber since the metallic mercury is present in great excess. The droplet

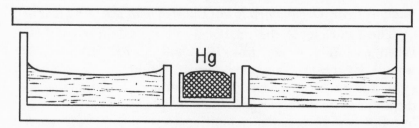

FIGURE 13-2. The Conway Microdiffusion Unit with a small glass cup used to measure isotope exchange.

is removed and counted in a well-shaped NaI crystal. Some self-absorption of the gamma rays takes place within the droplet and a correction for this change in counting efficiency must be made (Clarkson, Rothstein and Sutherland, 1964). The recoveries of $^{203}Hg^{++}$ from solutions containing known proportions of $HgCl_2$ and the organomercurial diuretic, chlormerodrin, are given in Figure 13-3. With chlormerodrin alone, no radioactivity was recovered in the metallic mercury. With $HgCl_2$ alone, all the radioactivity moved into the central compartment. When mixtures were present, the amount of radioactivity recovered was equal to the amount present in the solution as $HgCl_2$.

The same procedure may be used to determine inorganic mercury, Hg^{++}, present in tissue homogenates. The rate of exchange with the vapor is considerably slowed because the mercury is now complexed with tissue proteins. Nevertheless complete exchange takes place. Figure 13-4 shows typical results from $1\%(^w/v)$ homogenates of kidneys taken from rats injected with labelled $HgCl_2$ and killed at various times after injection. Exchange is virtually fraction of exchangeable radioactivity is small. The fraction incomplete in about 45 hours. In contrast when labelled chlormerodrin is given to rats, only part of the radioactivity exchanges (Fig. 13-5). When the animal is killed three hours after injection, the fraction of exchangeable radioactivity is small. The fraction in-

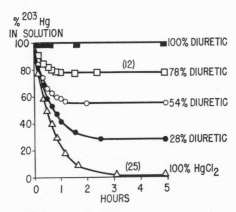

FIGURE 13-3. The recoveries of ^{203}Hg from solutions containing known proportions of $^{203}HgCl_2$ and ^{203}Hg-chlormerodrin.

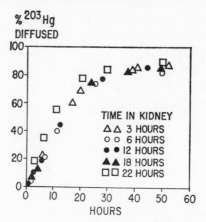

FIGURE 13-4. Recovery of ^{203}Hg from kidneys taken from rats injected with ^{203}HgCl$_2$ and killed at various times after injection.

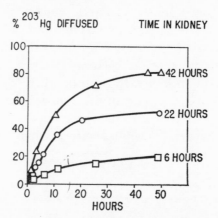

FIGURE 13-5. Recovery of ^{203}Hg from kidneys taken from rats injected with ^{203}Hg-chlormerodrin and killed at various times after injection.

creases with increasing time after injection, until at 45 hours after injection, all the radioactivity is exchangeable. The rate of exchange is identical to that observed when HgCl$_2$ was injected (Fig. 13-4). Evidently the exchangeable mercury observed when chlormerodrin is injected represents the amount of inorganic mercury which has been released from the chlormerodrin molecule. The addition of cysteine (0.2M) to the 1% homogenate accelerates the exchange process so that complete exchange is achieved in about 2 hours. This method has been successful with most of the labeled mercurials commercially available. However, it is desirable to have a method applicable to nonradioactive organomercurials. Many mercurials are not available in the radioactive form and, in most cases, human exposure is to the nonradioactive compounds.

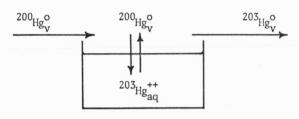

AT ISOTOPIC EQUILIBRIUM

$$[^{203}Hg_v^O] / [^{200}Hg_v^O] \; = \; [^{203}Hg_{aq}^{++}] / [^{200}Hg_{aq}^{++}]$$

FIGURE 13-6. A diagrammatic representation of isotope exchange where the vapor leaving the exchange vessel has attained isotopic equilibrium with the inorganic mercury in the solution.

Analysis Using Nonradioactive Organomercurial

The isotope exchange principle has been adapted to measure nonradioactive inorganic mercury in tissues (Clarkson and Greenwood, 1968). Moreover, in the presence of organomercurial compounds, the method is selective to inorganic mercury. The principle of the method is illustrated in Figure 13-6. Nonradioactive mercury vapor is passed over an aqueous solution containing $HgCl_2$ labeled with ^{203}Hg. As the vapor passes through the exchange vessel, it acquires radioactivity by isotope exchange with the radioactive mercury in the aqueous phase. The rate of loss of radioactivity from the solution is given by the equation

$$\frac{-d[^{203}Hg_{aq}].V}{dt} = \frac{a.[^{203}Hg_v]}{[^{200}Hg_v]} \tag{1}$$

where v ml is the volume of the aqueous phase, $[^{203}Hg_{aq}]$ and $[^{200}Hg_v]$ are the activities (in counts $min^{-1}ml^{-1}$) of the aqueous and vapor phases respectively, $[^{200}Hg_v]$ is the chemical concentration of mercury in the vapor phase (in $\mu g/ml$), and v is the vapor flow in $\mu g/min$. If the vapor achieves complete isotopic equilibrium with the solution, the specific activity of the vapor leaving the vessel will be equal to the specific activity of the mercury in the aqueous phase and the equation becomes

$$-d[^{203}Hg_{aq}].V = a.\frac{[^{203}Hg_{aq}]}{[^{200}Hg_{aq}]} dt \qquad (2)$$

where $[^{203}Hg_{aq}]$ is the concentration of the mercury in the aqueous phase, in $\mu g/ml$. Provided that V, a and $[^{200}Hg_{aq}]$ remain constant throughout the exchange process, Equation 2 may be integrated to give

$$\frac{-\ln[^{203}Hg_{aq}]_t}{[^{203}Hg_{aq}]_o} = \frac{at}{[^{200}Hg_{aq}].V} \qquad (3)$$

where the subscripts o and t refer to zero time and time t respectively. Equation 3 states that the rate of loss of radioactivity will be first order and the time for the loss of half the total activity from the aqueous phase, $t_{\frac{1}{2}}$, will be given by

$$t_{\frac{1}{2}} = 0.690 \frac{.V[^{200}Hg_{aq}]}{a} \qquad (4)$$

In other words the half-time will be directly proportional to the concentration of mercury in the aqueous phase. The slope of the line may be calculated from the known values of V and a.

The apparatus is shown schematically in Figure 13-7. Mercury vapor was generated by passing nitrogen through a midget impinger containing a small quantity of metallic mercury (vapor generator). The impinger was immersed in water at 45°C. The vapor was passed through a second impinger (not shown in Figure 13-7) containing a similar amount of metallic mercury but maintained at room temperature. The nitrogen gas emerging from the second impinger contained almost the saturation level of mercury (20 mg/m³). The vapor then passed to the exchange vessel where it bubbled through the aqueous phase containing radioactive mercury. The dimension of the vessel, length 52.5 cm and diameter 1.9 cm, allowed complete isotopic exchange. The issuing vapor was passed through a water trap and then through a tube containing activated hopcalite. The hopcalite absorbed all the vapor so that by placing the tube in NaI well-crystal, all the ^{203}Hg activity in the vapor stream could be recorded.

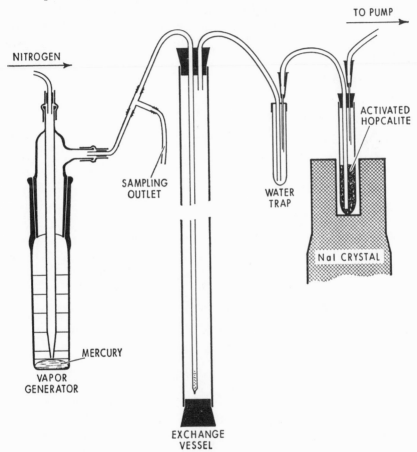

FIGURE 13-7. The exchange and counting assembly.

The results of a typical exchange run, using a solution ^{203}HgCl$_2$ in 0.2 M cysteine is shown in Figure 13-8. The buildup of radioactivity in the hopcalite is shown in Figure 13-8a as a tracing of the pen-recorder used as the read-out of the counting system. Eventually the activity in the hopcalite becomes steady. At this point all the radioactivity in the exchange vessel is completely transferred to the hopcalite so that the final trace of the recorder represents the initial radioactivity in the exchange vessel. The difference between the final pen recording and the pen recording at any given time during the exchange run is the amount of activity remaining in the solution in the exchange vessel. This value is plotted on semilog paper (Fig. 13-8b) against time and yields a straight line from which the exchange half-time ($t_{\frac{1}{2}}$ of Equation 4) may be calculated.

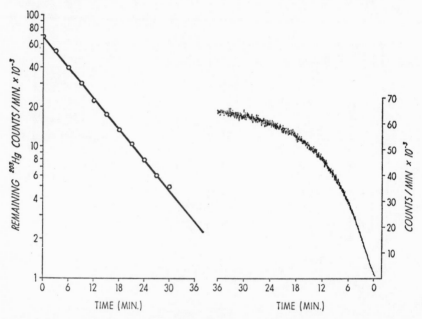

FIGURE 13-8. (a) Recorder tracing, (b) semilog plot. The radioactivity remaining in the solution at any time was calculated from the pen-recording as described in the text.

When $t_{\frac{1}{2}}$ was plotted against the concentration of Hg^{++} using solutions of $HgCl_2$ in 0.2 M cysteine, a straight line was attained (Fig. 13-9) as predicted by Equation 4. The broken line is the theoretical relationship calculated from Equation 4 from the known values of V and a. Reducing the vapor tension (Fig. 13-9b), as predicted by Equation 4 increases the slope of the standard curve so that lower concentrations of mercury give exchange half-times which can be measured accurately.

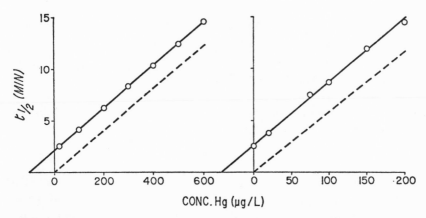

FIGURE 13-9. (a) Standard curve obtained with a vapor concentration of 17 mg Hg/m³. (b) Standard curve used for low concentrations of mercury with a vapor concentration of 5 mg Hg/m³.

The slope of the two standard curves corresponds closely to the theoretical slope indicating that isotopic exchange is close to 100 per cent. The intercept on the abscissae in both standard curves represent a blank reading due to mercury present in the standards in excess of that which was added.

To apply this technique to the determination of inorganic mercury in tissues of animals exposed to nonradioactive organomercurials requires that three conditions should be met: first, that the inorganic mercury is converted to a form which readily exchanges with the vapor. In the tissues, Hg^{++} is bound to tissue proteins and exchanges slowly (Clarkson *et al.*, 1964) . Second, the inorganic mercury must be labelled with the [203]Hg isotope without changing the chemical concentration to any appreciable extent, and third, that the conditions of the exchange run must not allow appreciable breakdown of the organomercurial molecules.

These conditions may be met by preparing at 1% (wet wt/vol) tissue homogenate in the presence of 0.2 M cysteine and adding a trace quantity of ^{203}Hg isotope as HgCl$_2$ to give at least 10,000 cycles per minute (cpm) and a final isotope concentration of 10^{-8} M or less. The radioactive homogenate is now added to the exchange vessel for analysis. Urine may be analyzed without dilution by adding cysteine and tracer ^{203}Hg.

Standard curves obtained by adding known amounts of HgCl$_2$ to urine or tissue homogenates have slopes identical to the curves of Figure 13-9. The intercept on the abscissae represent the inorganic mercury originally present in the samples. When known combinations of methylmercury and HgCl$_2$ and of chlormerodrin and HgCl$_2$, the method determined only the inorganic mercury present.

Results

All the organo-mercurials tested to date both in our laboratories and by other workers have been found to release inorganic mercury in animal tissues, Table 13-I. Otherwise the organic moieties vary widely in their chemical structure and the compounds possess different chemical and physical properties. Methylmercury having the simplest structure is volatile and lipid soluble. On the other hand the diuretic drug, Mersalyl, is nonvolatile and water soluble.

Different compounds are metabolized at different rates. Gage (1964) has reported that phenylmercury salts release inorganic mercury in rat tissues much more rapidly than methylmercury salts. The data given in Table 13-II indicate that there is a tenfold difference in the rates of deposition of inorganic mercury in kidney after dosing with different mercury compounds. Four hours after giving p-mercuribenzoate to rats, approximately 29 per cent of the injected dose was present in the kidneys as inorganic mercury and 3 per cent as the unchanged mercurial. In contrast, after chlormerodrin the inorganic mercury was down to 6 per cent and the organic mercury was up to 68 per cent of the injected dose.

The rate of deposition of inorganic mercury in kidneys varies with the species. With p-mercuribenzoate and p-mercuribenzenesulfonate, the greatest deposition was seen in the rat (Table 13-

TABLE 13-I
ORGANIC MERCURIAL COMPOUNDS RELEASING Hg^{++} IN MAMMALS

Name	Structure	Reference
Methylmercury	$Hg^+ - CH_3$	Webb
Ethylmercury	$Hg^+ - CH_2CH_3$	Webb
Phenylmercury	$Hg^+ -$	Webb
p–Mercuribenzoate	$Hg^+ -$ COO^-	This laboratory
p–Mercuribenzenesulfonate	$Hg^+ -$ SO_3^-	This laboratory
Chlormerodrin	$Hg^+ - CH_2CHCH_2NHCONH_2$ $\quad\quad\quad OCH_3$	This laboratory
Mersalyl	$Hg^+ - CH_2CHCH_2NHCO$ $\quad\quad OCH_3$ $\quad\quad \bar{O}OCCH_2O$	This laboratory

TABLE 13–II

RENAL METABOLISM OF ORGANOMERCURIALS IN THE MALE RAT

Compound 4 mg Hg/kg	Mercury in kidneys Inorganic	Organic % dose	L D₅₀ (seven day) mg Hg/kg
p–Mercuribenzoate	29	3	3.5
p–Mercuribenzenesulfonate	3.5	5	–
Chlormerodrin	4.3	68	24

III). The most outstanding species difference was seen with p-mercuribenzoate, where the amount of inorganic mercury deposited in rat kidneys was almost ten times the amount seen in the dog.

TABLE 13–III

SPECIES DIFFERENCES IN THE RENAL METABOLISM
OF ORGANOMERCURIALS

Compound	Inorganic Mercury In Kidneys	
4 mg Hg/kg (IV)	Dog	Rat
	% injected dose	
p–Mercuribenzoate	3.5	29
p–Mercuribenzenesulfonate	1.0	3.5
Chlormerodrin	0.9	4.3

Sex differences in renal metabolism of chlormerodrin in rats were small (Table 13-IV), and statistically insignificant. The levels of inorganic mercury at 4 hours after injection were identical. At 24 hours, both levels of inorganic mercury were small, with the female figures slightly less than the male.

The rate of deposition of inorganic mercury in various organs can only give an approximate idea of the true rate of conversion

TABLE 13–IV

SEX DIFFERENCES IN THE RENAL METABOLISM
OF CHLORMERODRIN IN RATS

Hours After Injection (SC) 4 mg Hg/kg	Inorganic Mercury in Kidneys			
	Males		Females	
		% dose		
	Total	Inorganic	Total	Inorganic
4	33.4	2.1	31.3	1.7
24	3.4	2.5	3.2	2.2

of these compounds to inorganic mercury. Following injection of an organomercurial, the level of inorganic mercury in the kidneys is determined by the following:

1. the rate of metabolism within the kidney,
2. the rates of renal uptake and excretion of the intact mercurial,
3. the uptake of inorganic mercury released from the mercurial in other parts of the body,
4. the excretion of inorganic mercury into the urine.

To compare rates of metabolism in different organs it is necessary to carry out studies with tissue slices. Preliminary studies, Table 13-V, indicate that the rate of metabolism of chlormerodrin is approximately the same in liver and kidney slices.

TABLE 13–V
THE METABOLISM OF CHLORMERODRIN IN RAT LIVER
AND KIDNEY SLICES

Kidney	Liver
ug HG/g. wet wt./hr.	
.027	.022

Following a single LD50 dose of chlormerodrin, the kidney levels of intact chlormerodrin and inorganic mercury released from chlormerodrin follow a time course outlined in Figure 13-10. Intact chlormerodrin and inorganic mercury reach maximal levels in about 4 hours after injection. The maximal level of chlormerodrin, 420 μgHg/g wet wt, is very much greater than that of inorganic mercury, 47 μgHg/g wet wt. After 4 hours, the levels of chlormerodrin fall rapidly. At 48 hours, the compound has virtually disappeared. Inorganic mercury (Fig. 13-10 and 13-11) falls during the first day and then becomes steady at about 15 to 20 μgHg/g wet wt.

After a single LD50 injection of $HgCl_2$, the level of mercury in the kidney rises more slowly to reach a maximum level of 33 μg/g wet wt in approximately 12 hours (Fig. 13-11). Thereafter the level falls very slowly and is still at 25 μg Hg/g wet wt at 50 hours. The kidney levels of inorganic mercury (Fig. 13-11) are higher after chlormerodrin during the first 12 hours and lower

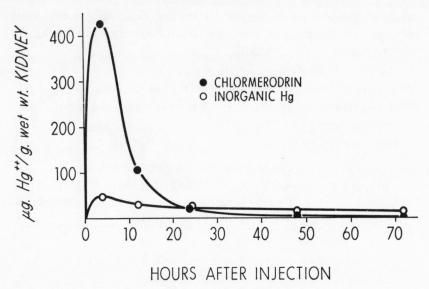

FIGURE 13-10. The kidney levels of chlormerodrin and inorganic mercury following an LD50 dose of chlormerodrin.

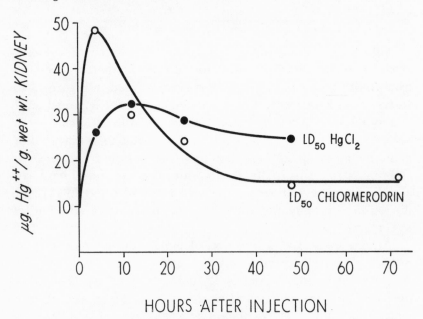

FIGURE 13-11. The kidney levels of inorganic mercury following LD 50 dose of chlormerodrin or mercuric chloride.

after that time. However, in view of the large difference in the LD50 doses, 20 mg Hg/kg of chlormerodrin and 3.4 mg Hg/kg of HgCl$_2$, the levels of inorganic mercury are remarkably similar.

Despite the very high levels of mercury seen in the kidney during the first 12 hours after injection of chlormerodrin, HgCl$_2$ produces a more rapid depression in kidney function. The creatinine clearance falls more rapidly and is practically zero 24 hours after HgCl$_2$ (Fig. 13-12). With chlormerodrin, the creatinine clearance falls more slowly and levels off at a low value.

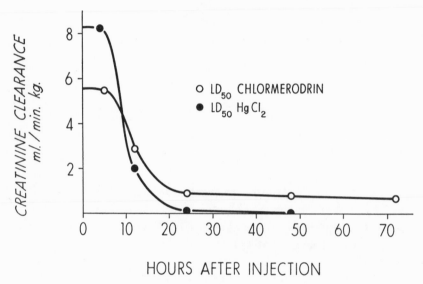

FIGURE 13-12. The endogenous creatinine clearances in rats following injections of LD50 doses of chlormerodrin or mercuric chloride.

Both compounds produce qualitatively the same effects on urine flow (Fig. 13-13). After an initial diuresis, the urine flows decline with the HgCl$_2$ treated rats showing the more rapid and more profound fall.

The urinary excretion of chloride falls dramatically after administration of either compound. In this case, chlormerodrin produced the more rapid fall but the effects of both compounds qualitatively and quantitatively are more notable for their similarities.

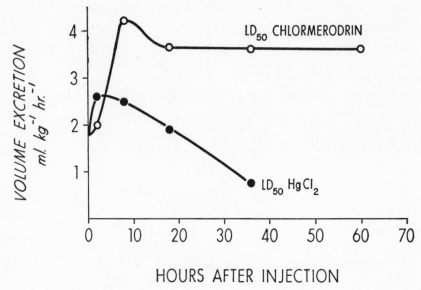

FIGURE 13-13. Urine flows in rats following injections of LD50 doses of chlormerodrin or mercuric chloride.

The profound depression in creatinine clearance, produced by $HgCl_2$, is not secondary to changes in blood pressure. The mean arterial blood pressure in rats given a LD50 dose of $HgCl_2$ remain within normal limits (Table 13-VI) during the first 48 hours after injection.

TABLE 13-VI
EFFECT OF LD50 DOSE OF $HgCl_2$ ON MEAN ARTERIAL
BLOOD PRESSURE

Hours After Injection	B. P. mm Hg	No. of Animals
0	120–130	2
12	120–130	1
20	145–160	1
24	99–110	1
24	157–165	1
48	116–121	1

Conclusion

Organomercury compounds release mercury slowly when allowed to stand in aqueous solutions (Mudge and Weiner, 1958).

This process is accelerated at low pH and in the presence of thiol containing compounds (Weiner, Levy and Mudge, 1962). The question arises as to whether the *in vivo* breakdown of the compounds is the same chemical process as *in vitro* or whether enzymes are involved.

The present results do not give decisive evidence on this question but suggest the participation of enzymes. First some species' differences exist in rates of breakdown, particularly in the case of p-mercuribenzoate. Second the relative rates of breakdown *in vivo* do not correspond with the *in vitro* rates. For example p-mercuribenzoate breaks down more slowly than chlormerodrin *in vitro* (Weiner *et al.*, 1962) but more rapidly in rat tissues. Third chlormerodrin releases inorganic mercury in the presence of tissue slices at pH 7.4 whereas much lower pH values are required in the absence of tissue slices.

It is possible that the toxicity of an organomercury compound is due to the inorganic mercury released from the compound. At present the evidence supporting this proposition is only suggestive. Mercuric chloride has the lowest acute LD50 of all the organic mercury compounds tested. The acute LD50 values in the rat for $HgCl_2$, p-mercuribenzoate and chlormerodrin are 3.0, 3.4 and 20 mg Hg/kg, respectively. This correlates with the rates of cleavage of these compounds; p-mercuribenzoate rapidly breaks down in rat tissue and has an LD50 close to that of $HgCl_2$ whereas chlormerodrin breaks down much more slowly and has a much higher LD50.

The changes in kidney function elicited by chlormerodrin are similar qualitatively to those produced by $HgCl_2$ (Fig. 13-12, 13-13 and 13-14). These changes occur at about the same level of inorganic mercury in the kidney. The concentration of intact chlormerodrin reaches kidney levels as high as 410 μg Hg/g wet wt. In contrast the highest level of inorganic mercury was 47 μg Hg/g wet wt as released from chlormerodrin and only 32 μg Hg/g wet wt following injection of the LD50 dose of $HgCl_2$. This high level of chlormerodrin has no detectable effect on kidney function. If anything, the depression in endogenous creatinine clearance (Fig. 13-12), volume excretion (Fig. 13-13) and chloride excretion (Fig. 13-14) is less after chlormerodrin than after $HgCl_2$.

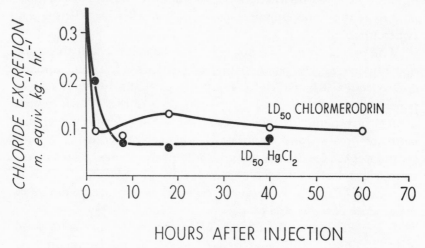

FIGURE 13-14. Urinary excretion of chloride following injections of LD50 doses of chlormerodrin or mercuric chloride.

Much more work will be required to test the general validity of the idea that the toxicity of organo-mercurials is due at least in part to inorganic mercury released from them. For example the alkylmercurials produces damage to the central nervous system only after a latency period of one to two weeks (Swensson, 1966). The intact mercurial is known to penetrate the brain (Gage, 1964). In view of the slow rate of breakdown of these compounds it may well take one to two weeks for sufficient inorganic mercury to accumulate in the brain. However, there have been no published accounts to date of attempts to discover if the intact mercurial or the released inorganic mercury is responsible for toxic symptoms.

REFERENCES

1. BRODIE, B.B.: Distribution and fate of drugs; therapeutic implications. In *Absorption and Distribution of Drugs.* Baltimore, Williams & Wilkins, 1964.
2. CLARKSON, T.W.; ROTHSTEIN, A., and SUTHERLAND, R.: The mechanism of

action of mercurial diuretics in rats; the metabolism of [203]Hg-labeled chlormerodrin. *Brit J Pharmacol, 24*:1, 1964.

3. CLARKSON, T.W., and GREENWOOD, M.: A simple and rapid determination of mercury in urine and tissues by isotope exchange. *Talanta*, 1968, vol. 15.

4. CONWAY, E.J.: *Microdiffusion Analysis and Volumetric Error*, revised ed. London, Crosby Lockwood & Son, 1947, p. 10.

5. GAGE, J.C.: Distribution and excretion of methyl and phenyl mercury salts. *Brit J Industr Med, 21*:197, 1964.

6. MUDGE, G.H., and WEINER, I.M.: The mechanism of action of mercurial and xanthine diuretics. *Ann NY Acad Sci, 71*:344, 1958.

7. SWENSSON, A.: Toxicology of different organo-mercurials used as fungicides. 15th International Congress of Occupational Medicine, 1966, vol. 3, p. 129.

8. WEBB, J.L.: *Enzyme and Metabolic Inhibitors*. New York, Academic, 1966, Vol. II, p. 961.

9. WEINER, I.M.; LEVY, R.I., and MUDGE, G.H.: Studies on mercurial diuresis; renal excretion, acid stability and structure — activity relationships of organic mercurials. *J Pharmacol Exp Ther, 138*:96, 1962.

DISCUSSION

SUZUKI: Is there any possibility in the analytical steps of your procedure for organic breakdown, especially with compounds that are suspected of being unstable?

CLARKSON: The evidence is simply that with the addition of an organic compound to an homogenate very little inorganic mercury is found. Organic compounds have inorganic impurities to begin with, amounting, in some cases to 1 per cent. Of the compounds we tested, only phenylmercury might break down slowly under test conditions. Some diuretics break down in acid, but not at pH 7.

BERGLUND: Mudge and Weiner [*Ann N Y Acad Sci, 71*: 344, 1958] proposed that the effect of diuretics actually depended on generation of mercuric ions within the kidney, and that this would be speeded up in the presence of sulfyhdryl groups in an acid environment.

CLARKSON: They (Mudge and Weiner) found certain compounds that would release inorganic mercury and certain compounds that would not; this was *in vitro* at pH 4 and with a high concentration of SH groups. They postulated that the same process took place in the animal, and they implied that those compounds that did not break down in the *in vitro* test would not break down in the animal. We showed that for certain compounds this in not true. Parachloromercurybenzoate which is quite stable *in vitro* is degraded in the rat and dog. Furthermore it is broken down at very different rates, rapidly in the rat and mouse, but slowly in the dog. There is also a large difference in its toxicity, it is extremely toxic to rats.

BERGLUND: What was the threshold level for the diuretic effect of mercury in your experiments on rats? [Clarkson, T. W.; Rothstein, A., and Sutherland, R.: *Brit J Pharmacol, 24:*1, 1965].

CLARKSON: It is very close to the toxic level in the rat, of the order of 20 $\mu g/g$. But I think that the diuretic effect in the rat is not the same as in the dog and could well be a first phase of the toxic effect.

BERGLUND: The mercury diuretics are a poor diuretics in the rat. Also the timing of diuretic action that you showed in the rat seemed to be slower than in the dog. I would like to see this experiment repeated in the dog.

CLARKSON: We have done this and, in fact, Dr. Vostal and I are continuing this work. These compounds do break down in the dog, and they do so at different rates. Even the nondiuretic mercurials, parachloromercurybenzoate for example, release inorganic mercury in the dog.

FASSETT: One of the oldest mercurial diuretics, Mercurophyllum was, as I recall, a derivative of camphoramic acid, with the Hg bound firmly to a carbon atom. The compound is interesting, because there are human data on this and other diuretics, with analyses of mercury levels in organs [Leff, W. and Nussbaum, H. E.: *Ann N Y Acad Sci, 65:*520, 1957].

WELCH: Is the diuresis one sees after a dose of chlormerodrin related to the inorganic mercury released or to the chlormerodrin itself?

CLARKSON: For the dog, there are not enough data to say yes or no. One can say that it is broken down and that there is inorganic mercury present in the kidney; then there's the job of correlating the extent of the diuresis with the levels of mercury.

Many factors are involved in estimating which fraction of the drug is active.

WELCH: How does the dose response curve of chlormerodrin as a diuretic compare to inorganic mercury?

BERGLUND: It may be difficult to get a complete dose-response curve with a mercurial diuretic. The diuresis is too heavy.

CLARKSON: Mudge and his group claim that on a mg/kg dose basis, mercury cysteine is the most potent diuretic among mercury compounds [*Ann N Y Acad Sci, 71:*344, 1958].

WELCH: Does this compound give very high levels of inorganic mercury in the kidney?

CLARKSON: Actually it does not, and much of the mercury in the kidneys is inertly bound. For example, one can reverse diuresis by giving (British Anti-Lewisite) but in doing this one has removed only about 20 per cent of the mercury from the kidney, so that the other 80 per cent is playing no role in diuresis [Weiner, I. M., *et al.: J Pharmacol Exp Ther, 127:*325, 1959].

WELCH: Does the breakdown of chlomerodrin take place in the kidney or is inorganic mercury transported from other organs to the kidney?

CLARKSON: It can occur in kidney because it is seen in kidney slices [Clarkson, T. W.: *unpublished observations*]. Kinetic considerations also indicate that kidneys are the important site of breakdown, because they hold up to 80 per cent of the injected dose. Chlomerodrin is so selectively accumulated in kidneys that its levels are one hundred times more than in almost any other tissue.

BERLIN: Inorganic mercury equilibrates with organic mercury, and we use this property to label organic compounds in our studies. Does this not interfere with your analytical method for inorganic mercury? Phenylmercury acetate is substantially labeled in minutes. Methylmercury takes hours.

CLARKSON: We did not use phenylmercury acetate. The isotope exchange in our analytical method is complete in 15 minutes, so the method is safe for organic compounds that give less than a 1 per cent exchange with inorganic mercury in 15 minutes.

BERLIN: It is possible to use your technique for determining total mercury content, by just waiting until the isotope is in equilibrium with all organic compounds?

CLARKSON: Yes! It is possible to do this with the acid-labile organomercurial diuretics. As first reported by Mudge and Weiner, the diuretics release inorganic mercury in acid solutions containing cysteine or other thiol compounds. Thus we can carry out the exchange reaction, after allowing breakdown to occur. The result will give the total mercury in the sample. We can then analyze another sample by making the exchange run at pH 7.0 and higher when the analysis gives only the inorganic mercury present in the sample. On the other hand, this approach will have to be modified if it is to be applied to the mercury pesticides. The compounds are generally stable at acid pH. I am presently searching for other conditions which will lead to the complete breakdown of such compounds as methylmercury chloride and phenylmercury acetate.

Chapter 14

PESTICIDE LEVELS IN HUMANS IN A VARIETY OF NATURAL AND EXPERIMENTAL CONDITIONS

J. L. RADOMSKI and WM. B. DEICHMANN

ABSTRACT

Pesticide levels were measured in five tissues in a stratified age and sex sample of the general population of Dade County, Florida. There was no significant increase in pesticide concentrations with age with the exception of the 0 to 5 age group which exhibited lower concentrations of p,p'-DDT and p,p'-DDE. Levels of p,p'-DDT and p,p'-DDE were almost twice as high in Negroes as in Whites. Pesticide concentrations were found to be approximately ten times less in the liver and one hundred times less in brain, kidney, and gonadal tissue than in adipose tissue. Pesticide levels approaching those found in adults were present in the tissues of six fetuses. Evidence that p,p'-DDT is metabolized to p,p'-DDD in the human was obtained.

A study of pesticide concentrations in the fat of Jackson Memorial Hospital autopsy cases exhibiting various pathological states of the liver, brain, and other tissues of the body has been carried out in 271 cases. No elevation of pesticide concentrations was observed in the presence of brain or primary liver tumors. A significantly elevated fat concentration of p,p'-DDE and dieldrin was found in portal cirrhosis and hypertension. Highly significant elevations of pesticide concentrations were present in cases of carcinoma of various tissues. An investigation of the home usage of pesticides by these individuals was made. It was found that individuals using pesticides extensively in the home had three to four times higher levels of p,p'-DDT and its metabolite p,p'-DDE than those who used minimal quantities.

REPORT

PRIOR TO THE INITIATION of our investigation in 1964, knowledge of the existence of pesticide levels in humans was confined to concentrations which were present in adipose tissue. In our project, we set out to determine pesticide concentrations in the five different tissues judged to be of greatest possible toxicological significance: fat, liver, kidney, brain, and gonad tissue. The pesticide concentrations in these tissues were measured in fifty-four people autopsied at the Medical Examiner's Office in Dade County, Florida. This was the pioneer study in a community studies program designed to monitor pesticide concentrations in people in a number of areas in the country.

This study was made feasible because of the recent successful application of gas chromatography to the determination of pesticides in biological material. No procedures were available at the time of the initiation of the project, however, for the gas-chromatographic analysis of pesticides in human tissues. Since this is not a paper on methodology, I will not go into these methods except to say that we were able to develop a universally applicable method which is simple, rapid and very sensitive, and remains to this day the standard, most widely used procedure in the field. The basic procedure consists of the extraction and direct injection of the extracts on a gas-chromatographic column without prior cleanup of the extract. Table 14-I shows the sensitivity achieved.

TABLE 14–I
SENSITIVITY LIMITS FOR THE DETERMINATION OF
PESTICIDES BY ELECTRON CAPTURE DETECTOR
WITHOUT PRIOR CLEANUP

	Standard picograms	Fat $\mu g/g$ (ppm)
Aldrin	1	0.01
Lindane	1	0.01
p,p'–DDE	2	0.02
Dieldrin	3	0.03
p,p'–DDT	20	0.20

The General Population

The pesticides found to be present in this early study were primarily p,p'-DDT, p,p'-DDE, p,p'-DDD and dieldrin. Evidence of the presence of aldrin was also obtained but was ignored at this stage because of the possibility that the peak believed to be this pesticide was an interference, and because of the evidence that aldrin is primarily converted to dieldrin in biological systems. The occurrence of the α, β and γ isomers of HCH and heptachlor epoxide were also observed occasionally as is discussed later.

Mean pesticide concentrations observed in the fat of our fifty-four cases in the various age groups collected in 1964 are shown in Table 14-II. These data were first examined to determine whether any significant difference existed between male and females. Rank tests were used for the detection of significance. Since no significant differences were observed, no further sex distinction was made in the experiment. The data was next examined for race effect. The mean Negro concentration of p,p′-DDT and p,p′-DDE can be seen to be approximately double the mean White concentration in the above thirty age group. Because of the small number of Negroes in this study, however, the difference was not significant. This difference is present in the other age groups also except for the 6 to 10 age group which was deficient in numbers. It is also rather interesting that the difference between the races does not extend to dieldrin.

TABLE 14–II

MEAN PESTICIDE CONCENTRATIONS IN VARIOUS AGE GROUPS
OF DECEASED DADE COUNTY, FLORIDA RESIDENTS IN 1964

Age	Race	No. of Cases	DDE	DDT	Dieldrin
0– 5	W	6	1.24	0.98	0.18
	N	6	4.18	2.07	0.13
6–10	W	4	8.00	4.36	0.10
	N	2	10.75	3.80	0.09
11–20	W	10	5.88	2.18	0.17
	N	2	6.65	2.65	0.20
21–30	W	8	3.60	1.58	0.16
	N	4	8.13	3.63	0.26
> 30	W	9	5.75	2.88	0.33
	N	3	14.17	3.66	0.32

Statistical analysis by rank test showed no significant increase in pesticide concentrations with age, except in the 0 to 5 age group which had lower p,p′-DDE and p,p′-DDT levels than the older age groups. This was particularly noticeable in the six cases of the White race which had a remarkably low mean level of only 1.24 ppm of p,p′-DDE and less than 1 ppm of p,p′-DDT. By six years of age, however, the level of p,p′-DDT and metabolites has already risen to the adult level.

This study in recent years has been continued by the Florida State Board of Health as part of the Community Studies Program of the United States Public Health Service, Bureau of State Services. Accumulation of a much larger number of cases has allowed the more exact examination of these effects (1). These results confirm the existence of the low pesticide levels in the 0 to 5 age group and the fact that there is no continued accumulation of pesticides with age. In an analysis of these results, Dr. John E. Davies of the Dade County Department of Public Health and the University of Miami concluded that there is definitely a race difference in pesticide concentrations (2). Again, the Negro race, both males and females, has roughly double the p,p'-DDT levels present in the white race. These differences are significant at less than the 1 per cent level. While he confirmed our observation that there is no difference in pesticide concentrations between white males and white females, he found a significant difference between Negro males and Negro females.

Pesticide concentrations of p,p'-DDT, p,p'-DDE, p,p'-DDD and dieldrin were also determined in liver, kidney, brain, and gonad tissue (Table 14-III). Here we see that, as might be expected, the pesticide concentrations in tissues other than the adipose tissue are markedly lower. Liver pesticide concentrations are approximately one-tenth those present in the adipose tissue, whereas pesticide concentrations present in kidney, brain and gonads are approximately 1/100 of those found in adipose tissue. At this point, an interesting observation was made of the existence of high concentrations of p,p'-DDD which were detected in the liver. This constituted the first evidence that the observation made in rats by Klein *et al.* (2), that p,p'-DDT is metabolized not only to p,p'-DDE, but also to p,p'-DDD, applies also to humans.

TABLE 14–III
THE AVERAGE PESTICIDE LEVELS IN 5 TISSUES OF
THE GENERAL POPULATION (ppm) −42 PEOPLE

Tissue	DDT	DDE	DDD	Total	Dieldrin
Adipose	2.81	6.67	0.28	10.56	0.215
Liver	0.12	0.35	0.34	0.889	0.035
Kidney	0.036	0.077	0.017	0.141	0.013
Brain	tr.	0.123	−	−	0.035
Gonads	tr.	0.059	−	−	0.035

It is likely that the low concentrations observed in the other tissues of the body reflect the low fat concentrations. Pesticide concentrations present in the brain were found to be extremely low and difficult to measure accurately. Every evidence indicates that the occurrence of pesticides in fat is very largely a physical chemical partitioning of these nonpolar materials in favor of the nonpolar neutral fat. The lipids present in the brain are, of course, phospholipids, which are a great deal more polar than neutral fat, and as such would not favor the partitioning of the pesticides into these tissues. The fact that essentially the same concentration of pesticide tissue exists in the brain of man as exists in other nonfatty tissues of the body suggests that pesticides penetrated the blood brain barrier but are not preferentially concentrated there because of the polarity of the lipids present. Higher concentrations in the liver may not only reflect the presence of neutral fat in the liver, but also the metabolic function of this tissue.

TABLE 14–IV
PERCENTAGE OF POSITIVE ANALYSES OF PESTICIDES
IN TISSUES OF 42 INDIVIDUALS

Pesticides	Fat %	Liver %	Kidney %
p,p′–DDE	100	100	96
Dieldrin	100	96	97
p,p′–DDD	90	97	74
p,p′–DDT	100	69	67
Heptachlor epoxide	53	23	19
o,p′–DDE	44	9	9
o,p′DDT	30	20	26
(γ+α) –HCH	19	6	3
β–HCH	13	4	4

Other pesticides were also detected in the tissues of these people (Table 14-IV). In addition to p,p′-DDE, dieldrin, p,p′-DDD and p,p′-DDT already described, o,p′-DDT, o,p′-DDE, heptachlor epoxide and α, β and γ hexachlorocyclohexane were detected in the fat. Heptachlor epoxide was detected in over 53 per cent of the cases. This is the metabolite of the insecticide heptachlor which was first discovered by Radomski and Davidow in 1953. In spite of the fact that these pesticide analyses were conducted without the use of a cleanup procedure, it is remarkable that in the series of 350 analyses only a few unidentified or unaccounted-for peaks were observed. This can be attributed to the considerable specificity of the electron capture detector.

We were extremely interested in the possible occurrence of pesticides in fetuses, which would answer the question of whether pesticides pass the placental barrier and may be transferred from mother to child prenatally. During the course of our study, a number of fetuses became available for analysis. Analysis of the adipose tissue indicated that considerable concentrations of pesticides were indeed present in the fetus (Fig. 14-1). These usually ran lower than the concentrations present in the adult, but clearly demonstrated that pesticides do possess the ability to penetrate the placental barrier. With present day concern about the mutagenicity and teratogenicity of compounds, this is a disquieting observation. Teratogenicity can occur with minute concentrations of compounds and their obvious existence during the first trimester of pregnancy should be a matter of considerable concern.

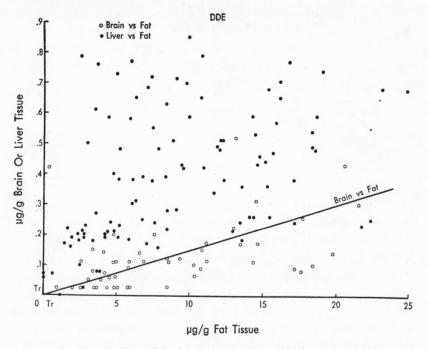

FIGURE 14-1. Correlation of brain concentration with fat concentration and liver concentration with fat concentration of pesticides in all disease groups combined.

Pesticide Concentration and its Relation to Certain Diseases

We next turned our attention to pesticide concentrations present in the liver, brain and adipose tissue of people who were suffering from certain diseases. Rather than undertake an experiment designed to investigate the relationship of pesticides to all possible pathological changes, attention was focused on the liver and brain. Extensive animal experimentation has shown that chronic exposure to the chlorinated hydrocarbon pesticides produces liver damage and that chronic exposure to aldrin and dieldrin have been shown to produce encephalographic changes and central nervous system toxicity in animals. If humans are developing liver or brain disease due to excessive exposure to pesticides, then it would seem reasonable to assume that those individuals so affected would bear higher concentrations of pesticides than normal. We therefore set out to determine the pesticide concentrations present in the tissues at autopsy following deaths from liver and brain disease at Jackson Memorial Hospital and to compare these concentrations with data obtained from the general population described above. Identical analytical procedures were applied to these analyses, and the analyses were carried out by the same analysts, in the same laboratories, and on the same gas chromatographs. Although these determinations were carried out a year or two later, there is no evidence that there is an increasing concentration of pesticides in the people of Dade County. Originally it was our intention to collect the tissues from three types of cases: those with liver disease, those with brain disease, and a third group made up of random cases not believed to be suffering from either liver or brain disease. During the analysis of the data at the completion of the experiment, however, it was discovered that many individuals believed to have liver disease on the autopsy table showed only, for instance, chronic passive congestion secondary to myocardial infarction. In addition, a large group of cases in the random group were found to have such diseases as atherosclerosis, hypertension, and various malignant diseases. We therefore reclassified all of the cases without regard to the original classification. Classification of these cases was carried out without knowledge of pesticide concentrations. In addition, a group of cases was discovered which we have grouped together as infectious diseases, with the idea that these may serve as a second control since all of these cases had a well recognized etiology not associated with pesticides. While the mean pesticide levels in these cases ran slightly higher than our control, the difference was not significant.

Liver Diseases

Mean levels of p,p'-DDE and dieldrin in the body fat in thirty-two cases of portal cirrhosis of the liver (alcoholic, nutritional or Laennec's) were significantly higher than those of the normal population (Table 14-V). Mean p,p'-DDT levels were also approximately double those of the normal population, but because of a high degree of variability, the observation was not significant. Levels of p,p'-DDE, p,p'-DDT, and dieldrin were all significantly elevated in the liver tissue itself; however, p,p'-DDD levels in the liver were lower than normal. Since we have seen that p,p'-DDD is a metabolite of p,p'-DDT which originates in the liver, impaired function would presumably lower the rate of this metabolic conversion.

TABLE 14–V

PESTICIDE CONCENTRATIONS IN THE FAT AND LIVER TISSUE
OF PATIENTS TERMINATING WITH LIVER DISEASE ppm (μg/g)

Disease Group	W/N	Tissue	DDE	DDD	DDT	Dieldrin
Normal	31/11	Fat	6.69	0.28	2.77	0.21
		Liver	0.35	0.36	0.10	0.03
Portal	25/ 7	Fat	11.49	0.34	5.19	0.50
Cirrhosis		Liver	1.19	0.20	0.53	0.10
Fatty	13/ 1	Fat	7.12	0.27	3.53	0.35
Metamorphosis		Liver	0.63	0.15	0.05	0.04
Malignancy,	20/10	Fat	19.27	0.37	5.87	0.49
Metastatic		Liver	2.44	0.22	0.08	0.07
Malignancy,	3/ 1	Fat	13.58	0.49	7.70	0.36
Primary		Liver	0.60	0.16	0.02	0.03
Amyloidosis	1/ 0	Fat	50.80	1.84	–	–
		Liver	1.39	0.14	0.50	tr.

The most striking elevations of pesticide concentrations were observed in the rather large series of cases of metastatic malignancy of the liver. The mean p,p'-DDE concentration in the fat was three times as high as normal and seven times as high in the liver tissue itself. P,p'-DDT concentrations were also elevated in the fat but, surprisingly, not in the liver tissue. Dieldrin and p,p'-DDD concentrations, on the other hand, were essentially normal. These would appear to be fairly specific alterations which may have some real, but at the moment, obscure meaning. It does not seem likely that these altered concentrations were related to increased fat content of the liver, for in this case, across-the-board increases should have been observed. Furthermore, similar elevations of pesticide concentrations were not observed where only fatty metamorphosis was present.

One of our initial primary interests in carrying out this study was to seek a relationship between chlorinated hydrocarbon pesticides and liver tumors. Unfortunately, only a small number of primary liver tumor cases were obtained. Three of the four cases had rather normal p,p'-DDE, p,p'-DDD, p,p'-DDT and dieldrin concentrations. The fourth case, a male Negro, had 35.04 ppm of p,p'-DDE in the fat and 1.25 ppm in the liver.

Of interest was one case of ideopathic amyloidosis, which had remarkably elevated levels of all the pesticides measured.

Brain and Neurological Disease

Significant elevations of the p,p'-DDE concentrations in the fat as well as in the brain tissue were observed in encephalomalacia (Table 14-VI). There was also a highly significant elevation of the dieldrin concentration in the fat. In this category were included

TABLE 14–VI
PESTICIDE CONCENTRATIONS IN THE FAT AND BRAIN TISSUE
OF PATIENTS TERMINATING WITH BRAIN DISEASE ppm (μg/g)

Disease Group	W/N	Tissue	DDE	DDD	DDT	Dieldrin
Normal	31/11	Fat	6.69	0.28	2.77	0.21
		Brain	0.12	–	0.04	0.04
Encephal omalacia	14/10	Fat	12.63	0.27	5.40	0.50
		Brain	0.20	0.00	0.00	0.01
Cerebral Hemorrhage	10/ 6	Fat	12.20	0.38	3.92	0.59
		Brain	0.12	0.00	0.00	0.01
Brain Tumor	4/ 2	Fat	7.33	0.27	3.91	0.32
		Brain	0.08	0.01	0.01	0.02
Degenerative Changes	9/ 1	Fat	10.44	0.28	2.00	0.30
		Brain	0.16	0.00	0.01	0.02

all cases of definite encephalomalacia, due to cerebral thrombosis and brain infarct, either old or recent; patients in this category were, of course, commonly aged; atherosclerosis and myocardial infarcts were also commonly present. Cases where there was distinct evidence of hemorrhage, either old or recent, were classified under cerebral hemorrhage. Dieldrin levels were significantly elevated in the fat but were essentially normal in the brain tissue (Table 14-VI).

One of the initial major interests of this investigation was the possibility of chlorinated hydrocarbon pesticides producing tumors of the brain. Primary brain tumors are rare and only four cases were found in this study. That normal pesticide levels were observed in both fat and brain tissue (Table 14-VI) would seem to suggest that chlorinated hydrocarbon pesticides are not involved in the etiology of the brain.

Miscellaneous Diseases

Moderately elevated levels of p,p'-DDE, p,p'-DDT and dieldrin were observed in the fat of many of the cases of atherosclerosis (Table 14-VII). However, mean pesticide concentrations were not significantly different from those of the normal population.

TABLE 14–VII
PESTICIDE CONCENTRATIONS IN THE FAT TISSUE
OF PATIENTS TERMINATING OF MISCELLANEOUS DISEASES ppm (μg/g)

Disease Group	W/N	Tissue	DDE	DDD	DDT	Dieldrin
Normal	31/11	Fat	6.69	0.28	2.77	0.21
Infectious Diseases	15/ 5	Fat	8.89	0.12	3.94	0.38
Atherosclerosis	41/13	Fat	12.01	0.27	5.10	0.37
Hypertension	1/ 7	Fat	17.91	0.40	6.54	0.73
Carcinoma	26/14	Fat	15.97	0.34	5.65	0.55
Leukemia	4/ 1	Fat	16.10	0.58	4.69	0.47
Chronic Renal Disease	5/ 3	Fat	8.11	0.21	2.11	0.23
Hodgkin's Disease	5/ 0	Fat	10.06	0.38	3.22	0.51

Although the p,p'-DDE and p,p'-DDT concentrations in fat in the presence of hypertension were markedly elevated compared to the control on a mixed race basis (Table 14-VII), it is necessary to point out that seven of the eight cases were Negroes. Dieldrin concentrations, however, were significantly elevated (p < 0.01) when compared to both the mixed race and Negro control levels.

Concentrations of all of the pesticides were remarkably high in the carcinoma group, averaging two to three times the normal concentration. Included in this group were carcinomas of the lungs, stomach, rectum, pancreas, prostate, and urinary bladder. Before grouping all of these various types of carcinomas together, each type was analyzed individually in an attempt to discover whether the elevated levels were associated with a particular neoplastic disease. However, no such association could be discerned.

Correlation of Pesticide Concentrations in Adipose Tissue with Liver Tissue and with Brain Tissue

As part of our statistical analysis of the data of this experiment, the correlation of pesticide concentrations in fat with liver tissue, and in fat with brain tissue was studied. Contrary to our expectations, virtually no correlation was observed between the fat and liver concentration of any of the pesticides measured (Fig. 14-2). On the other hand, fat concentrations of pesticides are correlated with pesticide concentrations in the brain.

When similar correlation studies were performed on our data from the general population, correlation was observed between liver and fat concentrations of p,p'-DDT and its metabolite. It would therefore appear that the lack of correlation between pesticide concentrations in liver and fat was associated with the presence of disease.

The Relationship of Pesticide Levels to the Home Usage of Pesticides

In the examination of these data, we were struck by the wide variance of pesticide levels in individuals, some being very high and others very low. Hospital records indicated no occupational exposure to pesticides in this study. Therefore, it was considered that perhaps these remarkable discrepancies could be related to the home usage of pesticides.

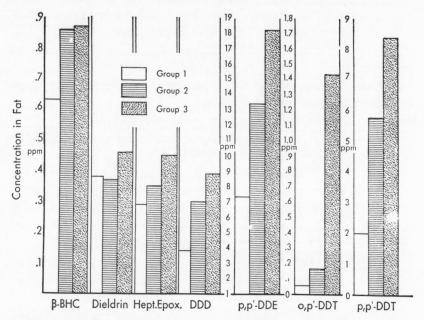

FIGURE 14-2. Mean concentration of pesticides as related to approximate extent of household usage. Group I: Little or no household usage of pesticides. Group II: Moderate or average household usage of pesticides. Group III: Heavy household usage of pesticides. Extent of household usage determined by interview of next-of-kin.

It has been our casual observation that there are two types of individuals: (a) those with an abhorrence of insects and a contempt for chemical intoxication, who are heavy users of household pesticides, and (b) those with an exaggerated fear of chemical intoxication who use pesticides seldom if at all. The difference in pesticide exposure between these two groups might be tremendous and might well explain the wide variation in adipose pesticide concentrations observed. The next-of-kin of as many as possible of these cases were interviewed and without knowledge of the results of the pesticide analyses, the cases were classified in three groups: Group III, heavy pesticide users; Group II, moderate or normal pesticide users; Group I, light or nonpesticide users. Mean adipose pesticide concentrations were calculated. (Fig. 14-3). The mean p,p'-DDT concentration in the high exposure group is over four times the mean p,p'-DDT concentration in the low exposure group (p = 0.001), while the mean p,p'-DDE concentration is almost three times greater in the high exposure group than the low exposure group. In the case of dieldrin and heptachlor epoxide, there were no significant differences in mean pesticide levels in the three exposure groups. β-HCH showed a significant increase (p = 0.01) from Exposure Group I to Exposure Group II but not between Exposure Group II and Exposure Group III. These results indicate that the home usage of pesticides is one, if not the major, factor in the occurrence of *elevated* pesticide body levels of p,p'-DDT and its metabolites. Similar results with dieldrin and heptachlor epoxide would not be expected since these pesticides are not normally present in pesticide mixtures intended for the home usage. It should be pointed out that there is a higher ratio of Negroes in Group III than in Group II and Group I, reflecting

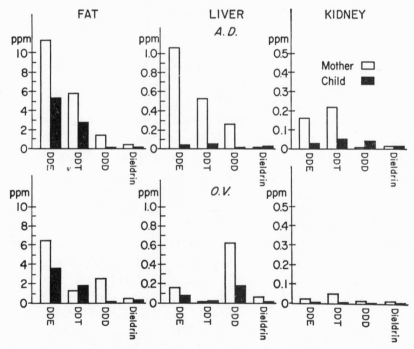

FIGURE 14-3. The concentration of pesticides in fat, liver and kidney of mother and fetus.

the fact that perhaps Negroes are heavier and perhaps more careless users than whites. These results offer a possible explanation for race differences observed by Dr. Davies described earlier.

While there is a great deal of evidence that occupational exposure to p,p'-DDT and its metabolites produces an elevation in pesticide concentration, there has been no evidence produced in the literature prior to this that nonoccupational household use of pesticides produces such an elevation.

REFERENCES

1. DAVIES, JOHN E.: *Public Health Rep, 83*:207, 1968, and *Community Studies on Pesticides,* vol. 11.
2. DATTA, P.R.; LAUG, E.P., and KLEIN, A.K.: Conversion of *p,p'*-DDT to *p,p'*-DDD in the liver of the rat. *Science, 145*:3635, 1964.
3. RADOMSKI, J.L., and DAVIDOW, B.: The metabolite of heptachlor, its estimation, storage and toxicity. *J Pharmacol Exp Ther, 107*:266, 1953.

DISCUSSION

BERG: I mailed a questionnaire to 230 Rochester families, asking them to look at the labels on their shelves, to count the cans, or boxes they have that contain this class of compounds, and to tell us what they will do with them. Perhaps this is the type of investigation you would like to see done.

RADOMSKI: I think this is a great neglected area in pesticide research and that we need much more of this type of information. Of course, the use of blood analyses of pesticides is now feasible and is perhaps a better approach.

FASSETT: Might I briefly comment on the blood analysis method? Long *et al.* [*Amer Industr Hyg Assn J, 29*:112, 1968 (abstr.)] reported recently on 155 farmers, some of whom used considerable amounts of pesticides in their work, and others none. An interesting result was that analysis of the blood showed only DDE. The amounts of DDE in the blood were in the order of 0.078 and 0.083 ppb, which really indicated that there wasn't much difference between "exposed" and "unexposed" individuals. The main intake of pesticides into blood was apparently from food, and not from exposure in the fields. With a situation like this I am not convinced that one can rely on blood anaylses.

HUNT: We were a little more fortunate, in that we were studying residue levels in blood of birds, rather than of mammals. We find levels of up to 25 and 30 ppm in whole blood of birds. This is ppm, not ppb.

GAGE: I am curious to know whether or not there might be correlation between race and pesticide use. Perhaps Negroes living in a situation where there may be more insects use more pesticides.

RADOMSKI: I believe this is so. The race difference may be simply a matter of usage, and of education about the use of these compounds. It is particularly striking in young children (0 to 5 age group), where the level was four times greater in the Negro population than in the white. Use of pesticides around the house is, I think, the determining factor.

BUTLER: One more comment on the relationship of residues to human habits. In our monitoring program, most of the stations in the State of Washington are negative, except for the one adjacent to the Indian reservation. This is the one place where the estuarine waters are typically polluted with DDT.

BERGLUND: I was surprised to see the low levels of the 0 to 5 age group [Quniby, *et al.*: *Nature, 207:*726, 1965] [Eagan, *et al.*: *Brit Med J, 2:*66, 1965.] Several groups have shown that the concentration of DDT metabolites in women's milk is much higher than in cow's milk, despite the fact that the cow ingests much more DDT in its food per gram body weight per day than a woman. This would mean that a breast-fed baby would get much more DDT supplied in its food per gram body weight than the adult. These values have caused concern in Sweden, where we have approximately the same level of DDT in woman's milk as in the United States according to analyses done by Dr. Westöö (unpublished). I should like to know how one should look upon this problem and what could be done about it.

RADOMSKI: I do not know what your figures actually are, but I do not think most American women nurse their children.

DEICHMANN: There is another explanation: low or high concentration of pesticides in the milk may be partially due to the use of pesticides in the home.

HERMAN: Was there any difference in municipal (county or state) use of pesticides in the different areas?

RADOMSKI: Well, in Dade County (Florida) we get sprayed fairly evenly, thoroughly, and periodically with insecticides.

WEST: I feel it would be very profitable to sample residue levels not only in emaciated patients with pathological conditions but also in presumably healthy segments of the population. I am aware of one case of a very emaciated woman who died, and less than 1 per cent of her body weight was fat. She had very high levels of DDT but there was no explanation for this in her history. When we recalculated the body burden on a "normal amount of fat" basis, she came out with a normal level.

RADOMSKI: Many of our patients were emaciated. They were dying of cancer and of other chronic debilitating diseases. We have looked into that, too, but can find no correlation between emaciation and concentration.

DIECHMANN: One of the first studies we carried out was to determine what pesticides were being used to what extent in Dade County. We found three areas of heavy use of pesticides. First came residential areas where trees and shrubs were being sprayed. The second area was the so-called citrus belt in Southern Florida. And the third group was composed of individuals who use pesticides heavily as a matter of preference.

HERMAN: I might also point out that persons with what you referred to as "abnormal fear" of pesticides tend to select and wash their food more carefully.

RISEBROUGH: Is it possible that blacks eat more fish than whites?

FASSETT: There is quite a bit of personal fishing in Dade County, particularly along the canal banks.

RADOMSKI: The Negroes do fish mullet out of the canal but I don't know of any data on pesticide levels in fish in our area.

BUTLER: We have analyzed fish from the west coast of Florida and find relatively high levels of DDT, e.g., 2 to 10 ppm whole body in fish such as mullet and marsh minnows.

HUNT: Some of our bird DDT residue data [Keith, J. O., and Hunt, E. G.: Residues. In *Calif Fish Wildlife Trans*, p. 150,

19ᴬ6] correlate with this finding. We find that the fish-eating birds have high levels of DDT in their eggs.

RISEBROUGH: Do you find peaks other than those which are caused by DDE? Sometimes it is necessary to concentrate and subject the material to rigorous cleanup before the peaks of polychlorinated biphenyls show separately.

RADOMSKI: We have been on the lookout for strange peaks and have been able to account for all except three "strange" ones. I have noted your data on this type of analysis and expect to apply your techniques to some of our tissue samples.

DEICHMANN: So far we have not said anything about the possible effect of one pesticide on the retention of another. Our work on dogs has shown that with a combined feeding of DDT and dieldrin there was an increase in retention of both compounds, particularly in fat and blood. Retention of DDE was also increased. Street's work on rats showed the opposite of this. Combined feeding of DDT and aldrin decreased the retention of DDT. There is a species difference and we do not know where man fits in.

WURSTER: I should like to question the validity of correlating toxic action with body burdens when this is done by comparing the burdens found in healthy and in sick individuals. The whole concept of L.D.$_{50}$ is that if an equal exposure is given to each individual of a population, half will survive and half will die. The same reasoning holds for Dr. Robinson's analyses of birds' eggs. He pointed out that birds' eggs from successful nests showed no less residue level than from unsuccessful nests, and therefore concluded that there was no causative relationship. This is not necessarily true; there need not be a greater residue level in the victims than in the survivors. Therefore, the increased levels found in some of the diseased cases may have no bearing on any potential relationship between pesticides and the disease.

RADOMSKI: As I stated in my paper, I cannot really believe that all of these diseases are caused by pesticides. At different times both ways of looking at the problem seem valid to me. My view is somewhat the following: There are people with "normal" and others with elevated levels of pesticides and it's not a normal distribution. Are the people who have elevated levels, e.g., 20 ppm, more likely to have a disease of the liver than the people with the normal burden? If one were to take people who have liver disease then it would seem to me that one ought to find elevated residue levels if that is the factor involved. But I am not sure of this. Your point is certainly well taken.

FASSETT: Dr. Spencer has asked for a few minutes during discussion in order to speak about government control of pesticides. This subject fits quite well here and I invite him to present it (see Appendix p. 501).

Chapter 15

ENZYME ACTION ON HALOGEN-CONTAINING PESTICIDES

R. G. YOUNG

ABSTRACT

The enzymatic dehydrohalogenation of DDT by a number of species of insects, mammals, and fish has been shown to occur. The rates are several hundred times lower than those for DDT-resistant houseflies. The enzyme action is enhanced by the presence of reduced glutathione and an indication of the presence of two dehydrohalogenases in houseflies was found. Some kinetic properties of the enzyme from rodents are illustrated.

Experiments with algae grown in the presence of DDT have been inconclusive with regard to dehydrohalogenation, but a presumed metabolite of unknown structure has been found by gas-liquid chromatography.

REPORT

Introduction

I T IS ALMOST TEN YEARS since Lipke and Kearns (7) purified and characterized one enzyme responsible for the degradation of DDT. Since then, relatively few reports have appeared which deal with the enzymatic processes of DDT detoxication. Increasing numbers of species have been investigated for their ability to degrade DDT and it can be stated with some confidence that most organisms probably have some mechanism for degrading the pesticide.

The low level of dehydrohalogenase activity has been investigated recently. Attention has been given to DDT and its analogues, as part of a larger interest in enzymatic degradation of halogen-containing compounds.

The biological degradation of DDT may occur in several ways.

1. *Dehydrohalogenation*

This process, discovered in DDT-resistant houseflies by Sternberg, Vinson and Kearns (10) entails removal of hydrogen and chlorine atoms from DDT, 2,2 bis (p-chlorophenyl) 1,1,1 tri-chloroethane to yield DDE, 2,2 bis (p-chlorophenyl) 1,1 dichloro-ethylene. That DDE has been found along with DDT in almost all living things which have been carefully examined is not evidence that it has been formed biologically, since DDE is an impurity in commercial DDT. The enzyme is quite labile and has been grouped with the "soluble" enzymes, i.e. those not sedimented by present centrifugation techniques.

2. *Reductive dehalogenation*

The product of this action is DDD, 2,2 bis (p-chlorophenyl) 1,1 dichloroethane. This appears to be as frequently detected as DDE as a metabolite of DDT. There are reports of its production by rat liver (4), yeast (6), and bacteria (12, 3). There is evidence that the conversion of DDT to DDD in some species is dependent on nicotinamide adenine dinucleotide phosphate (NADPH) which is characteristic of some microsomal enzymes (8). Available evidence denies its formation by dehydrogenase activity on DDE. DDD can serve as a substrate for dehydrohalogenase, the product being DDMU, 2,2 bis (p-chlorophenyl), 1 chloroethylene.

3. *Hydroxylation*

This process is now known to be carried out by the mixed-function oxidases associated with mammalian liver microsome preparations. The first report of hydroxylation on DDT was made by Tsukamato (11) who indicated that the end product was Kelthane, 2,2 bis (p-chlorophenyl) 1,1,1 trichloroethanol. Additional reports on other species (1, 2) have substantiated the reaction, but have not established the position in the molecule of the oxygen function, which may be on one of the rings.

Further degradation of DDT beyond the above mentioned slightly polar compounds is indicated by the finding of DDA, 2,2 bis (p-chlorophenyl) acetic acid production in rats by Judah (5).

Methods

The laboratory strains of white mice and rats used here were fed pellets and water *ad libitum*. Insects except cluster flies were reared in the laboratory. Fish were obtained from the Department of Conservation, Cornell University and National Fish Hatchery, Cortland, N. Y., through the courtesy of Professors A. Eipper and Arthur M. Phillips, respectively. Beef liver was provided by the Department of Animal Science.

Assays for *in vitro* breakdown of pesticides were done in 10-ml glass-stoppered bottles which contained 1.0 ml of tissue fraction (Potter Elvejhem homogenate or acetone powder), 1.0 ml of 0.2 M Tris. HCl, pH 7.5, containing 6 mg/ml reduced glutathione; 1.0 ml of DDT suspension, prepared by slowly adding 1.0 ml of 2% DDT in 2,2 ethoxyethoxyethanol (carbitol) to 99 ml of rapidly stirred distilled water. After flushing the flasks with nitrogen and incubating for appropriate times in a shaking bath (37°C for mammals, 25°C for fish, and 30°C for insects), the enzyme action was stopped by adding 1.0 ml acetonitrile. Hexane was then added, followed by 1 hour of vigorous shaking.

Analysis of DDE or other products was done by injection of sodium sulfate dried portions of the hexane layer into 4-foot Pyrex columns of Dow Corning 200 on Chromosorb W in an Aerograph 204. The column temperature was 175°C. The detector was an electron capture type. The inlet port was Teflon lined. Control bottles were run on the complete system by inactivation at the initial time.

The results were calculated from peak heights, or areas determined by planimeter, relative to standards of high purity.

Results and Discussion

Although it was reasonable to expect the liver of mammals and fish to be a major site for pesticide metabolism, a few *in vivo* experiments were done to test this expectation and to establish whether the level and duration of administration of DDT would affect the rate of metabolism. Table 15-I supports the idea of active metabolism in the liver by the finding of larger amounts of DDE and DDD than DDT. The results presented in Table 15-II show that the mouse liver, apparently, can metabolize more readily small doses of DDT than an equivalent total single dose. The data in the tables point to substantial *in vivo* reductive dechlorination by the liver. Our *in vitro* systems, however, have failed to demonstrate this.

TABLE 15-I
DISTRIBUTION OF DDT AND METABOLITIES IN THE MOUSE

Tissue	DDT	DDE	DDD
	ppm, wet weight		
Brain	1.6	nd	nd
Lung	2.1	nd	nd
Gonads	48.6	7.4	9.5
Gut	2.8	0.4	nd
Kidney	14.2	1.0	0.9
Liver	0.7	0.9	1.2

Dosage: 4 mg/kg in DMSO, on three consecutive days. Sacrificed on fifth day. nd: not detected. Ten male mice, organs pooled.

TABLE 15-II
IN VIVO METABOLISM BY MOUSE LIVER

DDT injected	DDT	DDE	DDD
	(ppm, wet weight found in liver)		
12 mg/kg*	7.2	0.1	0.6
12 mg/kg+	6.0	0.1	1.3
4 mg/kg, 3 days	1.6	0.9	5.1

* Sacrificed after first day.
+ Sacrificed after fifth day.

The rates of dehydrochlorination of DDT or DDD by all tissues so far tested are very low compared to those of DDT resistant houseflies. The values found are presented in Table 15-III. It is perhaps significant to note that the liver of all mammals and fish tested have levels of dehydrochlorinase within an order of magnitude of each other.

In view of the low levels of activity, a few attempts to purify the dehydrohalogenase were made. Overnight dialysis of mouse liver homogenates against thiomalate buffer, pH 7, resulted in precipitation of inactive protein and increased specific activity of the soluble enzyme. Evidence to date supports the view that dehydrohalogenase activity is not associated with cell particulates. Some characteristics of the dehydrohalogenating enzyme derived from rat and mouse liver are illustrated in Figures 15-1 to 15-3.

TABLE 15–III
DEHYDROHALOGENASE ACTIVITIES

Species	Organ	Activity (mg DDE $\times 10^6$/hr/mg protein)
Rat	Liver	4.0
Beef	Liver	2.0
Mouse	Liver	10.6
Yellow perch (*Perca flavescens*)	Liver	11.5
Brown trout (*Salmo trutta*)	Liver	11.2
Rock Bass (*Ambloplites rupestris*)	Liver	4.0
Milkweed bug (*Oncopeltus fasciatus*)	Whole	7.2
Wax moth larvae (*Galleria mellonella*)	Whole	37.6
Southern armyworm larvae (*Prodenia eridania*)	Whole	5.1
Cluster fly, adult (*Pollenia rudis*)	Whole	10.0
Housefly,* adult (*Musca domestica*)	Whole	3220
Housefly,* adult (*Musca domestica*)	Whole	1250

* Insecticide resistant strain.

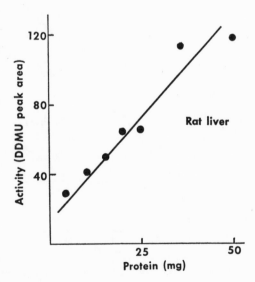

FIGURE 15-1. Dehydrohalogenation of DDD by rat liver as a function of protein concentration.

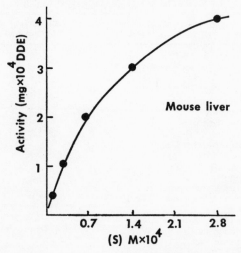

FIGURE 15-2. Dehydrohalogenase activity of mouse liver as a function of DDT concentration.

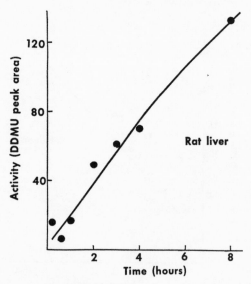

FIGURE 15-3. Dehydrohalogenation of DDD by rat liver as a function of time.

The time course of the dehydrohalogenation and its dependence on tissue concentration are typical of enzymatic reactions. The activity dependence on substrate concentration was not carried to the calculation of Michaelis-Menten constants because of the short range of concentration of substrate obtainable. In addition, the incubation system is presumably not a single phase.

Our experiments with mouse liver have consistently shown that reduced glutathione enhances dehydrohalogenase activity but is not obligatory, as found for resistant houseflies (7).

Experiments with preparative polyacrylamide gel and gel filtration of housefly tissues indicated the presence of two enzyme activities producing DDE from DDT.

There does not appear to be much information on metabolism of chlorinated insecticides by plants. In this laboratory, an indication for such has been found by Rice (9). Pure cultures containing 500 μg/liter DDT of *Chlorella vulgaris* and *Ankistrodesmus flacatus* yield upon extraction with acetone, a gas chromatographic peak with a retention time relative to DDT of 2.4 which is a significant amount relative to the DDT present. Isolation and characterization are in progress.

REFERENCES

1. AGASIN, M.; MICHAELI, D.; MISKUS, R.; NAGASAWA, W., and HOSKINS, W.M.: *J Econ Entomol, 54*:340, 1961.
2. AGASIN, M.; MORELLO, A., and SCAARAMELLI, N.: *J Econ Entomol, 57*:974, 1964.
3. BARKER, P.S.; MORRISON, F.O., and WHITAKER, R.S.: *Nature, 205*:621, 1965.
4. DATTA, P.R.; LAUG, E.P., and KLEIN, A.K.: *Science, 144*:1052, 1964.
5. JUDAH, J.D.: *Brit J Pharmacol, 4*:120, 1949.
6. KALLMAN, B.J., and ANDREWS, A.K.: *Science, 141*:1050, 1963.
7. LIPKE, H., and KEARNS, C.W.: *J Biol Chem, 234*:2123, 1959.
8. MORELLO, A.: *Canad J Biochem, 43*:1289, 1965.
9. RICE, C.: Unpublished data.
10. STERNBERG, J.G.; VINSON, E., and KEARNS, C.W.: *J Econ Entomol, 46*:513, 1954.
11. TSUKAMATO, M.: *Botyu-Kagaku, 24*:141, 1959.
12. WEDEMEYER, GARY: *Science, 152*:647, 1966.

DISCUSSION

RISEBROUGH: Do you know of any naturally occurring sub-strates for dehydrochlorinase?

YOUNG: There are no organochlorine compounds, and yet the enzyme is subject to selection pressure. I am curious, but I do not think it useful to assay purified dehydrochlorinase for phosphatase activity and every other enzyme activity I can think of.

RISEBROUGH: Is it not possible that the enzyme from flies has some other natural function than that from mammalian cells?

YOUNG: Possibly but not necessarily; the electrophoresis data we have indicate the two enzymes are approximately the same.

RISEBROUGH: Did not your data from the rat and the housefly indicate the presence of two different kinds of enzymes?

YOUNG: The bands appeared offset a little but these experiments were done some time apart, and the absolute distances from the origin are subject to that much variation.

RISEBROUGH: Would you say that these enzymes have a function of detoxifying foreign compounds?

YOUNG: From the electrophoresis data I would not expect them to be detoxifying enzymes.

WURSTER: DDE is almost always found in much larger quantities than DDT. I ask for comments on the statement that p,p'-DDE is the world's most widely distributed synthetic organic compound.

RISEBROUGH: The titer of DDT in the sea is very low but that of DDE is relatively much higher.

YOUNG: This, I would think, is a result of biological degradation.

WURSTER: Peterson and Robinson [*Toxic Appl Pharmacol,* 6:321, 1964], when they fed either DDT or DDD to rats, obtained various degradation products. But when they fed DDE, they recovered only DDE. Perhaps there is a shortage in nature of an enzyme to metabolize DDE; this might explain why DDE is worldwide in its distribution.

RISEBROUGH: Is there any enzyme that degrades DDE?

SPENCER: In some experiments with activated sewage sludge, DDE disappears but its rate of disappearance is slower than that for DDT [Hill, D. W., and McCarty, P. L.: *Water Pollution Control Federation Meeting.* Kansas City, 1966]. The DDE is degraded to a metabolite that was not included in Dr. Young's analyses.

YOUNG: We have some indications in the milkweed bug that injected DDE does disappear but as yet have no indication of its metabolic pathway (unpublished data).

WELCH: Is there any bacterial enzyme in the sewage sludge that will degrade DDE?

SPENCER: Biodegradation of DDT and its metabolites is an important way of eliminating this particular organochlorine, but the degradation is also catalyzed by reduced metals. Fe^{++}, for example, is very important for the reaction. In one experiment [Ecobichon, D. J., and Saschenbrecker, P. W.: *Science, 156*:663, 1967] blood was used as a medium. *p,p'-DDT* was introduced, and the rates of its disappearance and the formation of metabolites (TDE and DDE) were observed for 12 weeks. The percentages of TDE and DDE rose steadily for the first 7 weeks but then declined. DDT disappeared completely by the end of 12 weeks although the blood was kept at $-20°C$ during the experiment. Weekly thawing (for analytical purposes) probably released some catalytic activity from the blood.

DURHAM: In contrast to the animal species you have studied, we have some indirect information, based on *in vivo* rather that *in vitro* studies with the Rhesus monkey, which indicate that the Rhesus monkey is not able to make the conversion from DDT to DDE. In some long-term feeding studies we fed them DDT and, surprisingly, found very little DDE in the fat, where a lot of DDT was stored. When fed DDE, however, they did store it. Thus we concluded that they weren't able to make the conversion of DDT to DDE.

WELCH: Did you make cellular fractionations of mouse or rat livers to localize the enzymes?

YOUNG: It's a soluble enzyme, not brought down by 100,000 g's for 1 hour. This was also our finding with houseflies.

FASSETT: In Hayes' experiments [in Muller, P. (Ed).: *Pharmacology and Toxicology of DDT Insecticides.* Birkhauser-Basle, 1955, vol. 2] with human volunteers fed DDT, was there any analysis for DDE?

DURHAM: Yes. The volunteers built up quite a high level of DDT on DDT feeding but DDE accumulated to a much smaller degree. It appears that there is in man a limited capacity to convert DDT to DDE. For example, volunteers fed 35 mg DDT per day stored DDT at about 250 ppm in fat but their DDE storage was not much above that of a group fed DDT at 3.5 mg per day.

Chapter 16

GENERAL DISCUSSION, SESSION III

F ASSETT: Could we now consider what lessons might be learned from this elegant work on mercury, both from the point of view of the main topic of this discussion, which is the relation of the dosage to toxicity and the interpretation of this in terms of safety margins? Perhaps Dr. Berlin would be willing to comment on this topic of tissue levels, toxic effects and margins of safety.

BERLIN: My discussion will refer to a dose-response curve. One axis is the state of the living cell, which is a dynamic equilibrium. The other axis is the dose of mercury. In principle, the first molecule of any mercurial compound brought into the cell causes a change, alters the dynamic equilibrium. As the dose increases, there is a region where the change becomes detectable by the most sensitive biochemical methods. Next, a region where we begin to observe disturbances of function. Here other substances, such as chlorinated hydrocarbons, may make the cell more sensitive to mercury. This is still the physiological range, the changes can be reversed. If the whole body is considered, rather than one cell, the physiological range extends to higher doses. Next come pathological changes, and then irreversible and lethal effects.

In the case of mercury, I do not see a theoretical basis for expecting a threshold dose, such that there are no bad effects below the threshold. We are always exposed to lots of different kinds of agents. Obviously, the exposure in different parts of the world is different. The environment has different levels of metals. People have an individual capacity to compensate for exposure. It may be that a person living in Sweden has a better ability to counteract certain metals because of his surrounding than one living in South America. There are different routes and schedules of exposure, and the environment will add different kinds of stress to it.

It may be possible in the future to decide what is the optimal environment. For the present we have to face the facts and realize that we can't exclude exposure. We have to pay the price for industrial development, and we have to judge which kind of price we are willing to pay by trading useful gains for possible bad effects. However, we must protect people now from serious damage in the future and we certainly cannot permit exposures that cause changes in people which can be detected with present methods.

In the case of mercury I believe that we today have to accept a certain level of exposure for practical reasons. But it may be that in the future this level can be lowered. This amounts to setting a threshold in practice.

If the curve is for one single individual or one single cell, then to cover all individuals you have to draw a distribution of curves, ranging from most sensitive to least sensitive. We have to decide how large a part of the population we are going to protect. We would like to protect all the population but this is not always possible.

FASSETT: Toxicity of chemicals can be represented by another type of curve (Fig. 16-1), which is similar to an engineering stress-strain curve. At low exposures there is very low stress and the individual compensates for it and functions normally, and the effect of exposure is reversed when exposure ends. At higher doses there is progressively more strain, and the changes are irreversible and cumulative. Strain eventually reaches a breaking point, which represents the breakdown of regulation and death. I wish to give credit for this curve to Professor Hatch [*Am Industr Hyg Assn J*, *23*:1, 1962].

We know the levels of mercury in blood, brain, and other tissue. Where does this put us on this curve?

CLARKSON: We have some information that bears on this question. We were looking at mercury binding to the kidney in collaboration with Dr. Magos and we were curious why the kidney selectively accumulates mercury. The types of SH groups in the kidney were analyzed and it was possible to distinguish at least three main groups in terms of their affinity for mercury. The smallest group has the highest affinity. When this was translated back to the intact tissue and to the levels of mercury in tissue, it's at the point when enough mercury is bound in the kidney to saturate this first group of SH sites that we observe physiological or pharmacological effects [Clarkson, T. W., and Magos, L.: *Biochem J, 99*:62, 1966]. It might be that this first group of binding sites plays a protective role in the body, binding metals such as mercury in an inert complex. Then damage would appear only when this first group of sites was saturated.

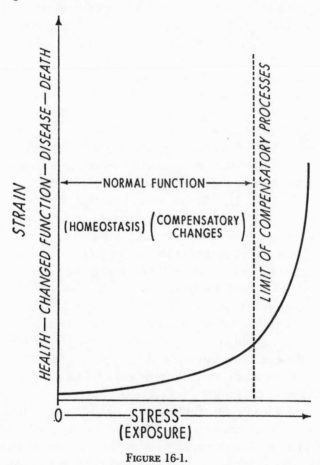

FIGURE 16-1.

BERG: It may be useful to distinguish naturally occurring molecules from "strange molecules" — a suggestive term I learned from Dr. Hodge. A living system which evolves in the presence of a heavy metal will be adapted to some range of exposures to the metal as to a normal environmental variable. One would expect to find some physiological regulating system, with damage occurring only when the limit of tolerance of that system is exceeded by the exposure. There will be individual variability but there will be a definite level below which heavy metals are tolerated — a physiological threshold in Dr. Berlin's terms.

When we shift to molecules which were not available for a couple of million years of evolution, one of two alternatives may be possible. Either the new molecules coincidentally happen to be handled by one of the available regulating systems, or there will not be, in Dr. Fasset's terms, any region of pure stress, there will be strain from the exposure to the first molecule, and no threshold. Can we arrive some time in our deliberations at a notion whether organochlorines, which are solely synthetic, are in any way like the familiar posions and are handled in the cell like the familiar poisons, or whether they are indeed strange and perhaps have no threshold for damage to an individual?

BERGLUND: Dr. Berlin considers that we don't have a threshold limit. Some pharmacologists claim we actually do have threshold limits. Peter Waser from Switzerland, has reported on the concentration of curare at the motor end plate, that is, at the target organ; he finds a sharp limit; above the limit there is an effect, and below the limit there is no effect.

I don't think one can characterize toxic agents as foreign and/or nonforeign substances. I'm inclined to accept the possibility of a threshold, which will vary depending on the way we measure the effect. For example, the thresholds for kidney failure due to mercury and for the diuretic effect of mercury were different but close to each other [*cf.* Clarkson, above].

FASSETT: The other thing that needs to be considered in studying any toxic material is that different metabolic pathways may be present with different dosages. This has been shown with phenols. The compounds occur normally and are handled by normal esterification reactions. The toxic effect seems to occur when the normal pathway for dealing with a compound of that class is overloaded. Then free phenol becomes available, and enters an alternate biochemical pathway which leads to damage. So little is known about the actual metabolic pathways of substances at very low doses that it is difficult to know whether or not metabolic experiments done at conveniently high levels are really giving relevant information.

JERNELÖV: I think we should keep in mind why pesticides are made. We use them because they are poisons, and for that reason cannot expect them to be harmless to us either.

FASSETT: Could we state any values, say of methylmercury in the brain, that would be relevant to mammalian toxicity? It seems to me from Dr. Suzuki's graph that 10 μg per gram of wet weight of brain produces symptoms in the central nervous system. In the case of Dr. Clarkson's rat kidneys with various mercury compounds, about 20 μg per gram of kidney were diuretic. We have a couple of points relating tissue levels to toxic effects. If we assume now that the whole population will be continuously exposed to mercurial compounds at a given level, can we estimate the hazard to the population? Where does the dose fit on the stress-strain curve?

RUDD: Perhaps we can extend this reasoning to ecological systems. Ecosystems also have homeostatic mechanisms which can be damaged by poisons. Biological control experts know, that it is not enough to speak only of the damage to individuals and of the damage to populations but we have to consider the interactions between populations in the ecological system.

BERG: A discussion of toxicity in ecological systems should begin with a reminder, that the stability of an ecosystem is a function of its complexity, and that complexity is always a critical variable. This is not the case in most of the studies of toxicity in organisms. The point at which an ecosystem breaks down depends on the number of species that coexist as normal members of the community. When you simplify the living community, the tolerance of the entire system to all kinds of damage goes down. In terms of the stress-strain curve, the boundary between reversible and irreversible damage moves down. So the introduction of a pesticide into an ecological system is not analogous to administering a mercurial diuretic to a patient, but it is quite analogous to administering an antimitotic drug to a cancer case; this kind of drug will knock out cancer cells, but it will also knock out intestinal epithelium and white cells, and thus "simplify" the patient and make him vulnerable to other damage. No toxicologist would recommend treating a patient with an antimitotic drug that persisted in an active form for years. Ecologists take the same view of persistent pesticides.

SUZUKI: I agree with Dr. Berg. For example, I don't think the dose of 20 μg of methylmercury per gram in the brain is a "fixed" point; it's a "floating" one, depending on the ecological setting, other exposures, etc.

RUDD: I agree completely that it is a "floating" point. We are not able to transfer our knowledge of toxicology to ecological systems and predict the ecological effect from the dose given. On the contrary, I have the impression, that the same chemical in presumably equivalent conditions gives widely different responses in different ecological systems.

HODGE: Ted Hatch applied this stress-strain curve to industrial posions, and their actions on people. He indicated that there was a rather wide area of stress or exposure in which the homeostatic mechanisms of the body were not overwhelmed, in which the stress exposure could be compensated, and in which at any moment it was impossible to know whether the individual was suffering from any effect. Then as the stress increased, there came a time when there were signs of illness, which is reversible. If the individual is then removed from exposure, he recovers since he hasn't passed the point of "no return." Only with increased exposure does the patient go into an area of irreversible illness and still later death.

RUDD: Aren't we straining systems analysis when we try to model ecological systems with such a curve? Would you put the chemical damage to a population of pink bollworm low on the curve and call it a compensated stress, and put the extermination of lacewings and other predaceous insects high on the curve and call it a damaging strain?

BERG: Agricultural chemists and ecologists disagree on the proper use of this model. Say, for instance, that DDT is sprayed over a crop, and removes a species of beetle, five species of spiders, and 1 species of insect-eating bird from that range. This can be modeled two ways, using the same analogy to a patient treated with a toxic chemical. The chemist would say that the treatment did not damage the crop, hence it damaged only vermin. The analogy is to a patient who lost his fleas and lice, and is better off for it. The ecologist would say that the treatment simplified the ecological system, and damaged its regulating powers. The analogy is to a patient who lost the function of one kidney as a result of exposure.

RUDD: The ecological system is most resistant to damage when it is managed with integrated controls, biological and chemical. The principle of such management is that none of the local species is vermin.

SESSION IV
MECHANISMS OF ACTION:
EFFECTS IN INDIVIDUALS

THE SIGNIFICANCE OF RESIDUES IN PHEASANT TISSUES RESULTING FROM CHRONIC EXPOSURES TO DDT

E. G. HUNT, J. A. AZEVEDO, JR., L. A. WOODS, JR., and W. T. CASTLE

ABSTRACT

The effects on pheasants (*Phasianus colchius torquatus*) of the chronic exposure to technical DDT were studied. Penned pheasants were placed on various dietary regimens containing 10, 100, 250, and 500 ppm of the insecticide. Residue data are presented for tissues of test birds, their eggs and feces. Results demonstrate that residue levels in whole blood, fat, reproductive organs and eggs reflected dietary levels of DDT.

Insecticide residue levels in brains have limited value in diagnosing cause of death of pheasants exposed to DDT. Maximum levels in excess of 1,000 ppm DDT were found in the eggs of some birds. One female voided approximately 60 mg of DDT in her eggs after receiving insecticide contaminated diets until the start of egg-laying.

REPORT

Introduction

THE EFFECTS OF DDT on pheasants have received considerable attention in California during the last seven years. The Department of Fish and Game has made investigations of chronic exposure to the insecticide of wild and penned pheasants (6).

A field study was made in 1962 of the effects of DDT on wild pheasants. In 1963 a statewide investigation was conducted to assess residue levels in eleven wild pheasant populations. Monitoring residues in pheasants has continued in two of the populations included in the statewide study.

An intensive study of the physiological effects of DDT on

Note: This study was supported in part by Public Helath Service research grant CC-00275 from the National Communicable Disease Center, Atlanta, Georgia. Support was also provided by Federal Aid in Wildlife Restoration Project FW-1-R, Pesticides Investigations.

penned pheasants has been underway since 1964. Various patterns of chronic exposure to technical DDT were used. Information was gathered on the following: (a) the toxic effects on the breeding of pheasants; (b) the influence of some stresses on survival of adults; and, (c) the effects of various feeding regimens on fertility, egg production, hatchability and survival of young. Early results were reported (1).

This paper emphasizes the major portion of the residue results obtained from penned pheasants and relates these data and others to biological effects. The use of various types of animal tissues for residue analysis is evaluated.

Methods

Test pheasants were maintained in large pens in heterosexual groups during fall and winter. DDT contaminated feed was provided for some groups, and other groups received uncontaminated feed. At the onset of the breeding season males and females were segregated — one male penned with each group of females for the breeding season. During breeding and egg-laying, birds were used in either individual or group trials.

Pheasants in individual trials were caged separately. This enabled accurate measurement of the amount of contaminated feed ingested. Eggs produced by these birds were used exclusively for residue analyses. Individual trial birds were sacrificed at the end of egg-laying and tissues taken for residue analysis. Correlations of intake of DDT with residues in eggs and tissues could be made accurately in samples from birds in these trials. Feces were collected from individually caged birds for residue analyses.

Each year from 180 to 350 pheasants were caged in groups and used primarily in production trials. Most of their eggs were incubated to evaluate the effects of contaminated diets on reproduction. Usually one cock and eight to fifteen hens comprised a unit in the group trials. For nine to twelve successive Sundays, most of the eggs from each group were saved for residue analysis. This provided a means for detecting possible major differences in residues in eggs from birds caged individually or in groups. The measurement of DDT ingestion per bird in group trials was based upon the amount of contaminated feed consumed by the group. Some of the group trial birds were sacrificed at intervals before, during, and following the breeding season. These periodic sacrifices permitted an assessment of the body load of insecticides during various intervals throughout the spring, and early summer.

Pellets, grain or a mixture of both comprised the diets of test and control pheasants. The dietary levels of DDT used were: 10, 100, and 500 ppm in 1964; 10, 100, and 250 ppm in 1965 and 1966; and 10 and 100 ppm in 1967. These diets were presented for various periods of time during the year to evaluate the biological effects and the effect of season and duration of exposure on the rate of insecticide accumulation. The detailed account of procedures used in feeding trials has been published (1).

Various types of tissue were used for residue analysis; i.e. body cavity and /or subcutaneous fat, whole brains, breast muscle and blood. Blood samples consisted of whole blood, serum, plasma or formed elements. Conventional methods of handling, preserving and fractioning blood were used. When necessary blood fractions were frozen for storage. Refrigerated whole blood was analyzed within 24 hours of collection. In most cases only the yolks of eggs were analyzed. Feces were collected on aluminum foil and dried prior to preparation for analyses.

Fresh tissue samples, weighed at time of autopsy, were ground with three or four times their weight of anhydrous sodium sulfate. Samples which were not analyzed immediately were stored at –10 to –20°C. Tissues such as blood, fat, brains, eggs, and gonads were ground in a beaker with the sodium sulfate and then extracted with nanograde hexane. Feces, flesh and whole chicks were blended with anhydrous sodium sulfate in an omni-mixer or Waring blender and then extracted with nanograde hexane. All samples were analyzed on a wet-weight basis except fecal samples which were dried to constant weight at 110°C before analysis.

Samples containing less than 10% fat were "cleaned up" by passing through a florisil column. The insecticide was eluted from the column with 250 ml of 6% diethyl ether in hexane. Samples with higher percentages of fat were partitioned after extraction with hexane. The partitioning process is essentially the same as

described by de Faubert Maunder *et al* (3) using dimethylformamide. Following partitioning the samples were passed through a
florisil column and the pesticides eluted from the column with
6% diethyl ether in hexane. All samples were analyzed by Elec-
tron-Capture gas chromatography using Wilkins Hy-Fi Model
600C gas chromatographs. The gas chromatographs were fitted
with $\frac{1}{8}$ inch by 5-foot spiral glass columns packed with a 1:1 mixture in tandem of 10% QF-1 silicone grease on 100/120 mesh Gas
Chrom Q solid support and 10% DC-200 silicone grease on the
same solid support. Operating temperature for the column was
200°C., injector temperature was 210°C., and nitrogen flow rate
of 110 ml per minute. During the first part of the study a column
consisting of 5% Dow-11 silicone grease on chromosorb w, 60/80,
was used. The peak height method was used to quantitate the residues found.

Percentage fat analysis was performed on adipose tissue and
eggs by the method described by Salwin *et al* (9) using a modified
Babcock procedure. Other tissues were analyzed for fat content by
evaporating an aliquot of hexane extract to dryness in a tared
aluminum planchet and weighing the residue. References to ppm
DDT in the text designates the total of all isomers and degradation products identified except where an isomer is specified.

Results

Residues in Brains

Residue levels have been determined in 268 analyses of brains
from 355 pheasants (Table 17-I). Results of these analyses have
been grouped according to the fate of the donor.

Pheasants that expired with symptoms typical of DDT intoxication and a few birds that died from other than mechanical or
identifiable natural causes were classified as being killed by DDT.
Birds that died from uterine failure or were killed by predators
were classified as sacrificed pheasants.

In brains of birds that died, residue values ranged from 0.5 to
108 ppm p,p'DDE and nondetected to 91 ppm p,p'DDT. In sacrificed pheasants the values were from nondetected to 35 ppm p,p'-
DDE and nondetected to 44 ppm p,p'DDT.

TABLE 17–I
RESIDUE LEVELS IN BRAINS OF TEST PHEASANTS

Diets	Fate of donor — Died ♂	Died ♀	Sac. ♂	Sac. ♀	Residues (ppm) — p,p′DDE	o,p′DDT	p,p′DDD	p,p′DDT	No. of analyses	No. of samples
Control			x		ND¶–1.54 (.469) §	ND–1.15 (.244)	ND–1.26 (.161)	ND–10.4 (1.20)	13	26
				x	ND– .321 (.184)	ND–.437 (.101)	ND–.047 (.009)	ND–3.83 (1.06)	5	5
	x				.690	ND	ND	ND	1	1
10 ppm			x		.169–5.67 (1.35)	ND–4.45 (.423)	ND–1.02 (.149)	ND–2.88 (1.18)	57	82
				x	.546–29.3 (5.06)	ND–4.74 (.806)	ND–.520 (.071)	ND–10.2) (2.38)	19	20
		x			.530–13.2 (6.85)	ND–.600 (.344)	ND–.422 (.084)	.410–7.20 (2.59)	5	5
	x				2.47–84.2 (32.8)	ND–7.40 (1.29)	ND–10.6 (2.82)	ND–25.2 (10.8)	15	15
100 ppm			x		.220–9.30 (2.26)	ND–4.93 (.640)	ND–1.45 (.213)	.690–7.60 (2.84)	34	70
				x	1.77–35.8 (10.2)	trace‡–3.23 (.980)	ND–5.00 (.779)	.644–44.0 (9.84)	18	19
		x			5.74–109.0 (47.1)	ND–25.6 (4.41)	ND–43.0 (9.74)	3.70–67.0 (23.7)	19	19
	x				1.60–92.2 (47.0)	ND–8.28 (1.48)	ND–32.8 (5.83)	3.35–79.4 (40.8)	50	50
250 ppm			x		.280–13.4 (3.30)	.140–2.80 (2.00)	ND	1.22–19.6 (6.32)	7	19
				x	3.00–4.94 (4.00)	.010–2.00 (.900)	ND	1.01–12.7 (6.91)	3	3
		x			30.2–66.2 (48.0)	6.00–17.9 (11.9)	ND	10.6–91.3 (52.6)	3	3
	x				16.1–28.3 (21.8)	.140–1.72 (1.16)	ND	31.6–51.0 (40.8)	4	4
500 ppm			x		26.0	ND	—	26.7	1	1†
			x		6.20	ND	—	2.80	1	1*
		x			7.30–12.3 (9.80)	3.20–4.10 (3.60)	—	15.1–24.5 (19.8)	2	2†
				x	9.10	ND	—	13.8	1	1*
	x				7.8–29.9 (16.0)	1.3–11.6 (4.12)	—	13.1–46.9 (25.7)	7	7*
	x				9.70–35.5 (18.1)	1.60–7.30 (4.12)	—	14.4–45.3 (28.4)	6	6†

* DDT through egg laying.
† DDT until egg laying.
‡ Less than .001 ppm.

§ Denotes average.
¶ Less than .0009 ppm.

The brain residue data were examined from the standpoint of using detected levels to determine the fate of the tissue donor. To determine the brain residue level that could be indicative of death a 95 per cent confidence interval was calculated for the mean brain residue values from male and female pheasants that died with symptoms of DDT intoxication. These mean residue levels were 49.6 to 88.8 ppm total insecticide in brains of females and 73.7 to 90.9 ppm total insecticide in brains of males.

Residues in Blood

Residue levels in blood were measured to obtain an indication of the role of this tissue in the distribution of DDT in the animal body. In general, blood residue levels reflected the levels of DDT in the diet; the higher the level of intake the higher the blood residue levels (Table 17-II). A maximum of 23.5 ppm p,p′-DDT occurred in a sample taken from birds with high exposure to DDT. Residue levels in the blood reflected exposure to DDT for at least 24 hours after ingestion of the insecticide. Maximum residue levels of 3.54 ppm p,p′DDE and 0.061 p,p′DDT were found in blood taken from pheasants sixty days after DDT was removed from their diet. Probably these residues resulted from the movement of insecticide from fat storage depots.

Under conditions of long-term continuous exposure to DDT

TABLE 17–II
RESIDUE IN WHOLE BLOOD OF TEST PHEASANTS

Diets	*Sex* ♂	♀	Residues (ppm) *pp′DDE*	op′DDT	pp′DDD	pp′DDT	*No. of Analyses*	*No. of Samples*
Control	x		ND‡–.178 (.039) †	ND–.035 (.007)	ND	ND–.416 (.094)	6	6
		x	ND–.065 (.007)	ND–.014 (.003)	ND–.005 (trace§)	ND–.130 (.014)	18	43
10 ppm	x		.017–1.30 (.447)	ND–.110 (.026)	ND–.113 (.018)	ND–4.28 (.475)	13	19
		x	.012–6.64 (.449)	ND–.285 (.022)	ND–.541 (.021)	.019–4.07 (.492)	26	104
100 ppm	x		.130–4.41 (.850)	ND–.131 (.051)	ND–.030 (.002)	.238–2.99 (1.36)	12	17
		x	.016–12.4 (1.74)	ND–1.50 (.132)	ND–.333 (.028)	.011–23.5 (4.49)	33	109
250 ppm*		x	.230–1.06 (.456)	.100–.560 (.340)	ND	.770–3.32 (1.88)	6	8

* Test birds received diets for much shorter period than other birds in this trial.
† Denotes average.
‡ Less than .0009 ppm.
§ Less than .001 ppm.

the blood residue levels in males and females were similar or dissimilar depending on the time of year in which the samples were taken. Residue levels in blood taken at the beginning of the breeding season tended to be about the same in both sexes. At this period residues were higher in the males than at the end of the breeding season. This may result from the hyperactivity and loss of weight by males prior to and during the breeding season. However, loss of weight (fat) per se did not in general correlate with increased residues in blood levels. In females during the breeding season, residue levels were comparatively low probably as a result of a drain in residues from the blood system into the ova. Differences in blood residue levels between sexes were greatest at the end of the egg-laying season. At this time, the high levels found in the blood of females may have resulted from the termination of egg-laying which also negated the facility to void or excrete residues in eggs. Levels in males at this period were generally lower than at the beginning of the breeding season, possibly reflecting an increase in input of residues into body fat and a reduction in the stress accompanying the breeding season. Maximum residue levels in the whole blood of males and females at the end of egg-laying season were 2.99 and 23. 5 ppm p,p'DDT, respectively.

Tests were made to determine the relative amount of residues occurring in blood fractions. Our principal intent was to select the blood fraction that would contain the highest residue levels and be the easiest to process. Analyses of the red blood cells and plasma were made from the same birds; most of the insecticide was in the plasma (Table 17-III). Similar tests comparing residue levels in serum and whole blood indicated slightly higher residues in whole blood. Comparatively higher residue levels, ease of obtaining large samples and reduced effort required to handle samples prior to chemical analysis support the use of whole blood for residue analysis.

Residues in Fat

Residue values in fat appeared to be influenced by the length and level of exposure to DDT, body condition and other physiological factors. Levels in excess of 2,000 ppm p,p'DDT were found

TABLE 17-III
RESIDUE IN BLOOD FRACTIONS OF TEST PHEASANTS

Diets	Tissue	Sex	pp'DDE	op'DDT	pp'DDD	pp'DDT	No. of Analyses	No. of Samples
			Residues (ppm)					
250 ppm	Serum	♀	.030–.480 (.197) §	trace†–.012 (.083)	ND¶	.103–2.05 (.831)	10	24
	Whole blood	♀	.230–1.06 (.456)	.100–.560 (.340)	ND	.770–3.32 (1.88)	6	8
100 ppm*	Plasma	♀	.470–.780 (.603)	.011–.140 (.071)	ND	1.11–1.75 (1.31)	3	18
	RBC‡	♀	.130–.178 (.155)	.027–.050 (.040)	ND	.018–.048 (.031)	3	18

* Plasma and RBC values from same birds.
† Less than .001 ppm.
‡ Formed element in blood including RBC, WBC, platelets, etc.
§ Denotes average.
¶ Less than .0009 ppm.

in sacrificed birds and birds that died from exposure to the insecticide (Table 17-IV). Contamination of this magnitude was usually associated with lengthy exposure to 100 or 250 ppm diets. Maximum levels in fat from birds on 10 ppm diets regardless of the length of exposure or their fate was about 350 ppm p,p'DDT.

The highest residue in a fat sample, 11,620 ppm p,p'DDT, was from a male that was fatally poisoned. The range of the five next highest levels was 2,098 to 4,590 ppm p,p'DDT. Three of these samples came from birds that died and two from sacrificed pheasants; four of the samples were from males and one from a female. The maximum level recorded in the fat of a female pheasant was 3,129 ppm p,p'DDT from a sacrificed bird. The highest DDT residue value we have found in wild pheasants is 3,339 ppm p,p'DDT in a nesting female shot near Richvale, California in 1963.

Tests were made to assess the comparability of residue data from muscle and adipose tissue from the same bird. Samples were taken from twenty-eight pheasants sacrificed on four dates during the breeding season. Analyses were made of these two tissues separately and residue results from both tissues recapitulated on a 100 per cent lipid basis. The fat in white muscle (breast) of pheasants ranged from .19 to 1.17 per cent. In comparison, aver-

TABLE 17–IV
PESTICIDE RESIDUES IN ADIPOSE TISSUES OF TEST PHEASANTS

Diets	Died ♂	Died ♀	Sac. ♂	Sac. ♀	p,p'DDE	Residues (ppm) o,p'DDT	p,p'DDD	p,p'DDT	No. of Analyses	No. of Samples
Control				x	.120–5.16 (1.12) *	ND†–2.00 (.400)	ND–.730 (.090)	ND–15.4 (3.22)	23	41
			x		2.70–4.65 (3.51)	.500–2.51 (1.44)	trace‡–1.40 (.620)	1.90–8.09 (4.33)	3	3
		x			.539–1.42 (1.06)	ND–.198 (.066)	ND	.922–2.49 (1.79)	3	3
	x				2.61	ND	ND	2.20	1	1
10 ppm				x	.85–100. (21.4)	ND–8.44 (2.14)	ND–2.72 (.150)	3.20–179. (58.7)	50	90
			x		28.8–137. (55.9)	ND–6.40 (2.23)	ND–8.31 (1.04)	4.60–265. (105.)	8	12
		x			18.1–41.6 (28.9)	ND–5.30 (1.29)	ND	89.8–122. (102.)	5	5
	x				23.6–122. (72.8)	ND	ND	133.–357. (245.)	2	2
100 ppm				x	17.9–717. (200.)	ND–82.1 (24.8)	ND–18.9 (1.30)	48.5–1200. (566.)	58	119
			x		111.–1598. (433.)	6.30–325. (60.3)	ND–122. (18.6)	383.–2098. (842.)	10	13
		x			22.3–1196. (363.)	ND–70.8 (28.8)	ND–35.4 (9.38)	45.8–1353. (862.)	5	5
	x				75.5–3380. (1097.)	ND–90.9 (27.9)	ND–333. (130.)	657.–11,620. (3800.)	6	6
250 ppm				x	29.0–672. (257.)	7.20–135. (61.8)	ND	226.–3129. (859.)	16	52
500 ppm				x	27.7–332. (187.)	55.0–233. (143.)	399.–1,733. (891.)	417.–2,269. (1,208.)	4	4
		x			80.4	94.5	—	1,185.	1	1

*Denotes average.
†Less than .0009 ppm
‡Less than .001 ppm

age fat levels in breast muscle of some other birds are: 4.21 per cent for the snowy egret (*Leucophoyx thula*), 4.78 per cent for the blue-winged teal (*Anas discors*) and 5.40 per cent for the barn owl (*Tyto alba*) (4).

The closest correlation of recapitulated values of adipose tissue and flesh occurred in some samples taken at the onset of the breeding season (Table 17-V). However, in general, recapitulated values from the two tissues were not directly comparable. Comparison of means of the recapitulated results showed the ratio in flesh samples to adipose samples ranged from .46 to 6.2, respectively, in samples taken from the same bird. On a 100 per cent lipid basis residue results from tissue samples were consistently higher than those from fat samples.

A series of residue analyses were made of body cavity and sub-cutaneous fat from the same birds. Each sample consisted of adipose tissue from several birds on the same dietary level of DDT. In comparing the percentage of fat in body cavity adipose tissue versus subcutaneous adipose tissue from the same female birds, the averages were identical, e.g., 87.7 per cent; the ranges being 77.5 to 98.9 per cent for subcutaneous and 81.0 to 96.0 per cent for body cavity tissue. The results indicate that a close correlation may exist between residues in these two types of adipose tissue (Table 17-VI). Neither tissue was consistently high or low in residue content. In a male pheasant the residue levels in the body cavity fat were 2.15 times the residue levels in the subcutaneous fat. This was the greatest variation found in residue values between body cavity and subcutaneous fat from the same bird. There was considerably less than twofold variation in most of the samples of fats from females. Similar comparisons of fat samples taken from a single bird suggest that there may be a greater variation in values than were expressed for the composited samples.

Residues in Reproductive Organs

Ovaries and testes were taken for analyses at the end of the breeding season. The residues were about proportional to dietary levels of the insecticide. The average level of p,p′DDT in ovaries was 10.4 ppm for females on the 10 ppm diets and 76.0 ppm for those on 100 ppm diets; 2.96 and 24.6 ppm p,p′DDT were in testes of males on 10 and 100 ppm diets, respectively (Table 17-VII).

Residue levels of the DDT complex were usually higher in ovaries than in testes. The maximum value of p,p′DDT in testes was 27.9 ppm in birds on 100 ppm diets; in ovaries of birds fed 100 ppm DDT the highest level was 161.16 of the same isomer. There was less variation in residues in testes from cocks on the same diets than in ovaries from females receiving similar exposure to DDT. The level of DDT reported in ovaries of control birds is

TABLE 17-V

COMPARATIVE RESIDUE VALUES IN FLESH AND FAT OF 56 FEMALE TEST PHEASANTS

	Sampled 2/23/66			Sampled 3/23/66			Sampled 5/3/66			Sampled 6/21/66		
	pp'DDE (ppm)	pp'DDT (ppm)	Total (ppm)	pp'DDE (ppm)	pp'DDT (ppm)	Total (ppm)	pp'DDE (ppm)	pp'DDT (ppm)	Total (ppm)	pp'DDE (ppm)	pp'DDT (ppm)	Total (ppm)
10 ppm diets												
Flesh — Range	8.30–30.9	31.2–72.7	50.0–220.	10.5–41.8	31.0–95.6	51.7–144.	34.3–97.4	114.–191.	177.–327.	11.8–297.	26.5–324.	143.–683.
Flesh — Average	21.1	56.6	107.	25.1	64.7	102.	68.0	140.	239.	119.	178.	371.
Flesh — Ratio	2.2*	1.2	1.6	3.2	1.9	2.4	3.0	1.5	1.9	6.2	2.6	4.0
Fat — Range	6.60–14.6	22.7–68.1	47.0–81.0	4.00–18.8	10.3–72.2	15.7–94.8	10.3–45.2	66.7–152.	79.7–204.	11.4–35.5	47.2–96.6	63.4–132.
Fat — Average	9.7	47.0	65.2	7.9	33.5	42.7	23.0	91.0	123.7	19.3	69.0	91.7
100 ppm diets												
Flesh — Range	161.–350.	557.–1026.	859.–1680.	72.7–863.	119.–440.	433.–1126.	348.–1170.	896.–1497.	1190.–3946.	424.–1060.	784.–1528.	1454.–2675.
Flesh — Average	268.	716.	1180.	356.	246.	808.	743.	1374.	2437.	621.	1141.	1927.
Flesh — Ratio	2.2*	1.0	1.4	6.4	0.46	2.7	3.2	1.6	2.1	2.5	1.3	1.6
Fat — Range	50.2–158.	624.–1025.	449.–1236.	29.3–96.8	300.–816.	207.–515.	116.–283.	639.–1048.	757.–1334.	136.–396.	624.–1176.	790.–1662.
Fat — Average	123.	679.	825.	55.9	536.	302.	226.	866.	1147.	249.	862.	1169.

*Indicates ratio of residues in the lipids in flesh vs. fat when recapitulated to 100 per cent lipid.

TABLE 17–VI

COMPARATIVE RESIDUE VALUES IN BODY CAVITY and SUBCUTANEOUS FAT OF TEST PHEASANTS

Diets	Sample No. and Sex	Body cavity fat Residues (ppm)			Subcutaneous fat Residues (ppm)			Ratio Body cavity/Subcutaneous fat		
		pp'DDE	pp'DDT	Total Pesticide	pp'DDE	pp'DDT	Total Pesticide	pp'DDE	pp'DDT	Total Pesticide
Control	–1–♀	.400	3.80	4.41	.910	2.90	3.81	.44	1.31	1.16
	–2–♀	.560	1.40	1.96	.660	.860	1.52	.85	1.63	1.29
10 ppm	–3–♀	45.5	121.	171.	47.6	126.	179.	.96	.96	.96
	–4–♀	34.5	98.7	138.	30.0	89.0	124.	1.15	1.11	1.11
	–5–♀	49.5	134.	188.	39.7	46.5	87.2	1.25	2.88	2.15
100 ppm	–6–♀	23.6	150.	203.	21.9	133.	181.	1.08	1.13	1.12
	–7–♀	29.8	159.	221.	30.5	159.	226.	.98	1.00	.98
	–8–♀	18.9	150.	190.	17.9	132.	169.	1.06	1.14	1.12
	–9–♀	35.1	223.	292.	39.0	268.	347.	.90	.83	.84
250 ppm	–10–♀	58.8	301.	437.	75.0	382.	565.	.78	.79	.77
	–11–♀	86.6	378.	553.	115.	542.	792.	.75	.70	.70
	–12–♀	59.2	380.	520.	44.0	287.	385.	1.34	1.32	1.35
	–13–♀	40.6	338.	438.	29.0	226.	305.	1.40	1.50	1.44

similar to that found in the fat of control birds. These residues probably resulted from contaminated ingredients in the commercially prepared diets fed to control pheasants. Small amounts of DDT and other pesticides may occur in commercial poultry feeds.

TABLE 17–VII
RESIDUES IN THE REPRODUCTIVE ORGANS OF TEST PHEASANTS
SACRIFICED JUNE 29, 1967

| Tissue | Diets | Residues (ppm) | | | | No. of Samples | No. of Analyses |
		pp'DDE	op'DDT	pp'DDD	pp'DDT		
Testes	Control	NDt–3.66 (1.26) *	ND–.230 (.076)	ND–.220 (.073)	.600–1.63 (1.26)	3	3
	10 ppm	3.18–7.13 (5.16)	.150–.540 (.345)	.180–.630 (.405)	2.74–3.13 (2.94)	2	2
	100 ppm	21.8–39.9 (30.8)	ND–2.04 (1.02)	2.61–4.88 (3.74)	21.3–27.8 (24.6)	2	2
Ovaries	Control	.240–.810 (.556)	ND	ND	1.54–4.83 (2.86)	3	3
	10 ppm	1.41–19.8 (5.24)	ND–4.32 (.800)	ND–1.56 (.628)	3.42–47.2 (10.4)	15	15
	100 ppm	6.60–68.6 (34.3)	.560–13.4 (5.25)	ND–12.0 (4.39)	29.1–161. (76.0)	23	16

*Denotes average.
†Less than .0009 ppm.

Residues in Eggs

Residue values were obtained from birds in individual and group trials. In general the values reflected the dietary levels of DDT. In most of the trials where 10 ppm diets were fed the average total pesticide level in egg yolk was approximately 30 ppm. In other 10 ppm DDT feeding trials, the average level in yolks was about 60 ppm. In these latter instances the samples came from trials where the adults were either stressed by starvation or were maintained on contaminated diets continuously throughout the 4-year study period. In trials where the diet contained 100, 250 and 500 ppm DDT, average egg yolk residues were about 260, 275 and 450 ppm, respectively. The highest values, in excess of 1,000 ppm, were in eggs from females fed 500 ppm DDT diets.

The maximum residue level in eggs from pheasants shot at

Richvale, California was 1,180 ppm. DDT at levels of 6,300 and 8,400 ppm were found on treated seed rice consumed by these birds.

DDT residues appeared in egg yolk within 1 day of the first exposure to DDT. Pheasants fed 100 ppm DDT diets 2 months prior to but not during egg-laying produced eggs at the end of a seventy-day laying period with about 50 ppm DDT. Intermittent exposure to DDT during egg-laying was generally reflected by large fluctuations in egg residue values. Samples from a bird on a 500 ppm DDT diet demonstrated the large amount of insecticide that can be voided in eggs. This pheasant ingested 480 mg of DDT during the twenty-three days just prior to egg-laying. She received clean feed during the next twenty-nine days. In this period fourteen eggs were laid which contained a total of 60.3 mg of DDT.

Two types of samples were used to measure residues in eggs laid by individually caged birds. Samples consisted of single eggs or composites of all eggs laid in a week. Results were similar regardless of the sampling method used except there was greater fluctuation in residue levels of successively laid eggs analyzed singly. Overall variations in residue values of approximately twentyfold occurred in eggs from one female. Generally eggs from one hen had a two- to fourfold range between high and low residue values. No definite pattern of egg residue buildup occurred in the eggs from birds maintained on contaminated diets throughout egg-laying. Insecticide residue levels were from 9.7 to 58.0 ppm DDT in eggs from birds on 10 ppm diets (fed throughout egg-laying) and from birds on 100 ppm diets (fed throughout egg-laying) 10.3 to 256.5 ppm DDT (Table 17-VIII).

The ovaries, seven ovum and an entire egg, were analyzed from each of two pheasants. DDT residues in ova reflected the inconsistencies reported for eggs. Ova residue levels ranged from 39.5 to 137 ppm DDT in samples from one bird and from 169 to 655 ppm DDT in samples from the other pheasant (Table 17-IX).

Information derived from eggs of birds in group trials suggests a similar pattern in residue data to that of individually caged birds. Two trials were conducted with grouped birds only. In one trial, birds were given uncontaminated diets the first half of egg-

TABLE 17–VIII

RESIDUES IN EGG YOLKS FROM TEST PHEASANTS IN INDIVIDUAL TRIALS

Diets	Exposure	No. of Analyses	Residue (ppm)			Total Pesticides
			pp'DDE	op'DDT	pp'DDT	
Control	No DDT	6	0.740	trace*	trace	0.740
10 ppm	DDT until egg-laying	14	0.150–3.60 (1.50) †	ND‡–4.30 (1.10)	1.90–6.80 (4.20)	3.53–10.8 (6.50)
	DDT through egg-laying	12	.910–11.8 (5.70)	tr.–240 (.120)	5.50–45.1 (19.2)	9.70–58.0 (30.4)
	DDT after egg-laying began	16	.013–2.00 (.780)	ND–.410 (.260)	ND–11.0 (3.19)	.013–13.4 (4.12)
100 ppm	DDT up to egg-laying	23	.920–32.5 (6.01)	ND–12.0 (3.00)	1.80–115. (43.9)	5.80–130. (57.9)
	DDT through egg-laying	21	2.04–42.5 (12.9)	.600–29.0 (11.8)	7.70–185. (86.1)	10.3–256. (113.)
	DDT after egg-laying began	18	ND–11.9 (4.18)	ND–4.90 (1.83)	ND–62.0 (23.2)	ND–78.8 (29.1)
250 ppm	DDT after egg-laying began	17	tr.–19.1 (7.50)	ND–8.80 (3.00)	ND–151. (59.8)	tr.–178 (70.2)
500 ppm	DDT up to egg-laying	18	3.60–300. (67.7)	ND–163. (30.6)	92.3–823. (336.)	123.–1020. (422.)
	DDT through egg-laying	21	4.80–175. (42.0)	7.50–67.7 (27.9)	70.3–585. (271.)	125.–825 (349.)

*Less than .001 ppm
†Denotes average.
‡Less than .0009 ppm

laying, and contaminated diets before and during the last half of egg-laying. In the other trial the pheasants were fed contaminated diets only during the last half of egg-laying. Residue levels approached 100 ppm DDT in egg yolks within a month after contaminated diets were offered in this latter trial. A summary of results of all group trials is presented in Table 17-X.

Residues in Liver

A series of analyses were conducted to determine the comparative residue values in liver and fat from the same bird. The pheasants sampled were wild birds shot near Richvale, California in 1963. Residues were higher in fat than in liver in the tissues of thirteen of the fifteen birds collected. The differences were greater

TABLE 17–IX
RESIDUES IN OVARIES AND OVUM OF TEST PHEASANTS
COLLECTED JUNE 29, 1967

Diets	Tissue	Sample Size (grams)	Residue (ppm) pp′DDE	op′DDT	pp′DDD	pp′DDT	Total μg Insecticide
10 ppm	Ovary	1.47	2.78	.810	.870	4.92	13.79
	Ovum #1	.12	12.5	9.09	ND*	34.1	6.68
	#2	.36	21.8	13.4	4.80	96.9	49.32
	#3	.97	17.9	3.66	2.72	32.3	54.82
	#4	2.01	19.5	2.43	1.06	40.0	126.63
	#5	3.73	21.9	8.60	ND	34.5	242.30
	#6	5.02	12.7	1.15	.550	25.1	198.24
	#7	8.17	20.8	.360	ND	41.3	510.30
	Whole egg	20.88	6.09	.810	.590	13.2	432.22
100 ppm	Ovary	1.98	17.2	3.36	1.86	29.1	102.11
	Ovum #1	.14	49.3	18.6	ND	101.	23.60
	#2	.60	173.	39.4	15.7	315.	325.91
	#3	1.29	160.	38.8	7.31	316.	672.77
	#4	2.86	183.	42.7	7.95	366.	1715.91
	#5	4.66	141.	33.0	4.10	471.	3024.81
	#6	6.92	174.	37.0	8.43	380.	4146.81
	#7	8.51	192.	29.8	ND	433.	5574.48
	Whole egg	28.07	25.9	.530	ND	51.1	2176.27

*Less than .0009 ppm.

than twofold in 73 per cent of the cases. The levels of total insecticide ranged from 5.6 to 3,339 ppm DDT in fat and from 1.5 to 128 ppm DDT in liver; means were 519.6 ppm DDT for fat and 33.0 ppm DDT for liver. Greatest variations were in samples in which the comparative values of total pesticide in fat to liver were 1,583 to 126; 1,204 to 16.9 and 3,339 to 102 ppm DDT.

Residues in Feces

The level of DDT in fecal samples was compared with the level ingested to estimate the amount of insecticide absorbed from the feed. Samples, comprised of five to seven days defecation, were taken from three birds receiving 250 ppm DDT diets and two samples from each pheasant analyzed for pesticide residues. Ranges in the levels of DDT absorbed by the birds from their feed were 96.5 to 98.8 per cent (Table 17-XI) .

Discussion

Residue levels in brain tissue have been looked upon as a key for determining the fate of birds exposed to persistent pesticides

TABLE 17–X
PESTICIDE RESIDUES IN EGGS OF GROUP TRIAL TEST PHEASANTS

| Diets | Exposure | No. of Analyses | Residues (ppm) | | | Total Pesticides |
			pp'DDE	op'DDT	pp'DDT	
Control	NO DDT	18	.034–.840 (.197) *	ND‡–.074 (.013)	ND–.042 (.003)	ND–2.848 (.331)
No DDT	DDT to parents only	3	12.7–82.0 (.226)	1.90–14.4 (.032)	ND–.147 (.083)	41.0–264. (.353)
10 ppm	DDT through egg-laying	68	1.70–88.7 (12.0)	ND–4.50 (.990)	4.17–121. (22.2)	5.55–212. (34.9)
100 ppm	DDT through egg-laying	63	24.6–205. (73.7)	trace–28.2 (9.25)	71.3–407. (178.)	106.–590. (242.)
	DDT up to & during latter half of egg laying	11	5.10–149. (75.8)	.041–26.0 (4.59)	8.6–255. (131.)	11.0–412. (216.)
	DDT latter half of egg-laying only	8	ND–20.1 (8.10)	ND–10.6 (5.60)	ND–89.5 (47.5)	ND–120. (69.9)
250 ppm	DDT through egg-laying	3	47.2–158. (110.)	28.2–39.2 (34.3)	260.–309. (281.)	355.–506. (418.)
	DDT up to & during latter half of egg-laying	11	.057–.442 (34.2)	trace†–.067 (6.90)	32.0–171. (92.9)	.262–.444 (124.)
500 ppm	DDT up to egg-laying	10	4.50–49.2 (25.6)	trace–39.6 (17.5)	63.0–258. (203.)	83.1–408. (260)

*Denotes average.
†Less than .001 ppm.
‡Less than .0009 ppm.

(2, 5, 9, 12). However, current knowledge of the relationship of residue values and pesticide intoxication has limited the use of brain residue data for this purpose. In wildlife monitoring programs, knowledge of the levels that represent the difference between background contamination and impending debility or mortality would help in assigning significance to residue information. If brain residue levels were a reliable diagnostic technique, they would be valuable in establishing the cause of death in losses of wildlife.

An attempt was made to assess the use of brain residue levels for this purpose by studying the relationships among DDT intake, the fate of test birds and the residues in the brain.

TABLE 17–XI

ABSORPTION OF DDT FROM PELLETED FEED WITH 250 ppm DDT BY INDIVIDUAL PHEASANTS

Pheasant #	Collection period	Feed consumed (grams)	DDT in feed (μg)*	DDE in feed (μg)†	Dry wt. fecal sample (grams)	DDT in feces‡ (ppm)	DDT in feces‡ (μg)	DDE in feces (ppm)	DDE in feces (μg)	Per cent absorbed DDT‡	Per cent absorbed DDE§	Per cent absorbed Total
6255	5/26–6/1/65	507	122,400	4,570	180.8	15.30	2,769	3.91	707	97.8	84.5	97.3
	6/2–6/6/65	402	97,400	3,630	159.6	11.74	1,874	7.50	1,197	98.1	67.1	97.0
6256	5/26–6/1/65	441	106,000	3,960	155.5	15.90	2,477	4.25	661	97.8	83.3	97.2
	6/2–6/6/65	344	83,000	3,100	114.1	6.23	711	3.1	354	99.2	88.6	98.8
6257	5/26–6/1/65	488	117,600	4,390	169.0	20.20	3,421	4.81	966	97.2	78.0	96.4
	6/2–6/6/65	367	88,700	3,310	146.7	17.10	2,509	5.2	763	97.3	77.0	96.5
Average		—	103,000	3,880	—	14.4	2,293	4.8	775	97.9	79.7	97.2

*Calculated amount of op and pp′DDT in feed consumed.
†Calculated amount of pp′DDE in feed consumed.
‡Total op and pp′DDT.
§pp′DDE.

In birds that died after comparable insecticide exposure, higher residues usually occurred in brains from males than in those from females. The mortality rate was also greater in males than in females on similar dietary levels of DDT. It was not established if the insecticide was absorbed in brain tissue more readily in males than in females. However, it is suspected that stresses occurring during the breeding season may have affected males making them more susceptible than females to the toxic effects of DDT.

The range in residue levels in brain tissue was less than in most other tissues taken from birds with similar exposure to DDT. Brain tissue took up DDT less readily than other tissues that contain appreciable amounts of fat. This was demonstrated by the comparatively low pesticide levels in the brain tissue of all test birds, and by a male pheasant given 10 ppm DDT diets continuously for three and a half years. This bird, killed by the other pheasants, had only 4.04 ppm DDT in his brain.

Among birds that died average brain DDT residues were less than the lower limit of the 95 per cent confidence level (mean brain residues of birds succumbing to DDT intoxication) for (a) males and females with short or long exposures to 10 ppm diets, (b) males (short exposure) on 250 ppm diets and (c) males and females with short exposures to 500 ppm diets. The lower limit of the 95 per cent confidence level for the mean was exceeded by the average brain residues from male and female pheasants with long exposure to 100 ppm diets, and by female birds with extensive exposure to 250 ppm diets. Brain residue values appear to be better indicators of death from chronic than acute poisoning from DDT.

Our data suggests that the whole brain residue values have limited application in diagnosing the cause of death in pheasants exposed to DDT. The results showed that residue values must be in the upper range of levels found in brains to be a reliable indicator of mortality. That birds on all dietary levels, but primarily on 10 ppm diets, died with symptoms of DDT intoxication but with brain residue levels considerably below those of many sacrificed pheasants supports this credence. These accounts, however, do not in any way imply that measurement of residue levels or changes in brain chemistry may not be applicable to diagnosing intoxication from this insecticide. We refer only to the measurement of total residue in whole brain samples as being limited for this purpose. There may be enzymatic changes in brain tissue caused by intake of DDT into the brain that could be used to measure the influence of the insecticide on vital body functions.

The measurement of residue levels in individual brain components rather than in the whole brain may be more valuable for diagnostic purposes. In white rat brains, the uptake of DDT is different in the white and gray matter and in the myelin layer (11). We had planned to analyze various components of pheasant brain but were not able to separate the brain fractions. Unfortunately, white and gray matter do not occur in distinguishable, mechanically separable layers in pheasant brains.

Additional study of the significance of isomers and degradation products of DDT in the brains of birds exposed to this insecticide may lead to the use of these residues in diagnosing debilitating effects. Current studies in which we are feeding components of the DDT complex to pheasants individually and collectively may provide practical information on this topic.

The residue levels reported in whole blood of pheasants are considerably higher than those reported for large animals (10). This may be due primarily to differences in proportions between the amount of pesticides ingested to the total blood volume of the test animals.

The data on blood residues from female pheasants indicated the major role of this tissue in carrying the insecticide to the developing ovum. The capacity of the blood to transport insecticides was aptly demonstrated by the large amount of DDT deposited in eggs.

Our data suggest that whole blood residue levels in pheasants are influenced primarily by four factors: the intake of insecticide, the turnover rate or mobilization of fat, the amount of residue in fat depots, and the insecticide excretion rate. Whole blood residue levels can be expected to be highest in the following instances:

1. when samples are taken within 24 hours of DDT ingestion;
2. when fat depots of the donor are depleted;

3. when the bird is under breeding stresses or reduced feed intake, and

4. for an indeterminate period after egg-laying when the molting female can no longer void DDT in her eggs.

In our trials self-contained egg yolk was the only source of insecticide exposure for chicks as they were fed noncontaminated diets exclusively. Although the amounts of insecticide laid down in yolks were undoubtedly an important cause of chick mortality other factors including stresses on the parents were involved. Chicks survived in instances where yolks were highly contaminated with DDT; chicks also died with symptoms of DDT poisoning, when yolks contained low residues. Our assessment of residue levels in the yolk in chicks was made on the basis of residue values in other eggs laid by the parents. We believe that the average residues in eggs from the same individuals or group of females provided a reasonable approximation of insecticide contamination of yolks in chicks.

The mortality level of chicks from parents on 10 ppm diets was similar to the levels of the control birds. However, in 1965 most of the adult pheasants partially rejected the 10 ppm diets, thereby drastically reducing total food intake. The stressing of these adults resulted in increased chick mortality comparable to the mortality of chicks from parents on 100 ppm DDT diets. In one 10 ppm group where food consumption was normal, the rate of chick mortality did not increase over that of the controls. This occurred even though the residues in eggs from the unstressed females (average 100 ppm) were higher than those from the stressed females (average 54.2 ppm).

The birds on higher dietary levels of DDT were not stressed by innutrition. Levels of chick mortality were six times as great in offspring from parents fed 100 ppm DDT diets as in offspring of controls. In chicks from parents on 250 ppm DDT diets the mortality rate was about 60 per cent in some cases. Adult pheasants could not be maintained for more than a few weeks on 500 ppm DDT diets. Even though the adults received this high DDT dietary level for short periods of time, the mortality rate of their chicks was about 80 per cent on some occasions.

Rice eaten by pheasants at Richvale contained 6,300 to 8,400 ppm DDT and residue levels in excess of 1,000 ppm occurred in some eggs from these birds. Of the chicks hatched from eggs that contained these high levels of insecticide about 80 per cent died or were crippled. However, there were always some chicks that survived regardless of the DDT exposure of the parents.

All factors that caused the differential mortality in chicks from parents with the same exposure to DDT were not established. In addition to parental stresses, factors that may have contributed to the differential mortality are: differential susceptibility to the toxic effects of DDT, and varying amounts of DDT present in the yolks of eggs laid by the same hen. There was no indication that exposure to DDT altered physical properties of the egg shell or increased egg-eating by adult birds. In some wildlife species the production of inferior eggshells and increased egg breakage has been reported as an indirect effect of exposure to persistent pesticides (7).

Results of our studies supported the findings of other workers that residue levels in adipose tissue are influenced by the history of exposure to DDT, the body condition of the birds sampled and the condition of fat deposits. High residue levels in adipose tissue usually were the result of (a) a lengthy chronic exposure to sublethal amounts of DDT and (b) a decreasing percentage of lipids in fat storage sites.

The mobility of insecticides in lipids was substantiated in several trials where DDT was fed intermittently. These situations involved extensive chronic exposure to DDT up to the start of, but not during, egg-laying. Some of the insecticide stored in fat depots was removed and deposited in eggs throughout egg-laying without significant reduction in the weight of the birds. Thus, insecticide transport is associated with constant turnover of lipids in fat depots.

Residue analysis of various tissues provided an indication of the comparative value of each tissue in determining body burden of the insecticide. Residue results from selected samples of either body cavity or subcutaneous fat did not vary appreciably. This was particularly true when composited samples from several birds were used.

Analysis of breast muscle tissue provided results that were much lower than those of fat tissues from the same bird when residue levels are expressed on a wet-weight basis. When values were recapitulated on 100 per cent lipid basis the levels in muscle tissue were considerably higher than those in fat. However, there was closer agreement in residue values when recapitulated on a 100 per cent lipid basis than when compared on a wet-weight basis. Residue results expressed on a 100 per cent lipid basis provide the best common denominator for comparing residue values when different tissues and different animals are being analyzed.

Those reporting residue data may be reluctant to express results on a 100 per cent lipid basis for fear of being accused of exaggerating true residue values. It would be helpful nevertheless, to include data on the fat content of samples, when appropriate, with results of residue analysis so that the reader may interpret the data on either a whole sample or 100 per cent lipid basis.

Residue data from liver samples did not reflect the body burden of insecticide in pheasant collected in the field. The wide variations found in residues in fat and liver from the same birds suggest that fat tissue analysis reflects the history of chronic exposure while residues in liver tend to reflect the more recent exposure to DDT. However, residue values of the two tissues were more comparable when results were expressed on a 100 per cent lipid basis.

The analysis of fecal samples indicated that the DDT was readily absorbed from the pelleted diets used in our studies. The levels of DDT absorbed from these studies was in excess of 95 per cent while absorption rates for other carriers such as corn oil did not exceed 25 per cent (5).

REFERENCES

1. AZEVEDO, J.A., JR.; HUNT, E.G., and WOODS, L.A., JR.: *Calif Fish Game,* *51* (4) :276, 1965.
2. BARKER, R.J.: *J Wildl Mgmt, 22* (3) :269, 1958.
3. DE FAUBERT MAUNDER, M.J., *et al.: Analyst, 89* (1056) :168, 1964.

4. GEORGE, J.C., and BERGER, A.J.: Avian myology. In *The Pectoralis: Bio-chemical Aspects.* New York, Academic, 1966, chapt. 4.
5. HARVEY, J.M.: *Canad J Zool, 45:*629, 1967.
6. HUNT, E.G.: *J Appl Ecol, 3* (suppl.) :113, 1966.
7. STICKEL, L.F., and STICKEL, W.H.: *Science, 151:*1549, 1966.
8. RATCLIFFE, D.A.: *Nature, 215* (5097) :208, 1967.
9. SALWIN, H.; BLOCK, I.K., and MITCHELL, J.H., JR.: *J Agr Food Chem, 3* (7) :588, 1955.
10. WITT, J.M., *et al.: Bull Environ Contam Toxic, 1* (5) :187, 1966.
11. WOOLLEY, D.E., and RUNNELLS, A.L.: *J Toxic Appl Pharm, 11:*389, 1967.
12. WURSTER, C.F., JR.; WURSTER, D.H., and STRICKLAND, W.N.: *Science, 148* (3666) :90, 1965.

DISCUSSION

RUDD: It is a temptation to say that DDT residues are very high, and that ring-necked pheasants in the central valleys of California are doing very well. Therefore, why should we concern ourselves with the occurrence of residues at very much lower levels in other species? What is wrong with this logic?

HUNT: I think you are trying to manuever me into a position where I would say that pheasants are not representatives of other bird species, and I think this is probably true. But by using a pheasant that occurs normally in the wild, we have a tool for studying events in nature. I would not say that such levels must be reached in other birds before they develop problems.

RISEBROUGH: Eggs of the California Murre now have about 60 ppm of total chlorinated hydrocarbons in their yolk. Were the pheasant chicks with the higher mortality with their parents?

HUNT: For parents on the low-level diets (10 ppm DDT diet), the levels in eggs were either 60 or 100 ppm, depending on the length of the diet. As long as the parents were not stressed, the mortalities of the chicks were the same as those of the controls. As soon as the female was stressed, owing to the fact that she would not eat enough food to maintain her body condition, the chick mortality increased sixfold, with no increase in the level of DDT in the egg yolk. It is therefore not wise to assume that there is a certain DDT level in eggs, below which mortality would not occur, because various stresses that cannot be measured and that occur in birds in the wild could induce this kind of mortality.

RISEBROUGH: Were these eggs incubated away from the hens?

HUNT: Yes; these penned birds are somewhat different from those found in the wild in that they produce about sixty eggs per hen per year. We collected eggs every day. Hens in the wild lay a clutch of up to twelve eggs. Some pheasant hens lay such a clutch two or three times a year, a fairly large number of eggs compared with most birds. This is correlated with careless nesting habits: they use dump nests.

WELCH: It is not surprising to me that there is a lack of correlation between the DDE residue levels in fat and the amount of exposure of the birds to DDT. Street and Blau have shown [*Toxic Appl Pharmacol, 8:*497, 1966] that feeding of one chlorinated insecticide will stimulate the metabolism of another and actually enhance its elimination from the fat depots. DDT can also stimulate detoxifying enzymes in the liver. It would be interesting to know how rapidly the liver can eliminate DDT from the body after chronic feeding of this insecticide. It is difficult at present to correlate mortality data with the residue levels of DDE in the fat. Does the black pigmentation you observed in pheasants fed DDT suggest an endocrine disturbance?

HUNT: I do not really know. We have an indication of metabolic change in stored DDT. Birds fed o,p'-DDT have residues consisting primarily of p,p'-DDT.

HERMAN: I should like to emphasize that too few studies have considered the relationship of per cent of lipids and ppm chlorinated hydrocarbons in tissue. We find a very close correlation in grebes between the per cent of extractable lipids and wet-weight concentrations of DDD in those tissues.

HUNT: In the analysis of a complete unit such as the brain or an egg yolk it is helpful to express residues in terms of total micrograms, which excludes the need for interpreting the amount of fat.

CLARKSON: Is the lipid content of plasma in birds higher than in mammals? Does not the plasma in laying hens, particularly at egg-laying time, have a high lipid content? If the blood of a laying hen is centrifuged, a layer of lipoprotein is seen, which, I think, goes into the egg. This might contain the extra DDT that one does not see in mammals.

Chapter 18

CHRONIC COLD EXPOSURE AND DDT TOXICITY

A. S. W. DEFREITAS, J. S. HART and H. V. MORLEY

ABSTRACT

Normal development of nonshivering heat producing mechanisms in cold (6°C) acclimated rats as measured by food consumption, growth rate and electromyographic activity is unaffected by up to 300 ppm DDT in the diet. Under conditions of extreme cold stress, sufficient to elicit maximal oxygen consumption, there is also no significant DDT effect. However, under less extreme stress conditions causing a slower but more exhaustive depletion of the animal's energy reserves, these dietary levels of DDT cause a significantly deleterious effect on survival.

REPORT

Introduction

CONTINUING WIDESPREAD AGRICULTURAL use of persistant organo-chlorine insecticides, such as DDT, has resulted in total biospheric contamination by these slowly degradable toxic compounds. The magnitude of this type of environmental contamination is demonstrated by the fact that over one hundred million pounds of DDT is used every year in the United States (3).

Considerable information is becoming available on the concentration of DDT usually found in animal and bird tissues (2) and the potentially lethal effect of DDT when present at high concentrations (3). However, there is little available information on the significance of the sublethal levels of DDT usually observed (0.1 to 10 ppm), either in terms of a direct or indirect effect on "survival potential" of the animal.

Since the *in vivo* administration of sublethal amounts of DDT alters the level of activity of a number of microsomal and soluble enzymes (4) and increases the metabolism of D-glucose by way of the pentose phosphate pathway (5), it seemed probable that metabolic changes induced by administering sublethal levels of

DDT to rats, could have measurable effects on their ability to adapt to a normal environmental stress such as cold (1).

The following report describes the results of a series of experiments to determine the effect of various dietary levels of p,p-DDT on the ability of rats to adapt to a cold (6°C) environment.

TABLE 18–I
COMPOSITION OF RAT DIET
(PELLET DIAMETER 3'/16)

Ingredients	(g/kg)
Wheat, Whole, Ground	455.00
Alfalfa Meal	50.00
Corn, Whole, Yellow, Ground	200.00
Soya Bean Meal (44% Protein)	100.00
Skim Milk Powder	100.00
Brewers Yeast	30.00
Mazola Oil	50.00
Iodized Salt (NaCI)	5.00
Calcium Phosphate Dibasic (CaHPO₄ 2H₂0)	10.00
Vitamin A, 12,500 Units/kg., Vitamin D₂, 4,140 Units/kg., Vitamin E Acetate, 5.5 Units/kg.	

Results and Discussion

Male Sprague-Dawley rats with an average body weight of 234g were divided into two groups, one group was placed in a thermal neutral environment at +28°C and the other group, in a cold environment at +6°C. Each experimental group of animals was subdivided into four separate dietary groups of twelve rats each, individually caged and fed ad libitum on pelleted diets (Table 18-I) containing 0, 50, 150, and 300 ppm DDT respectively for a period of 5 weeks. The presence of these DDT levels in the diet had no significant effect on either food consumption (19 g/rat/day at 28°C, 30 g/rat/day at 6°C) or on growth rate (4.4 g/rat/day at 28°C, 2.6 g/rat/day at 6°C) when compared with control animals on diets containing 0 ppm DDT (Table 18-II).

The presence of up to 300 ppm DDT in the diet did not prevent the normal development of nonshivering heat producing mechanisms in cold acclimated rats as measured by electromyographic activity at 0 to 2°C. Nor did these dietary amounts of DDT (Table 18-III) have any significant effect on the shivering response of warm-acclimated animals (Table 18-IV). Measurement of maximal oxygen consumption rates on acute cold exposure (−45°C for warm acclimated and −52°C for cold acclimated animals) was obtained from rats at each dietary level of DDT (Table 18-IV). The rate of maximal oxygen consumption for cold acclimated animals on 0 ppm DDT diet was 1.4 ml O_2/body wt.$^{0.5}$, approximately 35 per cent greater than the warm-acclimated controls. Feeding diets containing DDT during cold acclimation did not prevent this normal increase in maximal oxygen consumption. Nor did these dietary levels of DDT have any significant effect on the maximal O_2 consumption of warm-acclimated animals.

TABLE 18–III
FOOD CONSUMPTION AND GROWTH OF RATS*
LIVING IN WARM AND COLD ENVIRONMENTS†

Dietary	Acclimation Temperature			
Level of p, p–DDT µg DDT/g Diet	28°C Food Consumption g/day	6°C g/day	28°C Growth g/day	6°C g/day
J	19.0	31.2	4.4	2.8
50	20.1	30.9	4.5	2.9
150	18.6	28.7	4.5	2.4
300	20.0	31.5	4.1	2.6

*At start of acclimation, rats had an average body weight of 243 g (range 230 g–263 g).
†Data represent average values obtained from twelve rats in each case. During the 5-week experimental period, there was no statistically significant effect of dietary level of DDT at either acclimation temperature on growth or food consumption.

TABLE 18–III
DDT CONSUMPTION* OF RATS† LIVING
IN WARM AND COLD ENVIRONMENTS

Dietary	Acclimation Temperature			
Level of p, p–DDT µg DDT/g Diet	28°C Rate Consumed mg/rat/day	6°C mg/rat/day	28°C Total Consumed mg/rat	6°C mg/rat
0	—	—	—	—
50	1.01	1.41	35.26	52.2
150	2.81	3.88	98.18	144.6
300	5.99	8.61	209.48	319.1

*Total DDT consumed during a 35-day period.
†Data represent average values obtained from twelve rats in each case.

TABLE 18–IV
MAXIMAL O₂ CONSUMPTION AND MYOGRAPHIC
ACTIVITY OF WARM– AND COLD–ACCLIMATED RATS

Dietary *Level of*	*Acclimation Temperature*			
	28°C	6°C	28°C	6°C
p, p–DDT *µg/g Diet*	*Myographic Activity at O–2°C* mille volt x 10^3/rat		*Maximal O₂ Consumption†* ml O₂/Body Wt. ·⁵	
0	137 ± 9 (5) *	35 ± 3	1.06 ± .06 (7)	1.41 ± .05
50	138 ± 10 (4)	33 ± 3	—	1.32 ± .05
150	124 ± 23 (4)	36 ± 3	—	1.39 ± .08
300	172 ± 25	35 ± 2	1.08 ± .05	1.37 ± .07

*Figures in parenthesis refer to number of rats. All other observations were averaged over six rats in each case, and are reported with standard errors.
†Maximum O₂ consumption measured at –45°C in warm– acclimated and –52°C in cold–acclimated rats.

The effect of these nontoxic dietary levels of DDT on survival of cold-acclimated rats exposed to varying degrees of cold stress (Table 18-V) demonstrated that the presence of 300 µg DDT/g diet had no significant effect on survival time of rats exposed to an ambient temperature of either –36°C (survival time, 5 to 6 hours) or –28°C (survival time, 9 to 10 hours). Survival tests under less severe ambient temperatures of +6°C and +22°C demonstrated that rats fed the diet containing DDT had significantly shorter survival time than control animals on a diet containing no DDT. However, under none of these test conditions was there any significant difference between the weight loss experienced by rats which were fed the diet containing DDT and that of control animals fed a diet containing no DDT. Typical DDT-induced tremors were observed only in the group of animals fed DDT and exposed to a survival test temperature of +6°C. This observation tends to support the hypothesis that the rate of release of DDT from adipose tissue is directly related to (a) the concentration of DDT in adipose tissue, and (b) the rate of lipid mobilization from adipose tissue.

Analyses to determine the total amount of DDT and its major metabolites in the rat and their distribution in various tissues indicate that warm acclimated rats retain in their tissues a greater proportion (8.0%) of total ingested DDT than cold-acclimated rats (which retain only 4.0%). The total amount of DDT in eviscerated carcass tissues (excluding liver and brain tissue) of warm- and cold-acclimated animals was approximately 15.5 and 13.5 mg DDT respectively, corresponding to DDT concentrations of 560 and 880 μg DDT/g fat in warm- and cold-acclimated animals. Brain levels of DDT were estimated at below 0.5 μg DDT/g fat in both warm and cold-acclimated groups. The ratio of DDT to its metabolic breakdown product, DDE was approximately 10 in carcass tissue from both warm- and cold-acclimated animals.

TABLE 18–V

SURVIVAL* OF COLD (6°C) ACCLIMATED RATS EXPOSED TO VARYING DEGREES OF COLD STRESS

Dietary Of Level	Cold Stress Test Temperature			
	−36°C	−28°C	+6°C	+20°C
p, p–DDT μg/g	Survival Time (hours)	Survival Time (hours)	Survival Time (hours)	Survival Time (hours)
0	5.8 (25) p = .4‡	10.3 (41) p = .10	75.6 (110) p = .01	178.7 (131) p = .01
300	5.1 (25)	9.9 (42)	59.6† (97)	146.2 (133)

*During the survival test, the rat had access to water but not to food. Survival times represent antilog₁₀ of the average log₁₀ survival time. Values of "p" represent the probability of having "t" test of significance between sample means this large or larger by chance.
†At this test temperature (6°C) all rats previously fed a diet containing 300 μg DDT/g Diet exhibited typical DDT-induced tremors before death.
‡Figures in parenthesis refer to the weight loss in grams.

The results presented in this report indicate that relatively large (300 ppm) amounts of DDT in the rat's diet appear to have essentially no effect on its ability to adapt to a cold environment. Under conditions of extreme cold stress sufficient to elicit maximal oxygen consumption there is also no significant DDT effect. However, under less extreme stress conditions, conditions that cause a less rapid but more exhaustive depletion of an animal's energy reserves, there is a significant deleterious effect of these levels of DDT on survival.

REFERENCES

1. DEPOCAS, F.: *Brit Med Bull, 17* (1) :25, 1961.
2. DUSTMAN, E.H., and STICKEL, L.F.: Pesticide residues in the ecosystem. In *Pesticides and Their Effects on Soils and Water.* American Society of Agronomy Special Publication No. 8, pp. 109-121.
3. HAYES, W.J.: Pharmacology and toxicology of DDT. In Muller, P. (Ed.) : *DDT the Insecticide Dichlorodiphenyltrichloroethane and its Significance.* Basel, Switzerland, Birkhauser AG, 1959, Vol. II, chapt. VI.
4. SANCHEZ, E.: *Canad J Biochem, 45:*1809, 1967.
5. SIEKEVITZ, P.: *J Biol Chem, 195:*549, 1952.

DISCUSSION

HODGE: Were the DDT-fed animals at plus 6°C and plus 20°C actually losing weight faster than controls fed no DDT?

DeFREITAS: Yes. The higher rate of weight loss may have resulted from increased activity observed with the DDT fed animals.

BERG: Rats at minus 39°C and at minus 28°C lose weight at about 4 g per hour, but at a stress temperature of plus 6°C the weight loss is only about 1.75 g per hour.

DeFREITAS: During the initial stages of the cold stress, energy expenditure and hence rate of weight loss increases with decreasing temperature. However, at extreme cold test conditions such as minus 36°C, rats go into a moribund state for an hour or so before actual death; during this last hour their O_2 consumption is greatly reduced, and consequently utilization of body stores drops to a minimum.

SPENCER: Were these cold stress survival experiments carried out under conditions of complete or partial starvation?

DeFREITAS: Complete starvation; only water was available.

DEICHMANN: Why didn't the animals live longer than 178 hours at 20°C?

DeFREITAS: An ambient temperature of plus 20°C imposes a cold stress which results in an elevated energy expenditure by the rat in order to maintain homeothermy. If rats were starved at 28°C they would survive for approximately 48 to 72 hours longer than rats at 20°C.

DEICHMANN: Wasn't it starvation that killed the control animals after 178 hours?

DEFREITAS: Yes. However, the important point here is that the corresponding group of test animals fed DDT had a significantly shorter survival time.

WELCH: These results obviously indicate a stress. Several investigators have shown that *o,p'*-DDD stimulated degradation of hydrocortisone by the liver. [Bledsae, T., *et al.*: *J Clin Encocrinol, 24*:1303, 1964; Kupfer, D., and Peefs, L.: *Biochem Pharmacol, 15*: 573, 1966.] Others have shown that DDT also stimulates the activity of these enzymes in the liver. Therefore, one would expect a greater turnover of hydrocortisone in animals treated with DDT. A deficiency of hydrocortisone in the treated animals may partially explain the results of survival tests. It would be interesting to determine the activity of the enzymes and the breakdown of hydrocortisone in the liver of these animals.

SPENCER: How long were the rats fed the 300 ppm DDT diet before the start of the starvation period, and what increase above normal consumption of food did you observe?

DEFREITAS: Rats were fed the various test diets for five weeks which is well in excess of the normal time it takes to achieve cold adaptation. The food consumption increased from about 20 to 30 grams per day between the warm and cold acclimated groups respectively, on a pelleted diet containing 5% corn oil.

BERG: Is a chronic dose of 300 mg DDT/gram diet considered toxic?

DEFREITAS: No. We deliberately picked nontoxic dietary levels of DDT.

BERG: It is very significant that you found physiological impairment with this dose.

FASSETT: What was the age of the rats at the start of DDT feeding?

DEFREITAS: The rats used in this study were eight weeks old male Sprague Dawley animals with an average body weight of 230 g.

Chapter 19

CHLORINATED HYDROCARBON INSECTICIDES AND AVIAN REPRODUCTION: HOW ARE THEY RELATED?

C. F. WURSTER, JR.

ABSTRACT

The stability, solubility characteristics, and broad toxicity of certain chlorinated hydrocarbon insecticides are responsible for worldwide contamination of food chains associated with a potential for biological effects on many kinds of organisms. The compounds increase in concentration with increasing trophic level, generally reaching highest concentrations in carnivores, including carnivorous birds. On at least two continents, a pattern of diminished reproductive success of carnivorous birds coincides with the presence of chlorinated hydrocarbon residues, particularly those of DDT. Included are such birds as grebes, woodcock, various birds of prey, seabirds, and possibly others that have not been studied, although exceptions to these correlations exist. Additional information comes from various laboratory experiments and field observations including feeding experiments, egg injections, bird behavior, chick mortality, eggshell weight measurements, and enzyme induction. The data suggest that these chemicals interfere with calcium metabolism and hormonal mechanisms. Two mechanisms involving calcium metabolism are proposed to explain these observations:

1. Acute toxicity involves direct attack on the nerve axon with associated axon decalcification and symptoms of calcium deficiency. This mechanism can kill adult birds or chicks shortly after hatching.
2. Chlorinated hydrocarbons cause induction of hepatic enzymes that metabolize steroid hormones; estrogen breakdown is especially important, since estrogen influences calcium metabolism in birds. This would explain the widespread occurrence of calcium-deficient eggshells among carnivorous birds since introduction of DDT in the late 1940's. Current knowledge indicates, despite several anomalies, that certain chlorinated hydrocarbon insecticides are a probable cause of the widespread diminished reproduction of carnivorous birds.

REPORT

Introduction

THE CHLORINATED HYDROCARBON insecticides, including DDT, DDD, dieldrin, aldrin, endrin, heptachlor, and several others, as well as some of their metabolites, have three characteristics in common that make them potentially of unusual ecological importance. They are biologically very active and broadly toxic to a great variety of organisms, including vertebrates, rather than having an action restricted to the class Insecta. Chemically they are very stable materials, often persisting within the environment for years in their original, or only slightly modified, state. Finally, their nonpolar character gives them a very low solubility in water combined with a high lipid solubility. These solubility characteristics mean that the compounds consistently partition out of aqueous, inorganic media, and into lipid-containing organic and biological material, including living organisms. Being transported in various ways, these chemicals therefore may contaminate nontarget organisms that are remote from the site of insecticide application. Because of their stability, the original broad toxicity is retained, and exposure may therefore be associated with biological effects in these nontarget organisms. Without any one of these three characteristics, the ecological importance of the chlorinated hydrocarbons would be substantially diminished.

Dispersal of these insecticides through the environment is facilitated by a variety of mechanisms. Obviously they can be transported within the bodies of living organisms, and in spite of low water-solubilities and vapor pressures (10, 30), great quantities of water and air can carry large amounts of insecticides. The compounds readily form suspensions in both air and water (10, 30, 93), and they absorb to particulates that also are transported by these media (30, 53). Codistillation with water further facilitates escape into the air (4, 11, 12). The results of these mobility-imparting properties are shown by the appearance of chlorinated hydrocarbon insecticides in air (2, 6, 76, 77), in precipitation (1, 91), and in soil of untreated areas (21). Currents of air and water have distributed the chemicals to the remotest parts of the world, contaminating organisms in the Antarctic (35, 83), pelagic birds (75, 98), and apparently most other animals (53, 78, 94) including man (69).

These transport mechanisms predict a net transfer of chlorinated hydrocarbons from the Earth's treated land areas to its ocean basins, where accumulation and buildup would be expected. Increasing evidence indicates that this process is occurring (75, 77, 98).

Within a food chain, these compounds usually increase in concentration with increasing trophic level, reaching the highest concentrations in carnivores at the top (43, 46, 78, 94). The carnivores, primarily among fish and birds, therefore face special problems. These may include effects on reproduction.

Among the chlorinated hydrocarbons, DDT has been by far the most widely used and extensively studied, and has become one of the most widespread synthetic chemicals within the Earth's ecosystem. This paper is therefore particularly concerned with DDT and its metabolites, although most chlorinated hydrocarbons have many similar properties and should be expected to have comparable ecological effects.

Relationships between pesticides and avian reproduction have for years involved controversy and confusion, but recent pertinent data are rapidly accumulating from both field and laboratory investigations. The data involving these relationships is critically examined, first by summarizing observations from natural populations, and then by describing results from laboratory studies; finally, by offering hypotheses that attempt to integrate and explain some of these findings.

Observations on Natural Populations

On at least two continents, Europe and North America, there appears to be a pattern of diminished reproductive success, involving certain carnivorous birds at the ends of animal food chains, that coincides with the presence of residues of chlorinated hydrocarbon insecticides, particularly DDT. Herbivores generally are not involved.

WESTERN GREBE AT CLEAR LAKE. Additions of 14, 20, and 20

parts per billion (ppb) of DDT to the waters of Clear Lake, California, were followed by increased mortality of Western grebes (*Aechmophorus accidentalis*) and reduction of the nesting colony from 1,000 pairs before treatment to fifteen pairs in 1959, two years after the final DDD treatment (46, 47). By 1961 the colony approximated fifty pairs. These survivors averaged 656 to 723 ppm DDD in lipid tissue and had complete nesting failure for several years. Further details are given elsewhere in these proceedings by Herman Garrett, and Rudd.

WOODCOCK IN NEW BRUNSWICK. A statistically significant inverse correlation existed in New Brunswick between the feeding success of American woodcock (*Philohela minor*), as measured by the ratio of immatures to adult females, and the quantity of DDT applied annually to the habitat during a 6-year period (95). Eggs of these birds averaged 1.2 ppm DDT and 1.7 ppm heptachlor epoxide (from the winter range) in 1961, and adults carried somewhat higher amounts of both. Analyses were by the Schechter-Haller method and may be somewhat low.

OSPREY IN THE EASTERN UNITED STATES. The osprey (*Pandion haliaetus*) has been gradually declining in the northeastern United States since 1900, and for the past decade this decline has greatly accelerated to 30 per cent annually without obvious explanation (5). A colony in Connecticut, with the habitat and other factors apparently unchanged, declined from two hundred pairs in 1938 to twelve pairs in 1965. Whereas a normal osprey population produces 2.2 to 2.5 young per nest, a Maryland colony containing about 3.0 ppm DDT residues in its eggs fledged 1.1 young per nest, and the Connecticut colony carrying 5.1 ppm produced 0.5 young per nest. These low rates were due mainly to poor hatching. Analyses showed that the amount of DDT residues in the fish eaten by Connecticut ospreys was five to ten times higher than in the food of the Maryland birds. In Connecticut, Ames (5) saw no correlation between nesting success and degree of isolation of the nest site, nor was there evidence of a food shortage. Indications of egg-eating or other abnormal behavior were not apparent. Ospreys in other parts of the northeastern states have shown a similar pattern.

PEREGRINE FALCON. Extensive study of the peregrine falcon (*Falco peregrinus*) in Europe, especially in Britain, reveals a widespread, rapid population decline beginning during the early 1950's (70-73). British peregrines had been persecuted during the war because of their predation on homing pigeons, but had substantially recovered in the years immediately after the war. Recovery was interrupted and reversed by a sharp decline. The pattern of decline coincided both geographically and chronologically with the pattern of use of chlorinated hydrocarbon insecticides, and was characterized by breakage and devouring of eggs by parent birds, nest-abandonment, and other abnormal breeding behavior. Residues in eighteen eggs collected between 1963 and 1966 from unsuccessful nests averaged 17.4 ppm, while nine eggs from successful eyries averaged 12.7 ppm. Ratcliffe (73) considered the differences suggestive, but they were not statistically significant. Residues consisted mainly of p,p'-DDE with lesser amounts of dieldrin and heptachlor epoxide. Preliminary data for 1967 indicated a slight recover in breeding success in parts of Britain coinciding with a decrease in residue levels in eggs (73).

During the last fifteen years the peregrine has become extinct as a breeding bird in the eastern United States (42, 44). No residue analyses are available for these birds, but their decline was characterized by abnormal nesting behavior, including egg abandonment and breakage, and late nesting.

GOLDEN EAGLE IN SCOTLAND. Nesting success of golden eagles (*Aquilla chrysaëtos*) in western Scotland has declined in recent years, especially since 1960 (59, 72). Comparison of the periods 1937 to 1960 with 1961 to 1963 shows that the proportion of non-breeding eagles increased from 3 per cent to 41 per cent, nests with broken eggs increased from 14 per cent to 36 per cent, and pairs of eagles rearing young declined from 72 per cent to 29 per cent. Human disturbance, climatic change, and changes in food supply do not offer adequate explanation for this breeding failure. Abnormal nesting behavior characteristic of the peregrines also occurred with the eagles. Chlorinated hydrocarbon residues averaged 2.5 ppm in eggs, consisting mainly of dieldrin and DDE; sheep dipped in dieldrin later were eaten as carrion by the eagles, and the pattern of low breeding success correlated geographically and chronologically with the use of dieldrin in sheep dips.

BALD EAGLE. There have been numerous indications of declining reproductive success and population numbers of bald eagles (*Haliaeetus leucocephalus*) in the United States, with the proportion of immatures among the population gradually declining (3, 14, 24, 29, 84, 85, 87). Residues of DDT and dieldrin were found in all eagles analyzed, and nine eggs averaged 10.6 ppm (87). Data for the bald eagle are generally scattered and confusing, but declining reproductive success and chlorinated hydrocarbon residues clearly coincide.

SPARROW HAWK AND OTHER EUROPEAN FALCONIFORMES. In British sparrow hawks (*Accipiter nisus*), a decline in population and increased egg breakage and egg-eating since the mid-1950's correlated geographically and chronologically with the use of chlorinated hydrocarbons, and several analyses of eggs and tissues demonstrated exposure of the birds, particularly to DDE and dieldrin (7, 24, 25, 59). Comparable circumstances seemed apparent for the kestrel (*Falco tinnunculus*), barn owl (*Tyto alba*), buzzard (*Buteo buteo*), merlin (*Falco columbarius*), and hen harrier (*Circus cyaneus*), but adequate data are not available on these species (25, 66, 72).

SANDWICH TERN IN THE NETHERLANDS. During the past decade the sandwich tern (*Sterna sandvicensis*) population in The Netherlands has declined from 30,000 to less than 4,000 pairs (56). In 1964 and 1965 abnormal numbers of chicks died during the first week after hatching, exhibiting convulsions similar to the symptoms produced in laboratory chicks experimentally poisoned with dieldrin (55) ; residues in livers of both laboratory and field specimens were comparable. Dieldrin, telodrin, endrin, and DDE were present, and residues in tern eggs averaged 1 to 2 ppm. The residue concentrations seem low, but telodrin is extremely toxic. Lower mortality and no convulsions were seen in 1966, which coincided with reduced residues in eggs, and in fish on which sandwich terns feed. Lower residues in 1966 may have resulted from the elimination of a local source of cyclodiene contamination.

BERMUDA PETREL. Reproductive success of the Bermuda petrel

(*Pterodroma cahow*), a rare oceanic bird of the North Atlantic, shows a statistically significant decline averaging 3.25 per cent annually for the last 10 years (98). A linear extrapolation of the negative regression goes to zero reproduction in 1978, suggesting extinction of this species unless the cause of the decline is removed.

HERRING GULLS ON LAKE MICHIGAN. Herring gulls (*Larus argentatus*) on Lake Michigan have become heavily contaminated with residues of DDT; living adults averaged 99 ppm in breast muscle, 21 ppm in brain, and 2441 ppm in body fat (43). Abnormal behavior, including restlessness and aggressiveness, was noted among these birds (52, 61). When seven were starved, they developed tremors and died; subsequent analyses showed lethal concentrations (more than 200 ppm) of DDT residues in their brains (62). Eggs from several colonies averaged 120 to 227 ppm. These concentrations were much higher than any mentioned in the above studies, and there was an unusual tendency for eggshells to be cracked and chipped. Reproductive success was subnormal, but one slightly less, but still heavily contaminated colony showed apparently normal success (61).

PHEASANTS IN RICHVALE, CALIFORNIA. Residues in ring-necked pheasants (*Phasianus colchicus*) exposed to DDT and dieldrin in the field averaged 14 ppm in breast muscle and 2509 ppm in fat (47). Their eggs contained 568 ppm in the yolk; when incubated artificially, hatchability was unaffected, but mortality of chicks was 46.6 per cent from the study area and 27.0 per cent in controls. The study area produced nearly twice as many crippled young as did the control area (25% vs. 12.9%).

In each of these cases, contamination with residues of chlorinated hydrocarbon insecticides has coincided with some form of reduced reproductive success in carnivorous birds (pheasants are herbivores). It seems likely that other species would show a similar coincidence, but they have not been studied. Carnivorous birds such as the brown pelican (*Pelecanus occidentalis*), marsh hawk (*Circus cyaneus*), sharp-shinned hawk (*Accipiter striatus*), Cooper's hawk (*Accipiter cooperii*), belted kingfisher (*Megaceryle alcyon*), Eastern bluebird (*Sialia sialis*), Eastern phoebe (*Sayor-*

nis phoebe) , and red-eyed vireo *(Vireo olivaceus)* , to name a few, apparently have undergone population declines in recent years, but objective data are generally unavailable and they have not been investigated with regard to a possible involvement with chlorinated hydrocarbons.

The coincidental occurrence of residues and reduced reproduction does not establish a causal relationship, although the number of these correlations increases the likelihood that such a relationship exists. These correlations are not universal, however; not all contaminated populations show declines. Contamination of peregrines in northwest Canada and Alaska is as great as in Britain, and reduced reproduction is not evident so far (20, 32) . Sample numbers were small in these studies, with each study involving about a dozen pairs; these birds will need continued scrutiny for several years before a firm conclusion can be reached regarding either their reproductive success or population trends.

In Britain, eggs of the heron *(Ardea cinerea)* and great crested grebe *(Podiceps cristatus)* contained residues of DDT and dieldrin averaging 17.8 and 10.2 ppm, respectively (63) , but the heron population has been stable (68) and the great crested grebe has increased (67) . These data refer to population numbers rather than reproductive success, however, and the factors must be kept separate. That the habitat for the grebe has increased in Britain is a primary factor in the bird's expanding population. Herring gulls on Lake Michigan may suffer reduced reproductive success associated with insecticide contamination (52, 61) ; their population has exploded nevertheless, because increased production of human wastes has expanded their niche (60) , a factor that more than offsets the effects of insecticides. On the other hand, imposition of a declining rate of reproduction on a population structure simultaneously weakened by other factors obviously will cause a sharp decline in numbers. Clearly other ecological parameters must be considered in determining the relationship between cause and effect.

No matter how many correlations involving reduced reproduction coincidental with chlorinated hydrocarbon residues are discovered, a causal relationship will not be a certainty, even though other factors do not explain the reductions. Further evidence must come from laboratory studies and from certain qualitative observations in the field; some of these are described below.

LABORATORY FEEDING EXPERIMENTS. Sublethal amounts of chlorinated hydrocarbons, including DDT and dieldrin, reduced reproductive success when fed to ring-necked pheasants, bobwhites (*Colinus virginianus*), and Japanese quail (*Coturnix coturnix*) under laboratory conditions (8, 27, 28, 34, 51). The reduction resulted primarily from mortality of chicks during the first week after hatching; hatchability of eggs was hardly affected. Concentrations that caused chick mortality were much higher than those found in all field studies except the ones involving herring gulls (52, 61) and pheasants (47).

EGG INJECTION. When chicken eggs were injected with 17 ppm dieldrin, hatchability was unaffected, but starvation of the chicks for 2 to 4 days after hatching led to a rapid rise in the concentration of dieldrin in the blood, followed by convulsions and death (55). Control chicks, starved even longer, showed no ill effects.

FISH FRY MORTALITY. Concentrations of DDT residues from 0.4 to several ppm in trout and coho salmon eggs resulted in heavy mortality of fry after hatching and absorbing the yolk sac (16, 26, 50).

CALCIUM, DDT, AND DOGS. Dogs dosed with DDT developed tremors and died, but symptoms could be avoided by prior injection of calcium gluconate; even after development of tremors the dogs recovered quickly following injection (89).

CALCIUM, DDT, AND ARTHROPOD NERVES. In the nerves of several crustaceans, the action of DDT was suppressed by an increase and intensified by a decrease in calcium ion concentration (38).

MALE-FEMALE SUSCEPTIBILITY. Males outnumbered females among birds killed by DDT in Dutch elm disease control areas, and dead females contained higher DDT residues than did males (9, 48, 96, 97). Dieldrin evidently produces a similar effect (88).

BEHAVIOR OF CALCIUM-DEFICIENT BIRDS. Calcium-deficient chickens were much more active and aggressive than controls, and increased their preference for a diet enriched with calcium carbonate, including eggshells [41, 92, M. R. Kare, personal communication]. Calcium-deficient pheasants selected calcium-bearing grit when given a choice (80). Calcium-deficient hens lay eggs with shells that are thinner than normal (79).

DDT AND MALE CHICKENS. Injection of sublethal amounts of DDT into young male chickens inhibited development of comb and wattles, and resulted in testes weighing 19 per cent as much as those of controls (17).

DELAYED OVULATION. Ovulation in Bengalese finches (*Lonchura striata*) was significantly delayed by sublethal amounts of DDT in the diet (49). Abnormally late nesting was characteristic of peregrine falcons before their extinction in the New York City region (42).

CALCIUM-DEFICIENT EGGSHELLS. A highly significant, sudden, and widespread decrease in eggshell thickness and calcium carbonate content occurred during 1946 to 1950 in British peregrine falcons, sparrowhawks, and golden eagles (74). The data show stability of shell thicknesses from 1900 to 1946, then a decrease of 7 to 25 per cent within a few years, and no recovery since then. Similar changes occurred simultaneously in North America for the peregrine, osprey, and bald eagle (45).

CALCIUM EFFECTS IN MOLLUSCS. Growth of oyster shells, which are more than 90% calcium carbonate, was decreased by concentrations of DDT in the water as low as 1 ppb (18, 19). Clams exposed to 4.4 ppb methoxychlor for 4 days showed a loss of 20 to 40 per cent of their mantle and muscle calcium (31).

HEPATIC ENZYME INDUCTION. After feeding 10 ppm DDT and/or 2 ppm dieldrin to pigeons for a week, microsomal hepatic enzymes partially converted exogenous testosterone or progesterone into polar metabolites *in vitro* (65). The two insecticides produced different metabolites, suggesting the induction of different enzyme systems. Controls showed no such steroid metabolism. Similar hepatic enzyme induction occurs in mammals (see below).

These field and laboratory observations suggest that DDT and possibly other chlorinated hydrocarbons interfere with normal calcium metabolism. It seems likely that the interference is not restricted to birds but involves other vertebrates and possibly certain invertebrates as well.

The observations may be explainable by the operation of two mechanisms, both of which, curiously enough, involve calcium. Either one or both of them could diminish reproductive success in birds. One of these, acute toxicity, appears to involve the direct attack of the insecticide on the nervous system, probably at the nerve axon; at sufficient concentrations, death results from this action. The other mechanism is more subtle, operates at much lower, sublethal concentrations, and involves hepatic enzyme induction. The former mechanism is the more obvious in that it can produce direct mortality of individuals, and occasionally, spectacular kills. Enzyme induction, on the other hand, does not kill, but involves hormone balance and control mechanisms with resultant behavioral and reproductive effects, and may be of greater ecological importance.

Acute Toxicity

The direct, toxic action of DDT on the nervous system evidently involves formation of a complex between DDT and some component of the nerve axon, which at sufficient concentrations of DDT leads to hyperactivity of the nerve, repetitive firing, tremors, and death of the organism (64). Many of these same symptoms, both in isolated nerves and whole organisms, are induced by subnormal calcium concentrations (13, 33, 92). Furthermore, it was shown in crustaceans that increased calcium ion concentration rendered the nerve resistant to DDT intoxication, while decreased calcium increased susceptibility (38). Apparently by absorption or complex formation DDT competes with calcium ions for sites on the axon surface, thereby interfering with the normal calcification of the axon and decreasing its stabilization by calcium ion. Increased calcium ion concentration therefore increases calcification of the axon (Le Chatelier Principle) and decreases its instability, whereas decreased calcium aggravates it. The mechanism would explain why dogs were protected from DDT intoxication by calcium gluconate injections (89). One might suspect, then, that organisms carrying sublethal quantities of DDT might exhibit symptoms characteristic of calcium deficiency, including behavioral abnormalities. There is evidence to indicate that the cyclodienes have similar, but not identical, mechanisms of action (64). Their mechanisms have been studied much less than that of DDT, however.

Presumably the mechanism of acute toxicity operates to kill adult birds when lethal concentrations in the brain are reached (9, 86, 96, 97). This can include large carnivores like hawks and eagles (54, 87, 90). The involvement of calcium in this mechanism would explain why bird kills usually include more males than females, with males showing greater susceptibility and females succumbing only after accumulating higher residue concentrations in their tissues (9, 48, 88, 96, 97). Females would be more resistant than males because estrogen raises their blood calcium (82) thereby bathing nervous tissues in a higher concentration of calcium and affording females a degree of protection not shared by the males.

The effects of acute toxicity have been less thoroughly studied in young birds than in adults. Mortality of chicks shortly after hatch may be an important effect in reducing reproductive success among wild populations, however. During the several days after hatch when the final stage of yolk absorption occurs, concentrations of chlorinated hydrocarbons in the blood may rise rapidly, especially when accompanied by starvation, and high chick mortality may occur (55). Presumably this mechanism killed the laboratory pheasant and bobwhite chicks even without starvation (8, 27, 28, 34, 51); lower doses produced chick mortality when starvation was imposed, and starvation of chicks is a common occurrence in nature (55). Under these circumstances, the toxin cannot be stored in lipid tissue, but instead is circulated by the blood and contacts nervous tissue, leading to acutely toxic effects.

The high mortality of sandwich tern chicks from poisoning by chlorinated hydrocarbons shortly after hatching is clearly a major factor in the decline of this species in The Netherlands (56), and demonstrates that acute toxicity from these compounds can have substantial ecological impact. The acute toxicity of chlorinated hydrocarbon insecticides to newly hatched chicks, particularly under stress of starvation, has not been adequately studied either in the laboratory or in the field, and it is clearly of ecological importance to do so. Most studies have involved galliform species, but the applicability of these laboratory studies to field populations of falconiformes is questionable.

The mortality of fish fry during the period of final yolk sac absorption probably involves direct mortality by a mechanism comparable to that of the newly-hatched chicks (16, 26).

Hepatic Enzyme Induction

The induction of hepatic enzymes by chlorinated hydrocarbon insecticides was discovered accidentally in rats after treatment of the animal room with chlordane (39). The action of several drugs, including phenobarbital, was inhibited for weeks in these animals because hepatic drug-metabolizing enzymes had been stimulated by the insecticide. The phenomenon has since been found to be much more general (22). Other chlorinated hydrocarbons, including DDT and its metabolites, dieldrin, endrin, aldrin, methoxychlor, and heptachlor have also been found to stimulate these enzymes. Furthermore, the action of the enzymes is not restricted to drugs but includes several normal body substrates such as steroid hormones. These insecticide-induced, hepatic microsomal steroid hydroxylases cause hydroxylation of androgens, estrogens, and progestational steroids, producing polar metabolites with altered biological activity. The *in vivo* effect decreases the action of such steroids as testosterone, estradiol, estrone, and progesterone (23, 58).

Enzyme induction by chlorinated hydrocarbons apparently extends to a variety of vertebrates and occurs at insecticide concentrations well below those causing obvious toxic symptoms. Thus in rats, one injection of 1 mg/kg DDT, or feeding of 5 ppm DDT or DDE in the diet, significantly increased hepatic enzyme activity (36, 40); more recently, microsomal epoxidation enzymes were induced in rats by a dietary level of 2.5 ppm DDT (37). Daily injection of chlordane (10 mg/kg) into female rats for fourteen days increased metabolism of estradiol several-fold (57), weekly injection of 25 mg/kg for three weeks decreased fertility of female mice (22), and injection of young male rats with chlordane or DDT (50 mg/kg) daily for four days markedly increased metabolism of testosterone (22). Increased enzyme activity in rat liver was associated with storage of DDT and DDE in fat at concentrations as low as 10 ppm (36, 81), a level that is frequently exceeded in many animals, including man, under current environmental conditions. The average human in the United States stores about 12 ppm of DDT and its metabolites in his fat (69). Numerous other examples of enzyme induction in various mammals, in trout (15), and in pigeons (65) by various chlorinated hydrocarbon insecticides have been demonstrated (23, 58), indicating the widespread nature of the phenomenon among vertebrates.

Enzyme Induction in Nature

p,p'-DDE, the metabolite of DDT so common in the environment, has an acute toxicity about half that of DDT to birds (88), but its major importance in nature may be as an enzyme inducer.

The induction by chlorinated hydrocarbon insecticides of hepatic enzymes that metabolize estrogens may be the critical factor affecting the reproductive performance of adult female birds in nature. In birds, estrogen mediates increased absorption of calcium from the diet, decreased excretion of calcium, and deposition of calcium in the hollow parts of the skeleton (82). Just before calcification of the eggshell, part of this medullary bone calcium is transferred via the bloodstream from the skeleton to the oviduct where, along with dietary calcium, it forms the eggshell, some of which becomes incorporated into the skeleton of the developing embryo during incubation (82). A subnormal amount of estrogen interrupts this crucial chain of events in the reproductive cycle. Calcium excretion is undiminished, adequate medullary bone is not formed, and the bird would be expected to show a variety of symptoms of calcium deficiency, including the production of calcium-deficient eggshells (79).

Estrogen metabolism induced by DDT and/or its metabolites would explain the sudden and widespread decrease in eggshell content of various carnivorous birds that coincided with introduction of DDT into the world environment during the late 1940's (45, 74). It would explain the egg-chipping and breakage that is frequently reported. Breakdown of estrogen and progesterone might also explain the various abnormal behavior characteristics —egg-eating, nest-abandonment, nervousness, and aggressiveness— that have been observed within certain populations. Interference with steroid sex hormone balance would also explain the delay in ovulation of Bengalese finches (49), and the abnormally late nesting of peregrine falcons (42).

Induction by DDT of enzymes that metabolize testosterone (65) offers a plausible explanation for the failure of young roosters treated with DDT to develop the secondary sex characteristics that are mediated by testosterone (17).

It must be noted that large differences in degree of contamination exist in birds showing reduced reproductive success. Contamination of pheasants, quail, and herring gulls was one to two orders of magnitude greater than that of most of the other carnivores (8, 27, 28, 34, 43, 47, 51, 52, 61). If susceptibilities between species were equal, this would argue against the low levels in carnivores being the cause of reduced reproduction, since comparable quantities in pheasants and quail are without apparent effect. There is little reason to expect susceptibilities between species to be equal, however, especially between galliformes and falconiformes; wide variation between species is characteristic of enzyme induction (23).

Conclusion

The multiple cases of reduced reproduction among carnivorous birds correlated with the presence of residues of chlorinated hydrocarbon insecticides suggest that these pesticides are causative agents. The dual mechanisms of acute toxicity to chicks, and hepatic enzyme induction in adults, when correlated with both laboratory and field data, add strong additional support for the existence of causal relationship. Further investigations, such as the crucial experiments on kestrels in progress at the Patuxent Wildlife Research Center, are required for a greater understanding of the relationships and mechanisms involved. Based on existing knowledge, however, the probability appears high that residues of chlorinated hydrocarbon insecticides, particularly DDT, are a major causative factor in the widespread declines in reproductive success and population numbers among carnivorous birds. Many species appear to be involved, and if world usage of these chemicals is continued at present or expanded rates, we can anticipate suppression of these populations to very low levels followed in some cases by extinction of the species.

Properties of the Chlorinated Hydrocarbon

The chlorinated hydrocarbon insecticides have a combination of properties that predict a potential for widespread effects within the biosphere. It should not be surprising that we do find such effects.

1. They are compounds with great biological activity that is not restricted to insects, but includes most of the animal kingdom and even extends to some plants.
2. They are quite mobile materials, and do not stay where we put them. They form suspensions in water and in air. Their transport around the world in the air seems to be an important route of disperasl. Transport in both air and water is facilitated by their tendency to adsorb to particulate matter. They codistil from water, which is probably an important factor allowing the compounds to escape from a body of water or from wet soil into the air. Thus, they can circle the globe in a few weeks, and can come down in the precipitation, even in remote areas.
3. Chlorinated hydrocarbons are extremely stable compounds. They stay in a toxic form long enough to contaminate and affect nontarget organisms that may be far removed in both time and space from the place of application: months or years later, and thousands of miles away.
4. They are extremely insoluble in water; DDT, for example, has one of the lowest water solubilities of any organic chemical. At the same time they are highly lipid-soluble. The partition coefficient therefore favors their uptake and retention by living organisms. They are not returned to the environment, won't get diluted in the ocean, but will be accumulated in living

organisms. Once the materials get into food chains, they are further concentrated by transfer to higher trophic levels.

In summary, if we eliminated any one of these four characteristics, we would have a compound that did less damage to nontarget organisms than the chlorinated hydrocarbons. With selective toxicity, they would not affect nontarget organisms. Locally bound, immobile compounds would reach fewer nontarget areas. Unstable compounds would not be biologically active when they got there. Higher water solubility would not favor accumulation in living organisms. The four characteristics of the chlorinated hydrocarbons mean that these materials funnel into living organisms all over the world, work their way up through food chains and remain biologically active. Their properties explain why they have become such a serious problem.

REFERENCES

1. ABBOTT, D.C.; HARRISON, R.B.; TATTON, J.O'G., and THOMSON, J.: *Nature, 208:*1317, 1965.
2. ABBOTT, D.C.; HARRISON, R.B.; TATTON, J.O'G., and THOMSON, J.: *Nature, 211:*259, 1966.
3. ABBOTT, J.M.: *Atlantic Naturalist, 22:*20, 1967.
4. ACREE, F.; BEROZA, M., and BOWMAN, M.C.: *J Agr Food Chem, 11:*278, 1963.
5. AMES, P.L.: *J Appl Ecol, 3* (suppl.) :87, 1966.
6. ANTOMMARIA, P.; CORN, M., and DEMAIO, L.: *Science, 150:*1476, 1965.
7. ASH, J.S.: *Brit Birds, 53:*285, 1960.
8. AZEVEDO, J.A.; HUNT, E.G., and WOODS, L.A.: *Calif Fish and Game, 51:* 276, 1965.
9. BERNARD, R.F.: *Mich State Univ Mus Publ Biol, Ser 2* (3) :155, 1963.
10. BOWMAN, M.C.; ACREE, F., and CORBETT, M.K.: *J Agr Food Chem, 8:*406, 1960.
11. BOWMAN, M.C.; ACREE, F.; LOFGREN, C.S., and BEROZA, M.: *Science, 146:* 1480, 1964.
12. BOWMAN, M.C.; SCHECHTER, M.S., and CARTER, R.L.: *J Agr Food Chem, 13:*360, 1965.
13. BRINK, F.: *Pharmacol Rev, 6:*243, 1954.
14. BUCKLEY, J.L., and DEWITT, J.B.: *A Florida Notebook,* Natl. Audubon Soc. Annl. Conv., Miami, Fla., Nov. 9-13, 1963.
15. BUHLER, D.R.: *Fed Proc, 25:*343, 1966.

16. BURDICK, G.E., *et al.*: *Trans Am Fish Soc, 93*:127, 1964.
17. BURLINGTON, H., and LINDEMAN, V.F.: *Proc Soc Exp Biol Med, 74*:48, 1950.
18. BUTLER, P.A.: *Amer Fish Soc, Special Publ. No. 3*:110, 1966.
19. BUTLER, P.A.: *J Appl Ecol, 3* (suppl.) :253, 1966.
20. CADE, T.J.; WHITE, C.M., and HAUGH, J.R.: *Condor, 70*:170, 1968.
21. COLE, H.; BARRY, D.; FREAR, D.E.H., and BRADFORD, A.: *Bull Environ Contam Toxicol, 2*:127, 1967.
22. CONNEY, A.H.; WELCH, R.M.; KUNTZMAN, R., and BURNS, J.J.: *Clin Pharmacol Ther, 8*:2, 1966.
23. CONNEY, A.H.: *Pharmacol Rev, 19*:317, 1967.
24. CRAMP, S.: *Brit Birds, 56*:124, 1963.
25. CRAMP, S.; CONDER, P.J., and ASH, J.S.: Fourth Rept. Brit. Trust Ornithol., Royal Soc. Protection Birds, Toxic Chemicals, March 1964. 24 pp.
26. CUERRIER, J.-P.; KEITH, J.A., and STONE, E.: *Naturaliste Can, 94*:315, 1967.
27. DEWITT, J.B.: *J Agr Food Chem, 3*:672, 1955.
28. DEWITT, J.B.: *J Agr Food Chem, 4*:863, 1956.
29. DEWITT, J.B.: *Audubon Mag, 65*:30, 1963.
30. EDWARDS, C.A.: *Residue Rev, 13*:83, 1966.
31. EISLER, R., and WEINSTEIN, M.P.: *Chesapeake Sci, 8*:253, 1967.
32. ENDERSON, J.H., and BERGER, D.D.: *Condor, 70*:149, 1968.
33. FRANKENHAEUSER, B., and HODGKIN, A.L.: *J Physiol, 137*:218, 1957.
34. GENELLY, R.E., and RUDD, R.L.: *Auk, 73*:529, 1956.
35. GEORGE, J.L., and FREAR, D.E.H.: *J Appl Ecol, 3* (suppl.) :155, 1966.
36. GERBOTH, G., and SCHWABE, U.: *Arch Exp Path Pharm, 246*:469, 1964.
37. GILLETT, J.W.: *J Agr Food Chem, 16*:295, 1968.
38. GORDON, H.T., and WELSH, J.H.: *J Cell Physiol, 31*:395, 1948.
39. HART, L.G.; SHULTICE, R.W., and FOUTS, J.R.: *Toxic Appl Pharmacol, 5*: 371, 1963.
40. HART, L.G., and FOUTZ, J.R.: *Naunyn-Schmiedeberg Arch Exp Path Pharm, 249*:486, 1965.
41. HELLWALD, H.: *Z Psychol, 123*:94, 1931.
42. HERBERT, R.A., and HERBERT, K.G.S.: *Auk, 82*:62, 1965.
43. HICKEY, J.J.; KEITH, J.A., and COON, F.B.: *J Appl Ecol, 3* (suppl.) :141, 1966.
44. HICKEY, J.J. (Ed) :*Peregrine Falcon Populations, Their Biology and Decline.* Madison, U. of Wisc., 1968.
45. HICKEY, J.J., and ANDERSON, D.W.: *Science* (in press) .
46. HUNT, E.G., and BISCHOFF, A.I.: *Calif Fish Game, 46*:91, 1960.
47. HUNT, E.G., and KEITH, J.O.: Proc. 2nd Annl. Conf. on Use Agric. Chemicals in Calif., Univ. Calif. Davis, 1962. 29 pp.

48. HUNT, L.B.: Songbirds and Insecticides in a Suburban Elm Environment. Ph. D. thesis, U. of Wisc., 1968, 146 pp.
49. JEFFERIES, D.J.: *Ibis, 109*:266, 1967.
50. JOHNSON, H.: Press release, Michigan State Univ., March 7, 1968.
51. JONES, F.J.S., and SUMMERS, D.D.B.: *Nature, 217*:1162, 1968.
52. KEITH, J.A.: *J Appl Ecol*, 3 (suppl.) : 57, 1966.
53. KEITH, J.O., and HUNT, E.G.: Trans. 31st North Amer. Wildl. Natural Resources Conf., March 14-16, 1966, p. 150.
54. KOEMAN, J.H., and VAN GENDEREN, H.: *Mendel Landbouwhogesch Opsoekingssta Staat Gent, 30* (3) :1879, 1965.
55. KOEMAN, J.H.; OUDEJANS, R.C.H.M., and HUISMAN, E.A.: *Nature, 215:* 1094, 1967.
56. KOEMAN, J.H., et al.: *Mededel Rijksfaculteit Landbouwwetenschappen Gent, 32*:841, 1967.
57. KUNTZMAN, R.; SCHNEIDMAN, K., and CONNEY, A.H.: *J Pharmacol Exp Ther, 146*:280, 1964.
58. KUPFER, D.: *Residue Rev, 19*:11, 1967.
59. LOCKIE, J.D., and RATCLIFFE, D.A.: *Brit Birds, 57*:89, 1964.
60. LUDWIG, J.P.: Great Lakes Research Div., Univ. of Mich., Publ. No. 15, 1966, p. 80.
61. LUDWIG, J.P., and TOMOFF, C.S.: *Jack-Pine Warbler, 44*:77, 1966.
62. LUDWIG, J.P.: Personal communication, 1968.
63. MOORE, N.W., and WALKER, C.H.: *Nature, 201*:1072, 1964.
64. O'BRIEN, R.D.: *Insecticides, Action and Metabolism.* New York, Academic, 1967. 332 pp.
65. PEAKALL, D.B.: *Nature, 216*:505, 1967.
66. PRESST, I.: *Bird Study, 12*:66, 196, 1965.
67. PRESST, I., and MILLS, D.H.: *Bird Study, 13*:163, 1966.
68. PRESST, I.: *J Appl Ecol, 3* (suppl) .:107, 1966.
69. QUINBY, G.E.; HAYES, W.J.; ARMSTRONG, J.F., and DURHAM, W.F.: *JAMA, 191*:175, 1965.
70. RATCLIFFE, D.A.: *Bird Study, 10*:56, 1963.
71. RATCLIFFE, D.A.: *Bird Study, 12*:66, 1965.
72. RATCLIFFE, D.A.: *Brit Birds, 58*:65, 1965.
73. RATCLIFFE, D.A.: *Bird Study, 14*:238, 1967.
74. RATCLIFFE, D.A.: *Nature, 215*:208, 1967.
75. RISEBROUGH, R.W.; MENZEL, D.B.; MARTIN, D.J., and OLCOTT, H.S.: *Nature, 216*:589, 1967.
76. RISEBROUGH, R.W.; HUGGETT, R.J.; GRIFFIN, J.J., and GOLDBERG, E.D.: *Science, 159*:1233, 1968.
77. RISEBROUGH, R.W.: This Conference, 1968.
78. ROBINSON, J., et al.: *Nature 214*:1307, 1967.
79. ROMANOFF, A.L., and ROMANOFF, A.J.: *The Avian Egg.* New York, Wiley, 1949.

80. SADLER, K.C.: *J Wildl Mgmt, 25*:339, 1961.
81. SCHWABE, U.: *Arzneimittel-Forsch, 14*:1265, 1964.
82. SIMKISS, K.: *Biol Rev, 36*:321, 1961.
83. SLADEN, W.J.L.; MENZIE, C.M., and REICHEL, W.L.: *Nature, 210*:670, 1966.
84. SPRUNT, A.: *Audubon Mag, 65*:32, 1963.
85. SPRUNT, A., and LIGAS, F.J.: A Florida Notebook. Proc. Natl. Audubon Soc. Conv., Miami, Fla., Nov. 9, 1963, p. 2.
86. STICKEL, L.F.; STICKEL, W.H., and CHRISTENSEN, R.: *Science, 151*:1549, 1966.
87. STICKEL, L.F., *et al.*: Trans. 31st North Amer. Wildl. Nat. Res. Conf. March 14-16, 1966, p. 190.
88. U.S. Fish and Wildlife Service: Resource Publ. 43, May 1967.
89. VAZ, Z.; PEREIRA, R.S., and MALHEIRO, D.M.: *Science, 101*:434, 1945.
90. WALLACE, G.J.; NICKELL, W.P., and BERNARD, R.F.: *Cranbrook Inst. of Sci Bull, 41*:1, 1961. (Bloomfield Hills, Mich. 44 pp.).
91. WHEATLEY, G.A., and HARDMAN, J.A.: *Nature, 207*:486, 1965.
92. WOOD-GUSH, D.G.M., and KARE, M.R.: *Brit Poultry Sci, 7*:285, 1966.
93. WOODWELL, G.M.: *Forest Sci, 7*:194, 1961.
94. WOODWELL, G.M.; WURSTER, C.F., and ISAACSON, P.A.: *Science, 156*:821, 1967.
95. WRIGHT, B.S.: *J Wildl Mgmt, 29*:172, 1965.
96. WURSTER, C.F.; WURSTER, D.H., and STRICKLAND, W.N.: *Science, 148*:90, 1965.
97. WURSTER, D.H.; WURSTER, C.F., and STRICKLAND, W.N.: *Ecology, 46*:488, 1965.
98. WURSTER, C.F., and WINGATE, D.B.: *Science, 159*:979, 1968.

DISCUSSION

PEAKALL: Using polychlorinated biphenyls, we demonstrated a considerable increase in metabolism of estrogen K *in vitro* with pigeon liver. Temple showed the same effect *in vivo* by putting labeled estrogen into the hepatic portal vein of birds that have been treated with DDT, collecting immediately afterwards from the right auricle and comparing with controls; there was a threefold increase in the ratio of metabolites to unchanged steroids. *In vitro* we also found induction of enzymes which degrade estradiol in kestrels.

RISEBROUGH: Polychlorinated biphenyls were already in the environment in the late 1930's. Dr. Peakall has just told us that they, like DDT, dieldrin and other chlorinated hydrocarbons, have enzyme-inducing capacities. In considering, therefore, the eggshell thickness, we should keep in mind that probably all chlorinated hydrocarbons and perhaps other environmental pollutants as well have contributed to the effect.

HUNT: Let us talk about the apparent birth-control effect. Do you feel that DDT would affect estrogen levels, thereby inhibiting egg-laying?

RISEBROUGH: We can discuss this mechanism better after we hear Dr. Welch's paper.

HUNT: Did you say there were only one hundred petrels?

WURSTER: There are about one hundred in the world, and about half of them are at sea. As immatures they spend four years at sea before returning to the islet on which they were born to search for a burrow. They may wait another year or two before breeding.

BERLIN: How do you know that the observation of these birds isn't causing the decline in their numbers?

WURSTER: Wingate is extremely careful in his observations of these birds and takes great pains not to disturb them in any way. I doubt the birds even know they are being watched.

RISEBROUGH: Wingate's efforts, particularly in the early years of his work, have significantly increased the population [personal observation].

WURSTER: In fact, the population is still increasing. Although the data show a decline in reproductive success, the number of birds is increasing because predation by tropic birds has been eliminated. Thus, two influences are simultaneously at work, but the curve of population numbers must eventually turn down if reproductive success continues to decline.

BERLIN: Perhaps the downward trend is caused by over-population?

WURSTER: These birds once lived on this island by the millions.

JOHNELS: This is a symptom of something of very general importance. Observations of the petrel, peregrine, osprey, and sea eagle are certainly in accord with one another, namely there's a decline in reproductive success. However, this decline is not seen in populations in the remote areas of Sweden and adjacent countries. Along the west coast of Norway, in north Sweden and in east Finland the breeding results are — for the present at least — observed at the east coast of Sweden south of Stockholm and across the Baltic to the west part of Finland. The birds contain DDT as well as mercury: breast muscle from a dead sea eagle had 80 ppm Hg and 140 ppm DDT and metabolites. The PCB's are present also and in the same quantities. Generally, chlorinated hydrocarbons were not thought to be very important in Sweden, but we will have to look at them more carefully.

LATIES: What concentration will produce effects in mammals?

WURSTER: The storage of 10 ppm of either DDT or DDE in fat is associated with increased hepatic enzyme activity in mice. Storage at that level is common in nature, even humans average 12 ppm in their fat. The importance of microsomal enzyme induction is that it occurs at extremely low concentrations of the inducing chemical. It may not kill a single individual but it may wipe out a species.

CLARKSON: The Bermuda petrel feeds entirely at sea and is a surface bird. With respect to the surface, one reads of accounts of tankers spilling oil, contaminating vast areas of the sea. Does this have any influence on the concentration of hydrocarbons in or on the bird?

WURSTER: Wingate reports that he has never seen an oiled petrel. Once fledged, they have very long lives. Presumably the birds receive a much greater exposure through their food chain than from patches of oil, which they apparently avoid anyway.

CLARKSON: My question was directed not to the danger of direct damage to the bird by contamination with oil, but to the possibility that surface layers of oil would selectively concentrate such oil-soluble compounds, as the chlorinated hydrocarbon pesticides. This process would keep high concentration of pesticides at the ocean surface where the organism constituting the birds' food chain are to be found.

Chapter 20

EFFECT OF CHLORINATED INSECTICIDES ON STEROID METABOLISM

R. M. Welch, W. Levin and A. H. Conney

ABSTRACT

The metabolism of testosterone *in vitro* by rat liver microsomes was inhibited by organic phosphorothionate insecticides such as parathion, malathion and Chlorthion. Studies on the hydroxylation of testosterone in specific positions revealed that the addition of Chlorthion *in vitro* had a more marked inhibitory effect on the 16α-hydroxylation reaction than on the 6β- or 7α-hydroxylation reaction. Chronic treatment of rats with Chlorthion inhibited the metabolism of estradiol-17β, testosterone, progesterone and deoxycorticosterone by rat liver microsomes, while the chronic administration of the chlorinated hydrocarbon insecticides, DDT or chlordane, stimulated the metabolism of these steroids by rat liver microsomes. Treatment of immature female rats with purified γ-chlordane, dieldrin, hepatochlordane, dieldrin, heptachlor or lindane for four days stimulated the activity of liver microsomal enzymes that metabolize estrone, inhibited the estrone-induced increase in uterine wet weight and decreased the amount of tritiated estrogen found in the uterus after an injection of tritiated estrone. Enhanced total body metabolism of estrone *in vivo* was also observed in rats pretreated with chlordane. During the course of these studies with halogenated hydrocarbon insecticides, it was found that DDT and some of its analogs possess estrogenic acivity. The I.P. injection of 50 mg/kg of technical grade DDT, purified o,p'-DDT, purified methoxychlor, or purified p,p'-DDT increased the uterine wet weight by 46 64, 38, and 18 per cent, respectively at 6 hours after the injection. The injection of technical grade DDT or o,p'-DDT to ovariectomized adult rats also increased uterine wet weight, indicating that the uterotropic effect of DDT was not dependent on the ovaries.

REPORT

THE CHRONIC ADMINISTRATION of DDT or chlordane to several species of animals increases the metabolism of drugs *in vitro* by enzymes localized in liver microsomes (15, 16). The effect of

Note: This investigation was supported in part by Research Contract No. PH 43-65-1066 from the Pharmacology-Toxicology Programs, National Institute of General Medical Sciences, National Institutes of Health.

390

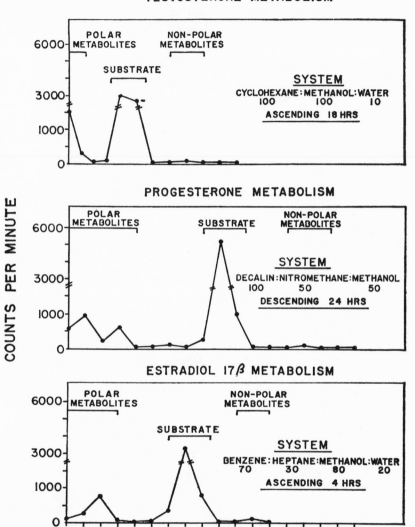

FIGURE 20-1. Metabolism of testosterone, progesterone and estradiol-17β to polar metabolites by rat liver. Seven hundred mμmol of steroid was incubated with microsomes from 165 mg of adult male rat liver in the presence of an NADPH-generating system. Extraction and chromatography were described previously [22].

low levels of DDT on drug metabolism in rats was investigated, and it was found that feeding as little as 5 ppm of DDT in the diet for 3 months resulted in elevated levels of liver microsomal enzymes that metabolize hexobarbital and aminopyrine (14). These results were paralleled by a decreased hypnotic effect of hexobarbital in the rats treated with DDT. Further studies (20) have shown that rats fed a diet containing as little as 5 ppm of toxaphene or 1 ppm of DDT had an increased level of liver enzymes which metabolize foreign compounds. Unlike the stimulatory effect that chlorinated hydrocarbon insecticides exert on liver microsomal enzymes, several organic phosphorothionate insecticides such as malathion and Chlorthion inhibit liver microsomal enzymes and prolong the duration of drug action (30, 40). Recent studies have shown that several factors which regulate the activity of drug metabolizing enzymes in liver microsomes also regulate the activity of hydroxylases in liver microsomes that metabolize naturally occurring steroids (21), and several reports have appeared that describe the effects of insecticides and closely related analogs on steroid metabolism and action in animals (7, 23, 25, 28, 38).

Inhibitory Effect of Organic Phosphorothionate Insecticides on Steroid Metabolism

Incubation of testosterone, estradiol-17β or progesterone with liver microsomes from adult male rats resulted in the formation of polar hydroxylated metabolites (Fig. 20-1) (22). The *in vitro* addition of organic phosphorothionate insecticides such as Chlorthion*, parathion and malathion, or chlorinated hydrocarbon

---

*Suppliers of the insecticides used in this study and the purity are as follows: *chlorthion* (p-nitro-m-chlorophenylthionophosphate) of 97% purity was furnished by Farbenfabriken Bayer A. G. Leverkusen-Bayerwerk, Germany; *Malathion* (S-(1,2-discarbethoxyethyl) O, O-dimethyl dithionophosphate) of 95% purity was furnished by American Cyanamid Company, Pearl River, New York; *Parathion* (p-nitrophenyldiethylthionophosphate) of 99% purity was purchased from K and K Laboratories, Plainview, New York; *Paraoxon* (p-nitrophenyldiethylphosphate) of 99% purity was purchased from K and K Laboratories, Plainview, New York; *Heptachlor* (1, 4, 5, 6, 7, 8, 8-heptachloro-3a, 4, 7, 7a-tetrahydro-4, 7-methanoindan), *dieldrin* (1, 2, 3, 4, 10, 10-hexachloro-6, 7-epoxy-1, 4, 4a, 5, 6, 7, 8, 8a-octa-hydro-1, 4-endoexo-5, 8-dimethanonaphthalene) and *lindane* (1, 2, 3, 4,

——————————→

insecticides, such as chlordane and DDT, inhibited the metabolism of testosterone to polar metabolites by liver microsomes (Table 20-I). The organic phosphorothionates at concentrations of 10^{-5}M and 10^{-4}M inhibited the formation of metabolites more polar than the substrate by about 30 and 70 per cent, respectively. Chlordane and DDT were less inhibitory since a concentration of 10^{-4}M of these substances produce about 35 per cent inhibition. Organic phosphorothionate insecticides become potent inhibitors of cholinesterase after oxidative metabolism to their oxygen analogs by rat liver (13). Paraoxon, the oxygen analog and major meta-

TABLE 20–I

In vitro effect of various insecticides on the metabolism of testosterone to polar metabolites by rat liver microsomes. Testosterone-4-C^{14} (700 mμmol) was incubated for 7 minutes with an NADPH-generating system and liver microsomes equivalent to 333 mg of liver obtained from adult male rats. Insecticides were added in 0.1 ml of ethanol immediately before the addition of substrate. Each value represents the average of three experiments.

Insecticide	*Concentration (M)*	*Inhibition of Polar Metabolite Formation (%)*
Chlorthion	10^{-4}	71
	10^{-5}	35
Parathion	10^{-4}	66
	10^{-5}	24
Paraoxon	10^{-4}	15
	10^{-5}	6
Malathion	10^{-4}	68
	10^{-5}	33
DDT	10^{-4}	38
Chlordane	10^{-4}	35

5, 6-hexachlorocyclohexane) with purities greater than 99% were obtained from Beckman Instruments, Inc., Fullerton, California; *p,p'–DDE* (1, 1-dichloro-2, 2-bis (p–chlorophenyl) ethylene) with a purity greater than 99% was obtained from Geigy Chemicals Corporation, Ardsley, New York; *p,p'–DDT* (1, 1-bis- (p–chlorophenyl) -2, 2, 2-trichloroethane), o,p'–DDT (1- (o–chlorophenyl) -1- (p–chlorophenyl) -2, 2, 2-trichloroethane), p,p–DDD 2, 2-bis- (P–chlorophenyl) (1, 1–dichloroethane) with purities greater than 99% were obtained from Aldrich Chemical Company, Inc., Milwaukee, Wisconsin; *DDT or DDT (tech)* refers to technical grade DDT obtained from Eastman Organic Chemicals, Distillation Products Industries, Rochester, New York; *γ–Chlordane* (1, 2, 4, 5, 6, 7, 8, 8-octachloro-3a, 4, 5, 5a-tetrahydro-4, 7-methanoindane) with a purity greater than 99% was obtained from Velsicol Chemical Corporation, Chicago, Illinois; *Chlordane* refers to the technical grade material obtained from Velsicol Chemical Corporation, Chicago, Illinois.

bolite of parathion, had little or no inhibitory effect on testosterone metabolism. These results indicate that the ability of organic
phosphorothionate insecticides to inhibit steroid hydroxylase activity in liver microsomes was not mediated through oxygenated
metabolites of the insecticides and was not related to the inhibitory effect of the insecticides on cholinesterase activity. Since the
metabolism of testosterone to polar metabolites by liver microsomes was inhibited by the addition of Chlorthion *in vitro*, studies were intiated to determine the effect of this organic phosphorothionate on the hydroxylation of testosterone in the 6β-,
7α- and 16α-positions (Table 20-II). The addition of 10^{-4}M
Chlorthion *in vitro* completely inhibited the 16α-hydroxylation
of testosterone, inhibited the 6β-hydroxylation by 55 per cent and
did not inhibit the 7α-hydroxylation reaction. At a concentration
of 10^{-5}M, 40 per cent inhibition of 16α-hydroxylation was observed without any inhibition of the 6β- and 7α-hydroxylations.
The selective inhibitory effect of Chlorthion on testosterone
hydroxylation in the various positions suggest that separate enzyme systems catalyze the hydroxylation of testosterone in the
6β-, 7α- and 16α-positions. Further support for the concept of
separate enzyme systems for the hydroxylation of testosterone in
the various positions was obtained from *in vivo* studies. Various
doses of Chlorthion were administered to male rats 60 minutes
prior to sacrifice, and the ability of whole liver homogenate to
metabolize testosterone was measured. An injection of Chlorthion (50 to 100 mg/kg) inhibited the metabolism of testoste-

TABLE 20-II

Inhibitory effect of Chlorthion on the oxidation of testosterone to 6β-, 7α- and
16α-hydroxytestosterone by rat liver microsomes. Liver microsomes from adult male
rats were incubated with testosterone as described in Table 20-I. The formation
of 6β-, 7α- and 16α-hydroxytestosterone were measured as previously described
(8). Each value represents the average obtained from several experiments.

| | Concentration (M) | Hydroxylated Testosterone Formed ($m\mu mol$) | | |
		6β-OH	7α-OH	16α-OH
Control	—	91	27	70
Chlorthion	10^{-4}	41	33	3
Chlorthion	10^{-5}	91	31	41

rone on polar metabolites by liver homogenate (Table 20-III).
It was also observed that the formation of 16α-hydroxysterone by
liver homogenate was almost completely inhibited after a 50
mg/kg dose of Chlorthion, while the formation of 6β- and 7α-
hydroxytestosterone was only slightly inhibited after a dose of
100 mg/kg of Chlorthion.

TABLE 20–III

Metabolism of testosterone by rat liver following the administration of a single
dose of Chlorthion. Male rats weighing 300 g were killed 60 minutes after the
IP administration of Chlorthion. Whole homogenate from 333 mg of liver was
incubated with 700 mμmol of testosterone in the presence of an NADPH-generating
system.

Dose of Chlorthion (mg/kg)	Polar Metabolite Formed (mμmol)	Hydroxylated Testosterone Formed (mμmol)		
		6β–OH	7α–OH	16α–OH
0	298	53	24	44
50	181	62	21	3
100	98	38	14	4

Since many chemicals inhibit liver microsomal drug oxidation for several hours after they are administered, but stimulate
drug metabolism at later times or when given chronically for
several days (18, 31), we have studied the effect of chronic Chlorthion administration on the metabolism of several steroids. The
daily administration of Chlorthion to immature rats for ten days
markedly inhibited the metabolism of testosterone, estradiol-17β,
progesterone and deoxycorticosterone to polar metabolites (Fig.
20-2). It was found that the inhibitory effect of chronic Chlorthion administration on the metabolism of estradiol-17β was
more marked than the effect of this insecticide on the metabolism of the other steroids.

Effect of Halogenated Hydrocarbon Insecticides on Steroid Metabolism and Action

In contrast to the inhibitory effect of Chlorthion on steroid
hydroxylation, the administration of halogenated hydrocarbon
insecticides such as chlordane or DDT for 10 days, stimulated
the metabolism of testosterone, estradiol-17β, progesterone and

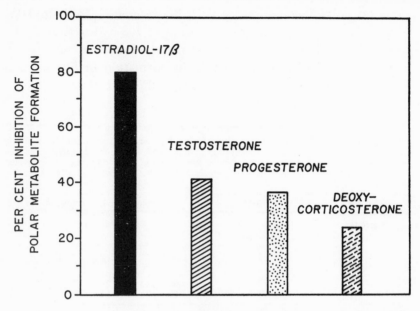

FIGURE 20-2. Inhibitory effect of chronic Chlorthion™ treatment on steroid metabolism by rat liver microsomes. Male rats weighing 50 g were injected I.P. with 25 mg/kg of Chlorthion twice daily for 10 days. Liver microsomes were incubated with 700 mμmol of steroid in the presence of an NADPH-generating system, and formation of metabolites more polar than the substrate was measured as previously described [38].

deoxycorticosterone to polar metabolites by liver microsomal enzymes (Fig. 20-3). The ability of halogenated hydrocarbon insecticide administration *in vivo* to stimulate the metabolism of estradiol-17β *in vitro* suggested that these insecticides might decrease the biological activity of estrogens. To investigate this possibility, adult ovariectomized rats were given 50 mg/kg of chlordane IP daily for six days. Tritiated estradiol-17β (0.3 μg) was administered 24 hours after the last dose of chlordane and the rats were killed 4 hours later. Pretreatment of rats with chlordane caused a marked inhibition in the uterotropic effect of estradiol-17β and decreased the concentration of estradiol-17β in the uterus (Table 20-IV). These results suggest that chlordane inhibits the action of estradiol-17β by increasing its metabolism

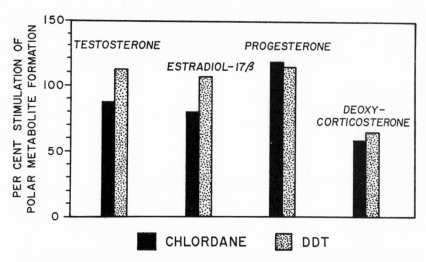

FIGURE 20-3. Stimulatory effect of chronic chlordane or DDT treatment on steroid metabolism by rat liver microsomes. Male rats weighing 50 g were injected IP with 25 mg/kg of chlordane or DDT twice daily for 10 days. Liver microsomes were incubated with 700 mμmol of the various steroids in the presence of an NADPH-generating system. Formation of metabolites more polar than the substrate was measured as previously described [38].

in vivo to nonestrogenic metabolites. Chronic treatment of immature female rats with γ-chlordane, dieldrin, heptachlor, or lindane, stimulated the activity of enzymes in liver microsomes that metabolize estrone (Fig. 20-4). This treatment also inhibited the action of tritiated estrone on the uterus and decreased the concentration of radioactivity in the uterus (Table 20-V). The effect of various doses of chlordane on the uterotropic action of estrone is shown in Table 20-VI. Pretreatment of rats with as little as 1 to 5 mg/kg of chlordane daily for seven days caused a significant inhibition in the uterotropic effect of estrone and decreased the concentration of estrogen in the uterus.

Studies were initiated to determine whether chlordane administration stimulates the total body metabolism of estrone in rats. Immature female rats were injected with chlordane once daily for seven days. Forty-eight hours after the last dose, the rats were injected with 0.2 μg of tritiated estrone and the whole body con-

TABLE 20–IV

Inhibitory effect of chlordane pretreatment on the action of estradiol–178β on the uterus of adult ovariectomized rats. Two weeks after ovariectomy, adult rats (190-210) gm) were given 50 mg/kg of chlordane IP daily for 6 days. Estradiol–17β-6, 7–H³ (0.3 μg) was administered 24 hours after the last dose of chlordane, and the rats were killed 4 hours later. For quantification of radioactivity the uteri were weighed and dissolved in 0.5 to 1 ml of hydroxide of hyamine at 50°C as previously described (25). More than 95 per cent of the radioactivity in the uterus was estradiol-17β. Values represent the mean ± SE from six-9 rats.

Pretreatment	Estradiol	Uterine Wet Weight (mg)	Estradiol in Uterus (pg)
Corn oil	—	86 ± 7	
Chlordane	—	82 ± 6	
Corn oil	+	133 ± 5	392 ± 10
Chlordane	+	104 ± 8	301 ± 18

TABLE 20–V

Effect of pretreatment with various chlorinated hydrocarbon insecticides on the action of estrone on the uterus of immature rats. Immature female rats were injected IP with insecticide once daily for 7 days. On the following day, rats were injected IP with 0.2 μg of estrone-6, 7H³ and killed 4 hours later. Uterine wet weight and radioactivity in the uterus was measured as previously described (25). Each value was obtained from six rats.

Pretreatment	Daily Dose (mg/kg)	Inhibition of Uterine Wet Weight Response (%)	Decrease in Radio activity in Uterus (%)
γ–Chlordane	25	92	57
Dieldrin	3	71	37
Heptachlor	10	62	48
Lindane	15	66	37

TABLE 20–VI

Inhibitory effect of pretreatment with various doses of chlordane on the action of estrone on the uterus of immature rats. Immature female rats were injected IP daily with chlordane for 7 days. Twenty-four hours after the last dose, the rats were injected IP with 0.2 μg of estrone–6, 7–H³ and killed 4 hours later. Values were taken from the mean response from six to eight rats.

Daily Dose of Chlordane (mg/kg)	Inhibition of Uterine Wet Weight Response (%)	Decrease in Radio activity in Uterus (%)
50	75	57
25	50	35
10	44	38
5	40	18
2	19	17
1	15	11

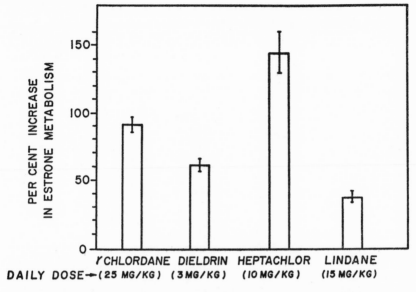

FIGURE 20-4. Effect of various halogenated hydrocarbon insecticides on the metabolism of estrone by rat liver microsomes. Immature female rats (19 days old) were injected IP daily with insecticide for 7 days. Rats were killed the following day, and microsomes equivalent to 333 mg of liver were incubated with 700 mμmol of estrone-6, 7-H³ for 15 minutes in the presence of an NADPH-generating system as previously described [25]. Values represent the mean ± SE of three determinations.

centration of estrone was determined 40 minutes later. Lower levels of estrone were present in rats treated with chlordane than in controls (Table 20-VII). The daily treatment of rats with 50 mg/kg of chlordane was more effective in enhancing the whole-body metabolism of estrone than was treatment with 10 mg/kg of the insecticide.

It is of interest that treatment of rats with phenobarbital and other drugs which stimulate the steroid hydroxylating enzymes in liver microsomes also antagonize the action of estrogens on the uterus, decrease the concentration of estrogens in the uterus and accelerate the whole-body metabolism of these hormones (25, 39). The data presented here show that halogenated hydrocarbon insecticides also antagonize the action of estrogens on the uterus and that this is related, at least in part, to enhanced estrogen metabolism. A stimulatory effect of halogenated hydrocarbon insecticides on hepatic steroid metabolism was recently observed in pigeons. Metabolism of testosterone and progesterone by liver microsomes was increased in pigeons that were treated with dieldrin or DDT (28).

TABLE 20–VII

Effect of chlordane pretreatment on the whole-body levels of estrone after administration of estrone–6, 7–H³. Immature female rats were injected with chlordane once daily for 7 days. Forty-eight hours after the last dose, the rats were injected IP with 0.2 μg of estrone–6, 7–H³. The rats were homogenized 40 minutes later, and the whole-body concentration of estrone was determined as previously described (39). Each value represents the mean $\pm$ SE from five rats.

Pretreatment	Daily Dose of Chlordane (mg/kg)	Estrone per Rat (% of dose)
Corn oil	—	21 $\pm$ 1
Chlordane	10	15 $\pm$ 1
Chlordane	50	9 $\pm$ 2

Studies in man indicate that the administration of phenobarbital or o,p-DDD, an analog of DDT, stimulates the extra-adrenal metabolism of cortisol *in vivo* to 6β-hydroxycortisol (4, 6, 34). This effect of phenobarbital and o,p'-DDD may be explained on the basis of enhanced liver microsomal hydroxylase activity, since treatment of animals with o,p'-DDD or phenobarbital stimulates the activity of liver microsomal enzymes that oxidize drugs (9, 33) and metabolize cortisol to polar metabolites (10, 24).

It is of interest to note that chlordane administration decreases fertility in rats (1) and that aldrin administration causes a disturbance of the estrus cycle in this species (3). Aldrin is a chlorinated hydrocarbon insecticide that is metabolized to dieldrin, and both compounds stimulate liver microsomal enzyme activity in rats (35). Additional studies are needed to determine whether the stimulatory effect of halogenated hydrocarbon insecticides on the metabolism of steroid hormones can explain the decreased fertility and altered estrus cycle observed in animals treated with these insecticides.

Uterotropic Effect of DDT and Its Analogs

During the course of studies with halogenated hydrocarbon insecticides, it was found that there was a marked increase in the uterine wet weight of immature female rats treated with technical grade DDT (DDT (tech)). Following the IP injection of 50 mg/kg of DDT (tech) into immature female rats, there was an increase in uterine wet weight which reached a maximum in 6 hours. Since DDT (tech) contains p,p'-DDT and o,p'-DDT, we have investigated the effect of these and related substances on the uterus. Immature female rats were given 50 mg/kg of o,p'-DDT, p,p'-DDT, p,p'-DDD, p,p'-DDE, methoxychlor or DDT (tech) and the uterine wet weight was determined 6 hours later. It was found that o,p'-DDT had marked uterotropic activity, methoxychlor and DDT (tech) were moderately active and p,p'-DDT and p,p'-DDD were weakly active (Fig. 20-5). The major metabolite of p,p'-DDT namely p,p'-DDE was inactive. The stimulatory effect of DDT (tech) or o,p'-DDT on uterine wet weight was also observed in ovariectomized adult rats, indicating that the uterotropic effect of DDT (tech) and o,p'-DDT was not caused by an indirect effect of the halogenated hydrocarbons on the ovaries or its hormones (Table 20-VIII). DDT (tech) and o,p'-DDT not only increase uterine wet weight, but like estradiol-17β, stimulate the incorporation of uniformly labeled glusose^{-14}C into lipid, protein, RNA and acid soluble constituents in the rat uterus*. An earlier study with p,p'-DDT indicated that this insecticide decreased testicular

TABLE 20–VIII

Stimulatory effect of DDT (tech) or o,p'–DDT on uterine wet weight in the ovariectomized rat.. Adult female rats were ovariectomized 2 weeks prior to treatment. Technical grade DDT or o,p'–DDT was injected IP in dimethylsulfoxide 6 hours prior to sacrifice. Each value represents the mean ± SE from seven rats.

Treatment	Dose (mg/kg)	Increase in Uterine-wet Weight (%)
DDT (tech)	5	34 ± 1
DDT (tech)	50	63 ± 3
o,p'–DDT	5	51 ± 2
o,p'–DDT	50	102 ± 6

*Unpublished observations.

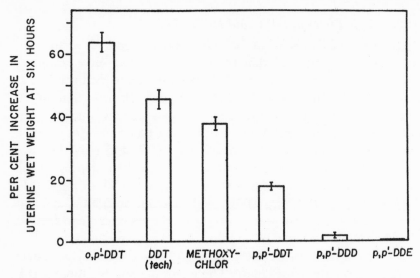

FIGURE 20-5. Effect of the various analogs of DDT on uterine wet weight in the rat. Immature female rats were injected IP with 50 mg/kg of analogs of DDT dissolved in corn oil. Six hours later, the rats were killed and the uterus was weighed. Results represent the mean ± SE from six rats.

growth and inhibited the development of secondary sex character-istics in the cockerel, an effect which suggested that this insecticide may possess estrogenic activity (5). However, in another study, it was concluded that p,ṕ-DDT did not have estrogenic activity, since it failed to maintain estrus in ovariectomized rats (12).

The widespread use of chemicals for the control of insects has raised concern about the possible harmful effects of these chem-icals on the public health. Reports have appeared, indicating that the population of several species of birds has decreased during the past decade (2, 11, 17, 19, 26, 27, 29, 32, 36, 37). It was suggested that this decrease in bird population may be due to pesticide resi-dues, but this relationship has not been firmly established. The data given in the present report show that some insecticides possess estrogenic activity while others inhibit or stimulate the metabol-ism of steroid hormones in animals. It is possible that the antifer-tility effect of certain insecticides in animals may be explained, in part, by one or more of the above observations, but further studies are needed to determine whether insecticides can influence reproduction by an interaction with steroid hormones or by a direct estrogenic action.

REFERENCES

1. AMBROSE, A.M.; CHRISTENSEN, H.E.; ROBBINS, D.J., and RATHER, L.J.: *Arch Industr Hyg, 7:*197, 1953.
2. AMES, P.L., and MERSEREAU, G.S.: *Auk, 81:*173, 1964.
3. BALL, W.L.; KINGSLEY, K., and SINCLAIR, J.W.: *Arch Industr Hyg, 7:*292, 1953.
4. BLEDSOE, T.; ISLAND, D.P.; NEY, R.L., and LIDDLE, G.W.: *J Clin Endocr, 24:*1303, 1964.
5. BURLINGTON, H., and LINDEMAN, V.F.: *Proc Soc Exp Biol Med, 74:*48, 1950.
6. BURSTEIN, S., and KLAIBER, E.L.: *J Clin Endocr, 25:*293, 1965.
7. CONNEY, A.H.; WELCH, R.M.; KUNTZMAN, R., and BURNS, J.J.: *Clin Pharmacol Ther, 8:*2, 1967.
8. CONNEY, A.H., and SCHNEIDMAN, K.: *J Pharmacol Exp Ther, 146:*225, 1964.
9. CONNEY, A.H.: *Pharmacol Rev, 19:*317, 1967.
10. CONNEY, A.H.; JACOBSON, M.; SCHNEIDMAN, K., and KUNTZMAN, R.: *Life Sci, 4:*1091, 1965.
11. CRAMP, S.: *Brit Birds, 56:*124, 1963.
12. FISHER, A.L.; KEASLING, H.H., and SCHUELLER, F.W.: *Proc Soc Exp Biol Med, 81:*439, 1952.
13. GAGE, J.C.: *Biochem J, 54:*426, 1953.
14. HART, L.G., and FOUTS, J.R.: *Naunyn Schmiedeberg Arch Pharm Exp Path, 249:*486, 1965.
15. HART, L.G., and FOUTS, J.R.: *Proc Soc Exp Biol Med, 114:*388, 1963.
16. HART, L.G.; SHULTICE, R.W., and FOUTS, J.R.: *Toxic Appl Pharmacol, 5:*371, 1963.
17. JEFFERIES, D.J.: *Ibis, 109:*266, 1967.
18. KATO, R.; CHIESARA, E., and VASSANELLI, P.: *Biochem Pharmacol, 13:*69, 1964.
19. KEITH, J.A.: *J Appl Ecol, 3* (suppl.) :57, 1966.
20. KINOSHITA, F.K.; FRAWLEY, J.P., and DuBOIS, K.P.: *Toxic Appl Pharmacol, 9:*505, 1966 .
21. KUNTZMAN, R.; JACOBSON, M.; SCHNEIDMAN, K., and CONNEY, A.H.: *J Pharmacol Exp Ther, 146:*280, 1964.
22. KUNTZMAN, R.; WELCH, R., and CONNEY, A.H.: *Advances Enzym Regulat, 4:*149, 1966.
23. KUPFER, D.: *Arch Biochem Biophys* (in press) .
24. KUPFER, D., and PEETS, L.: *Biochem Pharmacol, 15:*573, 1966.

25. LEVIN, W.; WELCH, R.M., and CONNEY, A.H.: *J Pharmacol Exp Ther, 159:*362, 1968.
26. LOCKIE, J.D., and RATCLIFFE, D.A.: *Brit Birds, 57:*89, 1964.
27. O'BRIEN, R.D.: Insecticides action and metabolism. *Insecticides and Environmental Health.* New York, Academic, 1967, chapt. 17.
28. PEAKALL, D.B.: *Nature, 216:*505, 1967.
29. RATCLIFFE, D.A.: *Bird Study, 12:*66, 1965.
30. ROSENBERG, P., and COON, J.M.: *Proc Soc Exp Biol Med, 98:*650, 1958.
31. SERRONE, D.M., and FUJIMOTO, J.M.: *J Pharmacol Exp Ther, 133:*12, 1961.
32. SPRUNT, A.N.: *Audubon, 65:*32, 1963.
33. STRAW, J.A.; WALTERS, I.W., and FREGLY, M.J.: *Proc Soc Exp Biol Med, 118:*391, 1965.
34. SOUTHREN, A.L.; TOCHIMOTO, S.; ISURUGI, K.; GORDON, G.G.; KRIKUN, F., and STYPULKOWSKI, W.: *Steroids, 7:*11, 1966.
35. TRIOLO, A.J., and COON, J.M.: *J Agric Food Chem, 14:*549, 1966.
36. WRIGHT, B.S.: *J Wildl Mgmt, 29:*172, 1965.
37. WURSTER, C.F., and WINGATE, D.B.: *Science, 159:*979, 1968.
38. WELCH, R.M.; LEVIN, W., and CONNEY, A.H.: *J Pharmacol Exp Ther, 155:*167, 1967.
39. WELCH, R.M.; LEVIN, W., and CONNEY, A.H.: *J Pharmacol Exp Ther, 160:*171, 1968.
40. WELCH, R.M.; ROSENBERG, P., and COON, J.M.: *Pharmacologist, 1:*64, 1959.

DISCUSSION

RISEBROUGH: You are making the "parts per million" approach to pollutant ecology obsolete, since the toxicity of high levels of pesticides may be irrelevant to ecological damage.

WELCH: Species differences with respect to pathways of insecticide detoxication and elimination must be kept in mind when evaluating insecticide metabolite residues in the tissues. One bird may make metabolite "x" while another may make metabolite "y." Each metabolite has a different toxicity. Man metabolizes the anticoagulant tromexan by hydroxylation while the rabbit de-esterifies the compound [Burrs et al.: *J. Pharmacol Exp Ther, 108:*33, 1953]. It could be that one investigator observes an effect on the reproductive process in one species of bird after feeding an insecticide while a different investigator, working with the same insecticide but a different species of bird fails to observe this effect. This apparent discrepancy may be attributed to different metabolic products made by the two different species of bird from the same insecticide.

WURSTER: Wynne-Edwards and others assembled extensive evidence of a correlation between steriod hormone balance and the number of organisms in the population [Christian, J. J., and Davis, D. E.: *Science, 146*:1550, 1964] [Wynn-Edwards, V. C.: *Animal Dispersion in Relation to Social Behavior,* New York, Hafner, 1962]. Your findings plus those of Drs. Peakall [*Nature, 216*:505, 1967], Conney [*Pharmacol Rev, 19*:317, 1967], and Kupfer [*Residue Rev, 19*:11, 1967] show that chlorinated hydrocarbons are of major ecological significance to the world's biota.

WELCH: Our studies in rats suggest that there are two ways in which DDT may bring about an alteration of the reproductive process. First, DDT increases the metabolism of naturally occurring steriods by liver microsomal enzymes. Second, DDT has an estrogenic action and therefore may interfere with the natural feedback mechanism which regulates secretion of gonadotropin by the pituitary gland.

BERG: This may be a new mechanism by which pesticides favor pest species over the species that control pests. I will draw a model (Fig. 20-6) of the contrast between a population with reproductive potential, represented by the oil torch, and a population with a low reproductive potential, represented by the candle. The torch is stable in the wind because it has a high energy flow. The candle can apparently be made just as stable with suitable shielding, but this is deceiving. The wind represents the environmental variables that add to mortality, and this includes only the life shortening effects of toxic chemicals. Those environmental variables that depress natality are represented by a drop of water, and they are highly selective: if a drip of water can get to the flame, then the low-energy system is vulnerable, and the high-energy system is not. Toxicity that depresses the reproductive potential acts like the drop of water.

Pest species are generally primary consumers adapted to heavy predation, and pest populations remain stable under high mortality because they have a high reproductive potential. The hormonal control of reproduction under such conditions is coarse, and

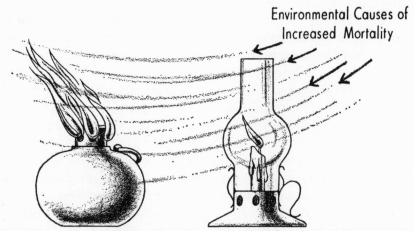

High Reproductive Potential,   Low Reproductive Potential,
        Stable Population   Stable Population

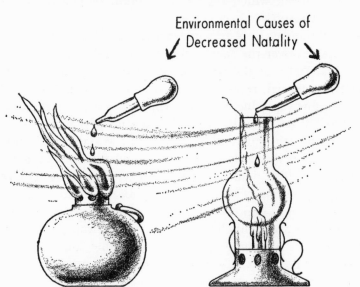

High Reproductive Potential,   Low Reproductive Potential,
        Stable Population   Extinct Population

FIGURE 20-6.

the effective concentrations of hormones in the body are high. In contrast to this, the predators that control pests are normally well shielded from environmental pressures in nature, as illustrated by the survival data on the petrels. As its mortality is low, such a species remains stable through a low and internally controlled natality. This means, that the control of reproduction by hormones is delicately poised [Christian, J. J.; Lloyd, J. A., and Davis, D. E.: The Role of endocrines in the self-regulation of mammalian populations, *Rec Progr Hormone Res, 21:*501-507, 1965]. and the concentrations of hormones in the body remain close to the effective thresholds. If organochlorine compounds interfere in small ways with the regulation of hormones in the body, and hence with reproduction, then this model predicts that low doses will do no damage whatever to pest populations while they exterminate the most stable and well-established predator populations.

Organochlorine pesticides are already known to favor pests over predators in three other ways:

1. At lethal exposures the larger population is more likely to leave breeding survivors;
2. at low exposures, the fast-breeding population is more likely to produce a resistant genotype; and
3. exposures are not equal, because the predator's intake of toxic compounds is amplified through the food chain. It should be clear, that the ecological damage of toxic chemicals cannot be evaluated from the usual studies of toxicity, which are done on the most commonly found primary consumers, including man.

Chapter 21

STUDIES OF INTRACELLULAR DISTRIBUTION OF MERCURY

T. NORSETH

ABSTRACT

The application of an analytical cell-fractionation method for studying intracellular distribution of mercury in rat liver cells is reported. The method is based on determination of marker enzyme activity in fractions and a subsequent statistical comparison with mercury distribution. When marker enzyme activity distribution is taken to represent the distribution of the corresponding cellular particle, the amount of mercury in the pure cell particle populations can be estimated. It is shown that misleading results are obtained if centrifugal fractions are taken to represent particle populations based on relative concentration of particles in fractions.

Typical results with an analytical approach to a centrifugal fractionation study of this kind are given. The model for treatment of the experimental results is outlined. The usefulness of the methods as compared to conventional methods without use of marker enzymes for analytical purposes is shown by the distribution of mercury at different times after a single injection and after exposure to different compounds.

Some general assumptions for the reality of the method are given. The possibility of applying a corresponding approach to distribution studies in other organs of interest in mercury intoxications is pointed out, and necessary precautions related to organ change are discussed.

REPORT

THE ASSUMPTION OF metabolic compartments within the cell is necessary to explain the dynamics of cell metabolism. The intracellular distribution of mercury may thus determine the toxic effects on the cell (6, 14, 17, 21, 24). Mercury metabolism at the cellular level must also be reflected in intracellular distribution. For these reasons an analytical centrifugal technique with

Note: This paper is based on work performed under contract with the United States Atomic Energy Commission at the University of Rochester Atomic Energy Project and is Report No. UR-49-953.

408

marker enzymes for cell fractionation was recently adopted for studying mercury distribution in rat liver cells (11, 13, 18).

The purpose of a cell fractionation may be either analytical or preparative. The technical procedures may be the same for both purposes, but the treatment of the results is different. With the preparative approach, the particle found in the highest concentration is taken to represent the fraction. For analytical purposes, the actual composition of the fraction with respect to all particles is taken into consideration. The composition of a fraction can be judged either morphologically or by use of marker enzymes. Although quantitative electron microscopy has been applied (1, 3, 4, 16), the use of marker enzymes is more useful for quantitative estimation of the different particle populations in a fraction. The marker enzyme principle has been thoroughly reviewed by de Duve (11, 13), and the technique has been successfully applied by his group for several years. When the distribution of enzyme activity is taken to represent the distribution of cell particles in a centrifugal fraction, it rests on the postulate of biochemical homogeneity. As outlined by de Duve, the postulation is that "granules of a given population are enzymically homogeneous, or at least cannot be separated into subgroups differing significantly in relative enzymic content" (11). It is thus assumed that the cytoplasmic particles are biochemically sufficiently homogeneous to allow meaningful extrapolation from enzymes to host particles.

Tables 21-I and 21-II taken from previous work on mercury distribution after methoxyethylmercury exposure show a typical

TABLE 21-I
ENZYMIC ACTIVITY AND CONTENT OF PROTEIN AND
MERCURY IN RAT LIVER

	No. of expts	Mean ± SD	
Protein	5	198.3 ± 40.8	mg
3-Hydroxybutyrate dehydrogenase	5	4.63 ± 0.81	units
Acid phosphatase	5	7.06 ± 1.34	units
Urate oxydase	5	2.88 ± 0.29	units
Glucose–6–phosphatase	5	21.30 ± 5.36	units
Mercury	5	0.83 ± 0.22	μg

Values are given per g wet-liver tissue.

TABLE 21-II

PERCENTAGE DISTRIBUTION OF ENZYMIC ACTIVITIES, (AND MERCURY AND PROTEIN
IN THE CENTRIFUGAL FRACTIONS

	Mean ± SD					
	N	M	L	P	S	Recovery
Protein	16.3 ± 3.1	26.3 ± 4.4	3.8 ± 1.0	15.7 ± 2.1	30.3 ± 4.4	92.5 ± 12.4
3-Hydroxybutyrate dehydrogenase	3.7 ± 3.1	80.5 ± 11.5	0.7 ± 0.4	2.7 ± 3.0	1.7 ± 1.0	89.2 ± 12.6
Acid phosphatase	5.4 ± 0.7	45.2 ± 2.4	13.2 ± 2.9	19.5 ± 6.6	9.1 ± 4.3	92.2 ± 11.8
Urate oxydase	2.6 ± 0.3	50.4 ± 8.3	12.3 ± 5.4	10.3 ± 2.0	0.3 ± 0.3	75.8 ± 12.4
Glucose-6-phosphatase	6.2 ± 1.8	18.2 ± 3.1	10.6 ± 5.2	43.7 ± 12.6	2.6 ± 2.8	82.2 ± 17.1
Mercury	11.2 ± 2.1	26.3 ± 4.1	5.0 ± 1.3	13.3 ± 4.2	41.4 ± 6.7	97.4 ± 12.7

Values are taken from the experiments in Table 21-I.
The fractions are, nuclear (N), mitochondrial (M), light mitochondrial (L), microsomal (P) and supernatant (S).

presentation of basic results from an analytical fractionation study
(18). 3-hydroxybutyrate dehydrogenase is used as marker enzyme
for mitochondria (5), acid phosphatase for lysosomes (9), urate
oxidase for peroxisomes (12), and glucose-6-phosphatase for
microsomes (10).

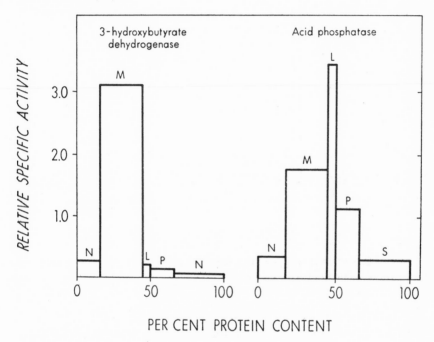

FIGURE 21-1. Distribution pattern of the marker enzymes 3-hydroxybutyrate
dehydrogenase and acid phosphatase. The results in Table 21-II are recalculated for variable recovery and expressed as percentage enzymic activity per
percentage protein in the different fractions (relative sp. act.).

Figures 21-1 and 21-2 give an impression of the particle distribution in the fractions (9). It is clearly seen that the same technical procedures for separating the homogenate into fractions can be
used for preparative purposes. N, M, L, P and S are related to the
centrifugal procedure and represent nuclear, mitochondrial, light
mitochondrial and microsomal fractions, and the supernatant,
respectively.

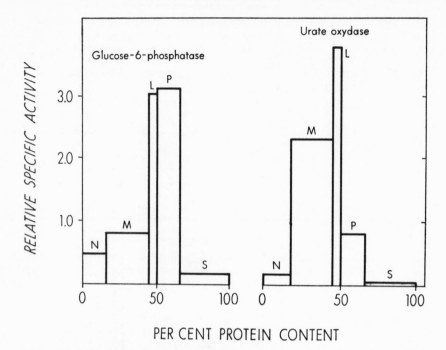

FIGURE 21-2. Distribution pattern of the marker enzymes glucose-6-phosphate and urate oxidase expressed as in Figure 21-1.

As seen from the Figures, the height of the blocks represents the degree of purity achieved for a certain particle population above that of the homogenate (9). The heterogeneous composition of the fractions is clearly visualized, the distribution patterns are similar to those published by others (2, 5, 9).

Because no marker enzymes are used for nuclei and supernatant, the analytical treatment of the results is restricted to mitochondria, lysosomes, peroxisomes, and microsomes. The following set of equations describes the model for estimating mercury content in the pure cellular particle populations.

$$Hg_M = \alpha_1 X_{1M} + \alpha_2 (X_{2M} + X_{3M}) + \alpha_3 X_{4M}$$
$$Hg_L = \alpha_1 X_{1L} + \alpha_2 (X_{2L} + X_{3L}) + \alpha_3 X_{4L}$$
$$Hg_P = \alpha_1 X_{1P} + \alpha_2 (X_{2P} + X_{3P}) + \alpha_3 X_{4P}$$

Hg represents the mean relative specific amount of mercury, X_1, X_2, X_3 and X_4 represent the mean relative specific activities of

3-hydroxybutyrate dehydrogenase, acid phosphatase, urate oxydase and glucose-6-phosphatase respectively, and α_1, α_2 and α_3 represent the corresponding unknown coefficients.

The equations thus express the mercury content in the mitochondrial (M), light mitochondrial (L) and microsomal (P) fractions as the sum of the mercury in mitochondria, lysosomes, peroxisomes and microsomes. The comments on the reality of the model and the assumptions for the statistical treatment of the results can be found elsewhere (18).

TABLE 21–III
MERCURY DISTRIBUTION AMONG CELLULAR COMPONENTS IN
CENTRIFUGAL FRACTIONS AND IN THE CELL

Cellular Component	Percentage (of total) Mercury Content in Fractions					
	N	M	L	P	S	Total
Mitochondria	5.6	12.1	0.1	0.3	0.2	13.3
Lysosomes and Peroximes	0.9	10.5	2.7	3.0	0.9	18.0
Microsomes	1.5	4.2	2.3	9.8	0.9	18.7

Table 21-III shows the percentage distribution of mercury in pure cellular particle populations estimated by this method from the values in Tables 21-I and 21-II. Without the use of marker enzymes the best estimates of mercury in these particle populations are the amount of mercury found in the M, L and P fractions. Without the analytical approach to the distribution these fractions represent mitochondria, lysosomes and peroxisomes, and microsomes respectively, because of the relative purity achieved for the different populations over the homogenate. The mercury distribution in Table 21-II thus represents only mercury distribution in fractions and cannot be taken to represent cellular particle populations.

This method has also been applied for evaluating the intracellular distribution of mercury at different time intervals after intravenous injection of mercuric chloride (19). The results (Table 21-IV) indicate some retention of mercury in the lysosomes-peroxisome population without statistical evaluation with marker enzyme distribution. The distribution patterns become, however, far more evident after an analytical analysis. The method is here extended with a preparative part to evaluate the distribution between lysosomes and peroxisomes (12, 19). The recovery of tissue in the purified lysosome and peroxisome populations related to starting material is very low. Because of the already known general distribution and the similarities between these two particle populations, meaningful conclusions can be drawn. No retention of mercury is found in peroxisomes (3).

TABLE 21–IV
MERCURY CONTENT IN FRACTIONS AND IN CELLULAR
PARTICLE POPULATIONS AT DIFFERENT TIMES AFTER
INJECTION OF MERCURIC CHLORIDE

	1 hr	1 day	4 days
Total mercury in liver	4.50	1.39	0.74
Mitochondria	13.2	13.2	17.7
M–fraction	16.0	21.6	29.7
Lysosomes and peroxisomes	5.4	13.4	25.0
L–fraction	2.9	3.6	6.8
Microsomes	13.6	14.9	12.5
P–fraction	9.9	12.2	10.8

Values for total amount are given in $\mu g/g$ wet-weight of liver.
Distribution figures are given as per cent of total.

Table 21-V shows the distribution of mercury after exposure to three different compounds (18, 19, 20). Mercury content in M, L and P fractions is given for comparison. The necessity of marker enzyme analysis for meaningful conclusions is clearly seen.

The general pattern in the reported distribution studies indicates the existence of mercury in different forms in the cell. Methods for separate determination of organic and inorganic mercury in homogenates are available (7, 8, 15, 22, 23). The same set of marker enzyme determinations can then be used for evaluating the distribution of the different forms of mercury in the cell. This may explain the differences in distribution already found and give valuable results concerning toxicity and metabolism of mercury compounds.

The liver was chosen for this analytical fractionation study because of the amount of previous work in this field on liver tissue (13). The relative homogeneity of the liver also facilitates the drawing of conclusions; morphological comparisons were likewise

TABLE 21-V

MERCURY CONTENT IN FRACTIONS AND IN CELLULAR PARTICLE POPULATION
AFTER EXPOSURE TO THREE DIFFERENT COMPOUNDS

	Mercuric Chloride	Methylmercury	Dicyandiamide	Methoxyethylmercury
Total amount in liver	0.74	0.82	0.82	0.83
Mitochondria	17.7	11.9	11.9	13.3
M-fraction	35.1	14.7	14.7	27.0
Lysosomes/peroxisomes	25.0	7.0	7.0	18.0
L–fraction	8.1	4.2	4.2	5.1
Microsomes	12.5	25.9	25.9	18.7
P–fraction	13.5	18.7	18.7	13.7

Values for total amount are given as μg/g wet weight of liver.
Distribution values are given as per cent of total.

found unnecessary. The method is tedious and time consuming but should give rewarding results when further applied to the liver and to other organs. The adoption of the method to target organs such as the kidney and brain for mercury intoxication is possible. The reality of the statistical model must, however, be established for different organs separately. The basic postulate of biochemical homogeneity must also be reevaluated if modifications of the centrifugal procedure are found necessary. All purification steps in a procedure will tend to make this postulate unreliable. Such steps are necessary for evaluating the reality of the model on the actual tissue, related if possible to morphological controls, but should not be part of the general analytical fractionation procedure.

REFERENCES

1. BAHR, G.F. and ZEITLER, E.J.: *J Cell Biol, 15*:489, 1962.
2. BAUDHUIN, P.; BEAUFAY, H.; RAHMAN-LI, Y.; SELLINGER, Z.; WATTIAUX, JACQUES, P., and DE DUVE, D.: *Biochem J, 92*:179, 1964.
3. BAUDHUIN, P., and BERTHET, J.: *J Cell Biol, 35*:631, 1967.
4. BAUDHUIN, P.; EVRARD, P., and BERTHET, J.: *J Cell Biol, 32*:181, 1967.
5. BEAUFAY, H.; BENDALL, D.S.; BAUDHUIN, P., and DE DUVE, C.: *Biochem J, 73*:623, 1959.
6. BERGSTRAND, A.; FRIBERG, L.; MENDEL, L., and ODEBLAD, J.: *Ultvastruct Res, 3*:234, 1959.
7. CLARKSON, T.W.; ROTHSTEIN, A., and SUTHERLAND, R.: *Brit J Pharmacol, 24*:1, 1965.
8. CLARKSON, T.W., and GREENWOOD, M.: *Talanta* (in press).
9. DE DUVE, C.; PRESSMAN, B.C.; GIANETTO, R.; WATTIAUX, R., and APPELMANS, F.: *Biochem J, 60*:604, 1955.
10. DE DUVE, C.; WATTIAUX, R., and BAUDHUIN, P.: In Nord, F.F. (Ed.): *Advances in Enzymology.* New York, Intersicence, 1962, vol. 24, p. 320.
11. DE DUVE, C.: *J Theor Biol, 6*:33, 1964.
12. DE DUVE, C., and BAUDHUIN, P.: *Physiol Rev, 46*:323, 1966.
13. DE DUVE, C.: *In Enzyme Cytology.* Roodyn, D.B. (Ed.): New York, Academic, 1967, Chapt. 1.
14. ELLIS, R.W., and FANG, S.C.: *Toxic Appl Pharmacol, 11*:104, 1967.
15. GAGE, J.C.: *Brit J Industr Med, 21*:197, 1964.
16. GLAS, U., and BAHR, G.F.: *J Cell Biol, 29*:507, 1966.
17. JONEK, J., and GRZYBEK,, H.: *Arch Gewerbepath Gewerbehyg, 20*:527, 1964.

18. NORSETH, T.: *Biochem Pharmacol, 16:*1645, 1967.
19. NORSETH, T.: *Biochem Pharmacol* (in press)
20. NORSETH, T.: Unpublished results.
21. TIMM, F.; NAUNDORF, C., and KRAFT, M.: *Arch Gewerbepath Gewerbehyg, 22:*236, 1966.
22. WESTOO, G.: *Acta Chem Scand, 20:*2131, 1966.
23. WESTOO, G.: *Acta Chem Scand, 21:*1790, 1967.
24. YOSHINO, Y.; MOZAI, T., and NAKAO, K.: *J Neurochem, 13:*397, 1966.

DISCUSSION

BERG: You and Dr. Berlin have traced mercury to the level of the cell, and you have now extended it to the subcellular localization in the organism. Do you expect this to influence our understanding of the mechanism of action?

NORSETH: Yes, I think so. I would, of course, like to apply this method to kidney and brain, which are organs more susceptible to mercury intoxication than the liver. It will be difficult to do this because the subcellular particles of these organs are very complex. The basic assumption of my method is that in the fractions I use for estimating mercury, there are no particles other than those characterized by a small set of enzymes.

It is possible to begin fractionation studies on subcellular particles with some purification steps, which remove unwanted particles.

The use of purification steps, however, is dangerous, because if the proteins and the total homogenates are not recovered to almost 100 per cent, the part of the homogenate containing the mercury may be lost.

BERG: Can mercury migrate from fraction to fraction in the centrifuge?

NORSETH: The probability of redistribution must always be considered in this kind of study. I controlled for it by measuring distribution at three concentrations of mercury. Even at the highest concentration of mercury, the distribution was far below the total capacity of the available SH-groups, which have the highest affinity for mercury, as shown by Clarkson and Magos in their work on liver homogenates [Clarkson, T. W., and Magos, L.: *Biochem J, 99:*62, 1966].

SUZUKI: Do you expect to be able to measure the distribution of mercury of such fractions in the brain?

NORSETH: Some results from fractionation studies on the brain have been obtained by Yoshino and associates from Japan *J Neurochem, 13*:397, 1966. I don't think their fractionation methods were good enough to draw any conclusions about the distribution of mercury in subcellular particles. They merely concluded that distribution is related to fractionation, and did not characterize their fraction. I feel no conclusions about mercury distribution in subcellular particles can be made if the particle content in the fractions one is working with is undefined. I would expect with a better method to find a subcellular distribution pattern of mercury in the brain different from that reported by Yoshino.

YOUNG: Did you resuspend and wash the particles?

NORSETH: Yes, twice.

YOUNG: And you still obtained 100 per cent recovery.

NORSETH: All of the recoveries were between 90 per cent and 100 per cent. With this method, however, recoveries are independent of resuspension and washing.

FASSETT: Did you do any electron microscope studies of the fractions?

NORSETH: No, but if this method is to applied to the brain and kidney, morphological studies of the fractions must also be done.

WELCH: Some time ago we had occasion to use glucose-6 phosphatase as a marker enzyme for microsomes. We could not completely clean the mitochondrial fraction of glucose-6 phosphatase. It appears from your experiments that some glucose-6 phosphatase activity still remained in the mitochondria after fractionation of the cell. Is it possible to clean the mitochondrial fraction completely of glucose-6 phosphatase?

NORSETH: It can be done if the centrifugation procedures are modified, but then too many of the mitochondria are removed from the mitochondrial fraction. Glucose-6 phosphatase also has an uneven distribution between the heavy and light mitochondrial fractions. It is not a good marker.

SUZUKI: In your method you must presume no change in the activity of enzymes, but if mercury accumulates in the subcellular fraction, there is a possibility that the enzyme activity would change. How would you take such a change into account?

NORSETH: I can only use amounts of mercury which do not affect the enzyme activity. I tested the effect of the highest amount of mercury that I had in my fractions on enzyme activity. There was one change of activity.

JOHNELS: In a study of the nervous system, would it not be useful to make autoradiographs of large nerve cells? These cells are not found in mammals but they are present for example, in squid, in the spinal cord of the lamprey, and in electric fish.

NORSETH: That is possible, but this procedure is not quantitative. A study done in kidney used a combined histochemical and centrifugal procedure, and demonstrated an accumulation of mercury in the lysosomes of the cell or of a population of cells could not be measured with any degree of precision [Timm, F., *et al.*: *Arch Gewerbepath Gewerbehyg, 22*:236, 1966].

FASSETT: Silver is another metal with an affinity for the SH-group. In a case of localized argyria, biopsies established that silver was present as silver sulphide all around the mitochondria [Buckley, W. R., *et al.*: *Arch Derm, 92*:697, 1965]. We did not detect it anywhere else. This may be an interesting marker.

NORSETH: Beryllium and gold are accumulated in the lysosomes like mercury.

FASSETT: Do you have any suggestions as to why methylmercurydicyandiamide did not accumulate in the lysosomes?

NORSETH: Perhaps because it is not degraded to inorganic mercury and the lysosome has a scavenger effect in the cell, taking out proteins which contain inorganic mercury and not those containing methylmercury, relative to the degree of denaturation.

SESSION V
ESTIMATION OF EFFECT ON HUMAN POPULATIONS AND EVALUATION OF CONTROL OF PESTICIDES

HUMAN RISK EVALUATION FOR VARIOUS POPULATIONS IN SWEDEN DUE TO METHYLMERCURY IN FISH

F. Berglund and M. Berlin

ABSTRACT

Two estimates of "allowable daily intake" (ADI) of methylmercury are presented. One ADI, equivalent to 0.7 mg Hg/week, was calculated from the linear relationship between intake and erythrocyte levels of mercury among fish-eating individuals in Sweden. The second ADI, equivalent to 0.42 mg Hg/week, equals the estimated equilibrium dose in a subject with a body burden of 60 mg Hg due to methylmercury. It is estimated that such a body burden would be associated with a toxic or near-toxic mercury level in the brain. These ADI's would result in approximate mercury levels in the erythrocytes of 120 and 70 mg/g, respectively, i.e. twelve and seven times as high as in a Swedish control population. Effects on the fetus or on the mitotic apparatus cannot be excluded at present. Only a small percentage of the Swedish population consumes enough freshwater fish to reach even the lower of the two ADI's proposed.

REPORT

VARIOUS METHODS MAY be used to protect the population from the hazards of toxic substances in the food, but the ideal way is to prevent the foreign substances from appearing in the food. For many chemicals (intentional food additives, pesticide residues, metals and metaloids) however, some degree of presence is tolerated. On the basis of toxicological data one may establish an "allowable daily intake" (ADI) of each chemical or group of chemicals. The total intake of the chemical through various foods should not exceed the ADI (5). When only incomplete toxicological data are available, however, a temporary ADI may be established (5). But if an ADI or a temporary ADI cannot be established, a "practical residue limit" may be used (5). "Practical residue limits" are based on levels found in "noncontaminated foods," and may thus differ between various food items and countries.

The sale of a certain kind of food may be prohibited due to contamination by a toxic (or nontoxic) substance, without any quantitative assessment of the dangers to the population. For example, the cranberry incident in the United States of America in 1959, resulted from the use of the pesticide aminotriazol on the plants and the deposition of residues on some of the marketed berries. This led to seizure of the cranberries by the Food and Drug Administration, because of the alleged potentially of amino-triazol for producing thyroid cancer. Since then it has been denied that animotriazol can produce a true cancer of the thyroid. The legal aspects of such an incident would probably vary in different countries.

The Problem of Methylmercury in Swedish Fish

High levels of mercury in fish from lakes, rivers and along the coasts of Sweden were reported by Westermark, Johnels and others since 1964. In general, fish from southern Sweden contain 0.02 to 10 mg Hg/kg, from northern Sweden 0.03 to 0.2 mg Hg/kg, and from the Atlantic Ocean 0.01 to 0.15 mg Hg/kg. In November 1965, Westöö (15) proved that the mercury in fish existed largely as a methylmercury compound. In view of the Japanese incidences of methylmercury intoxication in Minamata and Niigata, the Swedish health authorities in 1967 declared fish with mercury levels above 1 mg/kg to be unfit for human consumption, and banned nearly 1 per cent of Swedish fishing waters from commercial fishing. Thus most of the fish with 1 to 10 mg Hg/kg are excluded from commerce.

An attempt was then made to establish an ADI for mercury, and several alternative problems were considered. In 1966 two FAO and WHO expert committees proposed a "practical residue limit" of 0.02 to 0.05 mg/kg for mercury in foods (5). Clearly, this "practical residue limit" is not applicable to fish since, if strictly adhered to, it would prohibit fishing not only in most lakes in Sweden but also in many other countries; furthermore, some ocean fish would also be in the risk zone (e.g., tuna fish in the Indian Ocean). The Swedish health authorities have therefore taken it upon themselves to estimate the risk of toxic effects of methylmercury in various populations and the following factors have been under serious consideration:

1. the concentrations of methylmercury in different foods
2. the intake of methylmercury-contaminated foods by the population
3. the relationship between intake and body levels of methylmercury
4. toxic levels of methylmercury in man

1. *Concentration of methylmercury in different foods.* An average fish-free Swedish diet supplies somewhat less than 0.01 mg Hg per day. Fish, however, contain various amounts of mercury (*cf.* above), practically all of which is present as methylmercury. Since people who consume fish regularly are heavily exposed to methylmercury, it appeared possible to arrive at an estimate of maximum exposure to methylmercury by considering solely the intake of fish.

2. *The intake of methylmercury-contaminated fish by the population.* In 1967 the consumption of fish was investigated in three population groups in Sweden by the mailing of a questionnaire to individuals of these groups (9). One group consisted of 435 males, with ages ranging from 16 to 67 and randomly selected from the general populace. A second group consisted of 235 members of a Swedish lake fishermen's union; and a third group consisted of 230 members of any of the sea-fishermen's unions (ages 16 to 67 years). The two latter groups had a higher consumption of fish than the population in general. The response to the questionnaires averaged 81 per cent in the three groups. The groups were small when compared to the total population ($7\frac{1}{2}$ million). The consumption of fish among the population in general averaged 0.35 kg/week, but only $\frac{1}{4}$ of the fish originated from lakes or shore areas along the coasts. Only such fish were considered a toxicological problem, since they may have contained more than 0.2 mg Hg/kg. From the response to the questionnaire it appeared that not more than 15 per cent of the general population ate fish from these areas more than once a week.

3. *The relationship between intake and body levels of methyl-mercury.* On the basis of data published on the excretion of methylmercury under various experimental conditions one may assume a linear correlation at equilibrium between intake of methylmercury and concentration in each organ or tissue. [*cf.* Berglund and Berlin, contribution to Part III.] The mercury levels in the blood and hair are two readily accessible indices of the exposure to methylmercury. Methylmercury, in contrast to inorganic mercury, rapidly accumulates in the red blood cells and the mercury level in the red blood cells is therefore the best index of the total body burden of methylmercury at any time when equilibrium in the body can be postulated.

Even though methylmercury gives rise to much higher levels of mercury in the hair than in erythrocytes, the use of hair presents certain drawbacks. The concentration in hair varies on different parts of the body, possibly due to different growth rates, as has been shown, e.g., in the rat (6). Furthermore, external contamination may raise the mercury levels in hair. The relatively high mercury levels in hair of the Japanese (6.5 $\pm$ 3.2 μg/g compared to approximately 1.9 μg/g in nonresident Japanese) is supposedly due to local environmental conditions (7). Still, when hair samples are taken from the same part of the body in different subjects, e.g., the back of the head, there is often a good correlation between the levels of mercury in hair and erythrocytes over a wide range of exposure to methylmercury in the food (1). This applies to subjects on a relatively constant intake of methylmercury.

Birke *et al.* (1) studied mercury levels in blood and hair in seven subjects with a high intake of Swedish fish contaminated to varying degrees by methylmercury. One subject consumed approximately 150 g fish per day, with an average concentration of mercury of 6.7 mg/kg fresh weight, equivalent to 1.0 mg Hg/day or 7.0 mg Hg/week. The mercury level was 1 200 ng/g in his erythrocytes and 185 μg/g in his hair. The remaining subjects had a considerably lower intake of mercury, but the mercury levels in the erythrocytes seemed to be directly proportional to intake.

Tejning (14) investigated fifty-one fishermen living around Lake Vänern, the largest in Sweden. Their average intake of fish was 450 g/week. On the basis of published analyses on fish, Tejning calculated the average intake of mercury to be 0.39 mg/week. The average mercury level in the erythrocytes was 58 ng/g, i.e. almost six times as high as that of the control population (10 ng/g). Comparison of the data of Birke *et al.* (1) and Tejning (14) shows good agreement in the relationship between mercury intake and levels in erythrocytes.

Considering the obvious difficulties in obtaining exact figures (retrospectively) on fish intake and on the concentrations of mercury in the fish, the relationship between intake and mercury levels must be considered approximate.

4. *Toxic and nontoxic levels of methylmercury in man.* If hair and blood levels of methylmercury are proportional at equilibrium to the brain levels it should be of interest to look at available data in heavily exposed humans. In Niigata three patients with methylmercury intoxication had mercury levels in hair of 515, 565 and 763 μg/g, whereas in fifteen members of their families the levels varied between 15 and 412 μg/g (8) ; the mercury analyses were done by neutron activation technique. Diothizone analysis indicated that the patient with 515 μg Hg/g hair had 1,318 ng Hg/ml blood. The blood sample was taken one month later than the hair sample, and the mercury level in his erythrocytes was probably considerably higher than 2,600 ng/g at the onset of intoxication. In one patient 4,000 ng/ml blood was reported, equivalent to approximately 8,000 ng/g in the erythrocytes (10). Okinaka *et al.* (12) reported a level of 1,800 ng/ml blood three months after the onset of intoxication in one patient.

One of the subjects with no symptoms of intoxication reported by Birke *et al.* (1) seems to be of special interest, since data were available on both mercury intake and mercury levels in blood and hair. The reported intake of methylmercury through fish was equivalent to 1 mg Hg/day or 7 mg Hg/week. Mercury levels in erythrocytes (1,200 ng/g) and hair (185 μg/g) were less than half as high as in the cases of the toxicity reported above. Thorough clinical investigation did not reveal any signs of methylmercury intoxication, and currently it is suggested that this intake of mercury, i.e., 7 mg Hg/week, is a "no-effect dose" in man. This would require that the subject maintain a constant intake for at least one year. Calculation of a safety factor 10, which seems to be customary for "unintentional food additives," would yield an ADI of 0.1 mg Hg, equivalent to 0.7 mg Hg/week.

ESTIMATION OF ADI FROM BIOLOGICAL HALF-LIFE DATA IN MAN. The metabolism and retention of methylmercury 203 nitrate has recently been studied in three human subjects (2). Each took 3 μCi of the substance orally. The subjects were then measured daily, five days per week during the first two weeks, and then twice weekly in a whole-body counter. From the retention curves obtained a biological half-life of 65 to 74 days has been calculated, equivalent to a daily excretion of 1 per cent of the body burden. This approximately agress with the half-life data in erythrocytes (1). The mercury 203 was excreted mainly via the feces (25% of the dose in 29 days) and only a small fraction through the urine (1.3% in 29 days). Radioactivity was also measured over different parts of the body by scintillation scanning. Radioactivity was mainly located over the liver area, and 13 to 22 per cent was located in the head (3).

These data may also be used to calculate a toxic chronic dose of methylmercury. A critical point is that a concentration of mercury in the brain which causes toxicity, and we have roughly estimated this level to be 8 μg/g brain [Berglund and Berlin, session III], is equivalent to 12 mg Hg in the whole brain in the adult man. If this amount of mercury in the brain represents 20 per cent of the total body burden, the latter would be 60 mg. With a daily excretion of 1 per cent, the dose of methylmercury to maintain equilibrium would be equivalent to 0.6 mg Hg/day. If one accepts this as a minimum toxic chronic dose to adult man, and applies a safety factor 10, one obtains an ADI of 0.06 mg Hg/day, equivalent to 0.42 mg/week.

CONSIDERATION OF EFFECTS ON THE MITOTIC PROCESS AND ON THE FETUS. Ramel (13) studied the effect of organic mercury compounds on root cells of *Allium cepa*. The roots were grown in water solutions of the compounds for 4 to 72 hours. Disturbances of the mitotic spindle were caused by extremely low concentrations of the alkyl- and phenylmercury compounds, about 2.5×10^{-7} M or 50 ng Hg/g. The corresponding value for colchicine is almost a thousand times higher. This concentration may be compared with the mercury level in erythrocytes in man: in a Swedish control population around 10 ng/g, but frequently exceeding 50 ng/g in Swedes exposed to methylmercury by occupation (fertilizer factories) or by high consumption of freshwater fish (14), with a maximum level of 1,200 ng/g in one subject (1). An analysis is in progress concerning possible chromosomal aberrations in blood cultures from humans exposed to methylmercury. As yet no estimations of ADI can be made on the basis of the small amount of data concerning effects on mitosis.

Intrauterine toxic effects of newborns have been reported in Japan and Sweden. The Swedish case (4) occurred in family using seed dressed with methylmercury dicyandiamide. In the Swedish family and in two families in Japan (11) the detailed reports reveal that only one mother had any symptoms of toxicity which consisted of parestesia which disappeared after delivery. In the three families, several other members were severely affected by methylmercury poisoning, which clearly showed that the family exposure to methylmercury had been high. The two Japanese infants, which died at the age of $2\frac{1}{2}$ and 6 years, respectively, exhibited pathological changes in the central nervous system typical of methylmercury poisoning. The incidence of intrauterine methylmercury poisoning was alarmingly high in Minamata, Japan — twenty-two out of 110 cases. On the other hand, no cases of intrauterine intoxication have been reported from Niigata. The discrepancy between Minamata and Niigata in this respect is puzzling but may be due to different criteria of diagnosis. One must, at present, conclude that the fetus is more sensitive than its mother to methylmercury.

PRACTICAL CONSEQUENCES. ADI values for methylmercury have been calculated by two methods. The purpose has been to prevent the appearance of neurological damage. The ADI calculated from the relationship between the estimated intake of methylmercury via fish and the mercury content of the erythrocytes in a seemingly healthy subject were both equivalent to 0.1 mg Hg/day or 0.7 mg Hg/week. From the biological half-life and bodily distribution of methylmercury in man and a rough estimate of the lowest "toxic level" of methylmercury in the brain, an ADI equivalent to 0.06 mg Hg/day or 0.42 mg Hg/week has been calculated. These ADI's would result in mercury levels of 120 and 70 ng/g in the erythrocytes. The agreement between the two ADI's is fairly good. It should be emphasized that both ADI's are based on approximate data. Furthermore, no quantitative assessment of the sensitivity of the fetus or of the mitotic system has been made, even though a safety factor has been included. The estimations therefore have the character of a "temporary ADI," and the practical course to be taken has to be a matter of judgment.

In Sweden the regular diet, excluding fish, supplies about 10 μg Hg/day or 0.07 mg Hg/week. The two ADI's would therefore allow through the consumption of fish the addition of 0.63 mg Hg/week or 0.35 mg Hg/week, respectively. Assuming a maximal mercury level in fish of 1 mg/kg, the allowances would equal 0.63 kg fish or 0.35 kg fish per week, i.e., 2 to 4 meals of freshwater fish every week. The lower figure is equal to the average consumption of fish in Sweden, but most of the fish is ocean fish and contains less than 0.15 mg Hg/kg. Less than 10 per cent of the population eat more than 0.35 kg freshwater or coastal fish per week, and only in a few of these places does the mercury concentration approach 1 mg/kg fish. It is hoped that this small fraction of the population, as advised by the health authorities, limits its consumption of fish from potentially dangerous areas.

In 1967, a practical residue limit of 1 mg Hg/kg fish was established in Sweden, and commercial fishing was banned in almost 1 per cent of available fishing waters, thus presumably preventing the sale and consumption of fish with mercury concentrations above 1 mg/kg. At present, any lower residue limit for fish, e.g., 0.5 or 0.2 mg Hg/kg, would involve the banning of probably more than half or all fishing areas in Sweden, including lakes, rivers and coastal areas. Steps have therefore been taken to diminish industrial waste disposal of mercurials in an effort to prevent any further contamination of fishing areas with mercury compounds.

REFERENCES

1. BIRKE, G.; JOHNELS, A.G.; PLANTIN, L-O.; SJÖSTRAND, B., and WESTER-MARK, T.: *Läkartidningen, 64* (3) :628, 1967.
2. EKMAN, L.; GREITZ, U.; PERSSON, G., and ÅBERG, B.: *Nord Med, 79:*450, 1968.
3. EKMAN, L.; GREITZ, U.; SNIHS, J-O., and ÅBERG, B.: *Nord Med, 79:*456, 1968.
4. ENGLESON, G., and HERNER, T.: *Acta Paediat, 41:*289, 1952.
5. FAO and WHO: Pesticide Residues in Food. WHO Techn Report, Series No. 370, 1967.
6. GAGE, J.C.: *Brit J Industr Med, 21:*197, 1964.
7. HOSHINO, O.; TANZAWA, K.; HASEGAWA, Y., and UKITA, T.: *J Hyg Chem Sci, 12:*90, 1966.
8. HOSHINO, O.; TANZAWA, K.; TERAO, T., and UKITA, T.: *J Hyg Chem Sci, 12:*94, 1966.
9. JOHNSON, E.: To be published.
10. LUNDGREN, K-D., and SWENSSON, Å.: *J Industr Hyg Toxic, 31:*190, 1949.
11. MATSUMOTO, H.; GOYO, K., and TAKEUCHI, T.: *J. Neuropath Exp Neurol, 24:*563, 1965.
12. OKINAKA, S.; YSHIKAWA, M.; MOZAI, T.; MIZUNO,, Y.; TERAO, T.; WATANABE, H.; OGIHARA, K.; HIRAI, S.; YOSHINO, Y.; INOSE, T.; ANZAI, S., and TSUDA, M.: *Neurologia, 14:*69, 1964.
13. RAMEL, C.: *Hereditas, 57:*445, 1967.
14. TEJNING, S.: Report 670831, Dept. of Occupational Medicine, University of Lund, Sweden.
15. WESTÖÖ, G.: *Acta Chem Scand, 20:*2131, 1966.
16. WESTÖÖ, G.: *Vår Föda,* 1967, pp. 1-7 (in Swedish) .

DISCUSSION

VOSTAL: Are there any data about the effectiveness of the absorption of methylmercury from the intestinal lumen in people? How much of the material ingested in fish meat is absorbed, and how much is eliminated?

BERGLUND: According to Eichmann and co-workers, the total body burden measured over a seventy-day period corresponded to the total dose they took, which indicates complete absorption. There is always the question of how to extrapolate from a small dose to a large one, but our rat experiments indicate that a simple extrapolation is valid in this case. I should point out also that we have not taken into consideration any possible deleterious effects on the genetic mechanism and we have not been able to consider data on possible intrauterine damage.

SUZUKI: Why did you adopt a safety factor of 10?

BERGLUND: According to publications by the Food and Agriculture Organization, Rome, Italy; the World Health Organization, Geneva Switzerland, and others, the choice of a safety factor is arbitrary. The factor is usually chosen so that certain pesticides, certain food additives, etc. can still be used in practice. The choice of a safety factor is therefore dependent on a balance between damage and benefit, and there is no absolute answer to this. For example, for food additives (intentional or unintentional), safety factors first specified the mean intake; but now the ninth decile of intake is used, *i.e.,* that figure exceeded by no more than 15 per cent of the population. The safety factor for the intake of methylmercury in fish in Sweden calculated in this way is probably smaller than the safety factors for any other pesticide or food additives in Sweden.

Chapter 23

THE INFLUENCE OF OTHER FACTORS ON THE TOXICITY OF PESTICIDES

W. F. Durham

ABSTRACT

The primary factor which determines whether or not a chemical will have a toxic effect is dosage. However, the influence of other factors is of great importance in determining the effect of the pesticide on target or nontarget organisms. These interacting influences may be subdivided into factors which (a) primarily affect the response of the individual and (b) modify the pesticide before its absorption into the animal body. These factors may increase or decrease the inherent toxicity of the pesticide. The unifying concept which underlies many, but not all of these interactions involves the drug detoxifying enzyme in the liver microsomes. It is generally agreed that the net effect of microsomal induction is adaptive. The fact that such an adaptive change occurs is reassuring with regard to the capability of man to adapt to an environment which is increasingly contaminated with foreign chemicals.

REPORT

THE INFLUENCE OF OTHER factors on the toxicity of pesticides is of great importance in determining the effect of the pesticide on target or non-target organisms. Among the factors now known to influence the toxicity of pesticides are some which decrease as well as some which increase the toxic effect.

Since human health is the subject of our discussion in this session, I will review primarily, interactions which relate to the human population. The population at risk represents a wide range of individuals differing with regard to age, sex, nutrition, health status, and other factors. The environmental conditions under which exposure occurs vary, for example, with regard to temperature and light. The route of exposure may be oral, dermal, respiratory, or some combination of these three. The formulations in which a specific pesticide is dispersed can vary with regard to physical characteristics or content of added chemical compounds.

Also, in the environment of our complex, present day civilization, the population is continually in contact with a wide variety of foreign chemicals in addition to the pesticides.

These interacting influences may be subdivided into factors which (a) primarily affect the response of the individual and (b) modify the pesticide before its absorption into the animal body. Some of these factors may influence either the response of the individual or the chemical nature of the pesticide or both under different conditions.

The more important interacting factors influencing pesticide toxicity seem to be those which modify the response of the individual. These factors include age, sex, nutrition, disease, temperature, exposure and interaction with chemicals, drugs and other pesticides.

Age

The influence of age on toxicity is most evident in newborn animals which in several species have an almost complete lack of ability to metabolize certain drugs (18). However, this capacity increases quite rapidly during the first few weeks of life.

The lack of microsomal activity in the neonatal liver correlates well with the observed sensitivity of the newborn of several species to some pesticides. Thus, the toxicity of O-Ethyl O-(4-nitrophenyl) phenyl-phosphorothioate (EPN) (42) and of ethyl fenthion (8) was found to be much higher in 23-day-old male rats than in adult rats of the same sex.

In studies with drugs generally, Hoppe *et al.* (32) and Yeary and Benish (55) compared toxic levels in newborn and in adult mice and rats, respectively, and showed that adults were generally less susceptible.

As far as pesticides are concerned, Lu *et al.* (39) have studied the comparative oral toxicity of malathion, DDT, and dieldrin in rats of different ages. They found the toxicity of malathion increased in the following order: adult, preweaning, newborn. However, for both DDT and dieldrin, adult rats were more susceptible than newborn rats. With the exception of the newborn rats, only small differences were noted between the other age groups tested.

Brodeur and DuBois (2) compared the acute oral toxicity of sixteen anticholinesterase insecticides to weanling and to adult male rats. All of the phosphorothioates and phosphorodithioates tested were more toxic to weanlings than to adults. The greatest age differences were seen with EPN and carbophenothion (Trithion) to which weanlings were about five and four times, respectively, more susceptible than adult males. On the contrary, a phosphoroamide, schradan (OMPA), differed from all of the other compounds tested in that adult males were more susceptible than weanlings.

Some information on age differences in susceptibility is also available from cases of accidental human poisoning. In instances in which parathion-contaminated food has been eaten by people of different ages, death has occurred mainly or exclusively among children (34). Children five to six years old were killed by eating an estimated two mg. of parathion, a dosage of about 0.1 mg. parathion/kg body weight. In contrast, daily doses of 7.2 mg. (about 0.1 mg./kg.) given to adult volunteers for a period of 42 days produced no signs of poisoning and no symptoms other than a moderate decrease in blood cholinesterase level (15).

However, it appears that age has at most a small influence on the storage level of the chlorinated hydrocarbon pesticides in human fat. Thus, Laug *et al.* (36) did not detect an effect of age on storage of DDT. Read and McKinley (46) and Wassermann *et al.* (52) did not note significant differences in DDT or DDE storage levels in relation to age, nor did Robinson *et al.* (47) find an influence of age on the adipose tissue concentration of total DDT-derived material or of dieldrin. However, Egan *et al.* (16) found that tissue storage levels of benzene hexachlorine (BHC), dieldrin, and DDT were significantly higher for persons over forty years old than for those in the one- to thirty-nine-year age group.

Sex

The importance of sex in determining the toxicity of chemicals differs for different species of animals. The rat seems to show more variation between the sexes in its response to chemicals than any other species. This fact may have led to more concern than is justified regarding possible differences in the susceptibility of men and women to chemicals. Calculations made by Hayes (25) from data published by Gaines (23) for the oral toxicity of sixty-five pesticides showed that the difference in the oral lethal dose to 50 per cent of the animals (LD50) for male and female rats ranged from 0.21 (indicating greater susceptibility of the female) to 4.62 (indicating greater susceptibility of the male), and averaged 0.94. The corresponding factors for the dermal toxicity of thirty-seven pesticides were 0.11 to 2.93 with an average of 0.81. Sex differences seemed to be more important with regard to the organophosphorus than for the chlorinated hydrocarbon pesticides.

In the rat, these sex differences may be modified by hormones. Durham *et al.* (10) studied the influence of hormones and of gonadectomy on the storage of DDT and DDE in the rat. Testosterone propionate or oophorectomy decreased DDT storage in female rats while diethylstilbestrol or testectomy increased DDT storage in male rats. The effects on DDE storage were similar but of lesser magnitude. Administration of a male sex hormone to female rats and of female sex hormone to male rats tended to equalize their susceptibility to a single dose of parathion (9).

The evidence for or against a sex difference in DDT storage in species other than the rat is not so clear. Such differences were looked for but not found in relation to the storage of DDT in monkeys (11, 44). Robinson *et al.* (47) found somewhat more dieldrin, p,p'-DDE, and total DDT-derived material in fat from men than in fat from women. Zavon *et al.* (56) observed the same relationship for dieldrin, DDE, p,p'-DDT, o,p'-DDT, and heptachlor epoxide. Certain other authors failed to find a difference between men and women in the storage of DDT in fat (4, 16, 26, 37, 40, 46, 52). However, Dale *et al.* (5) found that plasma and whole blood levels of chlorinated hydrocarbon insecticides were significantly higher in men than in women.

Nutrition

Apparently only extremes of nutrition have produced significant alterations in the toxicity of pesticides (28). Various mammals and even fish are relatively resistant to poisoning by DDT if they are fat rather than thin (27). The same result has been produced with dieldrin under experimental conditions (1). Rats which had been given relatively large doses of DDT to produce significant fat storage levels of the compound developed characteristic DDT tremors during starvation (17). However, attempts to produce symptoms of poisoning in animals previously dosed with dieldrin by starvation have not been successful (28).

In addition to the protective mechanism provided by body fat stores to shield the sensitive nervous tissue from the chlorinated hydrocarbon pesticides, the reduced liver microsomal enzyme activity of starved animals may influence their susceptibility to poisoning. Both the mouse and the rat showed increased toxic effects from DDT when the percentage of fat in the basal diet was increased from 5 per cent to 15 per cent (48). A low protein diet accentuated the toxic effect of dieldrin as measured by increased mortality, increase in liver lipids, decrease in total liver content of vitamin A, and more marked histopathology (38). A dietary deficiency of riboflavin or nicotinic acid accentuated dieldrin toxicity in rats (51).

Disease

While in a diseased state resulting from an infectious agent, chemical intoxication, tumor, degenerative process, trauma, or other causes, the animal may, depending upon the severity of the disease, be spending a considerable fraction of its available physical resources dealing with the disease process (12). Thus, it might be expected that under certain conditions the response of a diseased animal to the added stress of exposure to a pesticide might differ from that of a healthy animal. However, very little study has been done to explore this hypothesis.

In a study of the interaction of trichinosis and DDT poisoning in rats (28), it was found that increase in mortality from DDT could be explained by the weight loss induced by the infection. However, in rats whose livers had been severely damaged by carbon tetrachloride, the toxicity of methoxychlor was increased markedly as was its propensity for storage in body fat (37). In an acute study there was little difference in DDT toxicity noted between normal rats and rats with carbon tetrachloride-induced liver damage (33).

With regard to the interaction between pesticides and disease in man, Maier-Bode (40) found no essential difference in storage of DDT or DDE between twenty-one persons who died of cancer and thirty-nine persons who died of other diseases. Robinson *et al.* (47) detected no differences in total DDT-derived material or dieldrin between fifty biopsy and fifty necropsy samples. Nor was there any correlation for the necropsy samples between storage level and cause of death, classified as neoplasm, cardiovascular disease, infection, or accident. More recently Hoffman and his colleagues analyzed fat storage levels of DDT, DDE, and lindane from 688 patients who died from a variety of diseases (31). They found no correlation between pesticide level and any disease studied.

The metabolism of parathion in man as measured by excretion levels of p-nitrophenol was influenced by intercurrent disease and skin and kidney pathology (6).

Ganelin and his colleagues (24) studied the possible health hazards of environmental exposure to parathion to a group of persons with bronchial asthma. There were no indications that this heavy exposure to parathion produced either clinical or laboratory signs of poisoning and the incidence of asthmatic attacks was similar for the exposed patients and for a control group who did not live near treated fields.

In studies with cell cultures of human tissues, Gabliks (22), noted that preparations treated with chlordane, dimethoate, dinocap, disulfoton, or trichlorfon were more susceptible to polio virus infection and that the infected cells degenerated more rapidly than controls. However, malathion treated cells did not show any change in susceptibility to the virus. In tests with diphtheria toxin, no changes in reactivity were noted after treatment with chlordane, dimethoate, disulfoton, dicofol, malathion, or trichlorfon. However, dinocap did sensitize the cells to the lethal action of diphtheria toxin.

Temperature

In contrast to the factors considered above, temperature is one variable which may have an influence on the reaction of the individual as well as on the chemical nature of the toxicant.

DDT, warfarin (35), and azinphosmethyl (53) are more toxic to rats at higher than at lower temperatures. It is well known that the nitrophenols, including those used as pesticides, are more toxic at higher environmental temperatures.

Limited information in man is also available both from studies on volunteers and from studies of exposed workers. Volunteers given dermal doses of parathion showed an increase in their excretion of paranitrophenol associated with an increase in ambient temperature (21). In studies of orchard workers engaged in spraying parathion, excretion rates of paranitrophenol for men with apparently similar exposures were generally higher during the hot days of July than during the cool days of May (13). In studies carried out in Israel, Wasserman *et al.* (36) noted that both respiratory and dermal exposure of spraymen were greater for day time than for nocturnal spraying and theorized that these differences were due to the diurnal variation in temperature.

Exposure

The route of exposure is important in determining the absorption and therefore the toxicity of pesticides. Compounds are usually more toxic by the oral than by the dermal route as noted for sixty-four of the sixty-seven pesticides studied in rats by Gaines (23). However, there were three exceptions involving compounds more toxic by the dermal route.

Techniques are available for measuring the respiratory and dermal contamination of exposed workers (14). One technique involves measurement of skin contamination by absorbent pads and estimation of potential respiratory exposure using a specially designed respirator equipped with an absorbent pad. Using these techniques in thirty-three different work activities involving ten different pesticides, potential dermal exposure was found to be greater than respiratory exposure in every situation studied.

Similar results were obtained by a different procedure involving estimation of total absorption of the pesticide from measurement of the compound or its biotransformation product in the urine or other excreta and estimating respiratory absorption from respiratory exposure or air concentration values (11). Dermal absorption was calculated from the difference between the total absorption and the potential respiratory absorption. The value obtained through use of this method should represent a minimal value for dermal absorption (11). Applying this procedure, the parathion absorption by orchard spraymen indicated that dermal exposure was greater than respirator exposure.

Interaction With Chemicals, Drugs and Other Pesticides

A number of chemicals, including chlorinated organic pesticides, carcinogenic hydrocarbons and drugs are capable of inducing microsomal enzyme activity. Early investigations in this regard with pesticides were made by Hart *et al.* (30), who noted that chlordane stimulated hepatic microsomal enzyme activity as measured by *in vitro* metabolism of hexabarbital, aminopyrene, and chloropromazine. Street (49) reported that storage of dieldrin in fat of female rats was markedly depressed when DDT and dieldrin were fed simultaneously. A similar effect on fat storage of dieldrin in the rat was also produced by certain drugs, including phenobarbital (3), aminopyrene, tolbulamide, phenylbutazone and heptabarbital (50).

With regard to the organo-phosphorus pesticides, interactions with chemicals which induce liver microsomal enzyme activity may produce either an increase or decrease in toxicity. Whether toxicity is increased or decreased seems to depend on the balance between the intoxication and detoxication reaction.

Also, the various combinations of the organo-phosphorus compounds have been shown to potentiate each other. Potentiation among these compounds was first shown to occur for malathion plus EPN (19). The maximum synergistic effect noted (88 to 134 times) has been observed with malathion plus triorthocreslphosphate (43).

Among the factors which may modify the toxic effect of a pesticide by influencing the chemical nature of the toxicant are formulation and light.

Formulation

The type of formulation in which a pesticide is dispersed can significantly effect the toxic hazards. For example, on dermal application to the rats, parathion is more toxic in xylene solution than as an aqueous suspension of water wettable powder (29). Dieldrin is absorbed almost as well in the dry powder form as in xylene solution, while DDT is absorbed through the skin to a negligible degree (27).

The addition of an emulsifier to the formulation has been shown to increase the dermal absorption of demeton (7). Also, addition of a surface active agent increased the oral toxicity to rats of DDT, parathion, and thiram, but not carbaryl (54).

Light

It has been shown that ultraviolet light can catalyze the oxidation of a number of organophosphorus pesticides. For example, parathion was converted to the more toxic analog paraoxon (20). It may be that this particular reaction may explain the sporadic outbreaks of parathion poisoning which have occurred in crop workers sometimes at relatively long intervals after application (45). The amounts of parathion present on the crop or foilage at the time of exposure do not seem to be great enough to account for the illness. However, the presence of paraoxon has been demonstrated on foilage in a peach orchard in which a residual poisoning incident occurred (41).

The primary factor which determines whether or not a chemical will have a toxic effect is dosage. However, as noted above, there are a number of other factors which may influence the response, including some factors which act on the animal organism and some which act on the toxicant molecule before absorption. The unifying concept which underlies many of the interactions of pesticides with other factors involves the drug detoxifying enzyme systems of the liver microsome. It is generally agreed that the net effect of microsomal induction is adaptive. The fact that such an adaptive change occurs is reassuring with regard to the capability of man to adapt to an environment which is increasingly contaminated with foreign chemicals.

REFERENCES

1. BARNES, J.M., and HEATH, D.F.: Some toxic effects of dieldrin in rats. *Brit J Indust Med, 21:*280, 1964.
2. BRODEUR, J., and DU BOIS, K.P.: Comparison of acute toxicity of anticholinesterase insecticides to weanling and adult male rats. *Proc Soc Expt Biol Med, 114:*509, 1963.
3. CUETO, C., JR., and HAYES, W.J., JR.: Effect of phenobarbital on the metabolism of dieldrin. *Toxic Appl Pharmacol, 7:*481, 1965.
4. DALE, W.E.; COPELAND, M.F., and HAYES, W.J., JR.: Chlorinated insecticides in the body fat of people in India. *Bull Wildl Health Organ, 33:*471, 1965.
5. DALE, W.E.; CURLEY, A., and CUETO, C., JR.: Hexane extractable chlorinated insecticides in human blood. *Life Sci, 5* (1) : 47, 1966.
6. DAVIES, J.E.; DAVIS, J.H.; FRAZIER, D.E.; MANN, J.B., and WELKE, J.O.: The use of P-nitrophenol in the Differential Diagnosis of Acute Parathion Intoxication and in the Daily Surveillance of the Occupationally-exposed Individual. Abstr. 150th Meeting Amer. Chem Soc., Atlantic City, N.J., 1965.
7. DEICHMANN, W.B.; BROWN, P., and DOWNING, C.: Unusual protective action of a new emulsifier for the handling of organic phosphates. *Science, 116:*221, 1952.
8. DU BOIS, K.P., and PUCHALA, E.: Studies on the sex differences in toxicity of a cholinergic phosphorothioate, *Proc Soc Exp Biol Med, 107:*908, 1961.
9. DU BOIS, K.P.; SALERNO, P.R., and COON, J.M.: Studies on the toxicity and mechanism of action of p-nitrophenyl diethyl thionophosphate (parathion) , *J Pharmacol Exp Ther, 95:*79, 1949.
10. DURHAM, W.F.; CUETO, C., JR., and HAYES, W.J., JR.: Hormonal influences on DDT metabolism in the white rat. *Amer J Physiol, 187:*373, 1956.
11. DURHAM, W.F., and WOLFE, H.R.: An additional note regarding measurement of the exposure of workers to pesticides. *Bull WHO, 29:*279, 1963.
12. DURHAM, W.F.: The interaction of pesticides with other factors. *Residue Rev,* 1967, vol. 18.
13. DURHAM, W.F.; WOLFE, H.R.; ELLIOTT, J.W., and ARTERBERRY, J.D.: Manuscript in preparation.
14. DURHAM, W.F., and WOLFE, H.R.: Measurement of the exposure of workers to pesticides. *Bull WHO, 26:*75, 1962.

15. EDSON, E.F.: Fisons Pest Control Limited. Mimeographed report from Medical Department, 1956.

16. EGAN, H.; GOULDING, R.; ROBURN, J., and TATTON, J.O.G.: Organochlorine pesticide residues in human fat and human milk. *Brit Med J, 2:*66, 1965.

17. FITZHUGH, O.G., and NELSON, A.A.: Chronic oral toxicity of DDT (2,2-bis-(p-chlorophenyl)-1,1,1-trichloroethane), *J Pharmacol Exp Ther, 89:*18, 1947.

18. FOUTS, J.R., and ADAMSON, R.H.: Drug metabolism in the newborn rabbit. *Science, 129:*897, 1959.

19. FRAWLEY, J.P.; FUYAT, H.N.; HAGAN, E.C.; BLAKE, J.R., and FITZHUGH, O.G.: Marked potentiation in mammalian toxicity from simultaneous administration of two anticholinesterase compounds. *J Pharmacol Exp Ther, 121:*96, 1957.

20. FRAWLEY, J.P.; COOK, J.W.; BLAKE, J.R., and FITZHUGH, O.G.: Insecticide stablity: Effect of light on chemical and biological properties of parathion. *J Agr Food Chem, 6:*28, 1958.

21. FUNCKES, A.J.; HAYES, G.R., JR., and HARTWELL, W.V.: Insecticide activity in man. Urinary excretion of paranitrophenol by volunteers following dermal exposure to parathion at different ambient temperatures. *J Agr Food Chem, 11:*455, 1963.

22. GABLIKS, J.: Responses of cell cultures to insecticides. II. Chronic toxicity and induced resistance. *Proc Soc Exp Biol Med, 120:*168, 1965.

23. GAINES, T.B.: Unpublished data.

24. GANELIN, R.S.; CUETO, C., JR., and MAIL, G.A.: Exposure to parathion: Effect on general population and asthmatics. *J Amer Assoc, 188:*807, 1964.

25. HAYES, W.J., JR.: Toxicity of pesticides to man: risks from present levels. *Proc Roy Soc [Biol], 167:*101, 1967.

26. HAYES, W.J., JR., QUINBY, G.E.; WALKER, K.C.; ELLIOTT, J.W., and UPHOLT, W.M.: Storage of DDT and DDE in people with different degrees of exposure to DDT. *Arch Ind Health, 18:*398, 1958.

27. HAYES, W.J., JR.: Pharmacology and toxicology of DDT. In Muller, P. (Ed.): *DDT the Insecticide Dichlorodiphenyltrichloroethane and its Significance.* Basel, Birkhauser, (1959, vol. 2, pp. 9-247.

28. HAYES, W.J., JR.: Unpublished data.

29. HAYES, W.J., JR., and PEARCE, G.W.: Pesticides formulation. Relation to safety in use. *J Agr Food Chem, 1:*466, 1953.

30. HART, L.G.; SCHULTICE, R.W., and FOUTS, J.R.: Stimulatory effects of chlordane on hepatic microsomal drug metabolism in the rat. *Toxic Appl Pharmacol, 5:*371, 1963.

31. HOFFMAN, W.S.; ADLER, H.; FISHBEIN, W.I., and BAUER, F.C.: Relation of pesticide concentrations in fat to pathological changes in tissues. *Arch Environ Health, 1967, vol. 15.

32. Hoppe, J.O.; Duprey, L.P., and Dennis, E.W.: Acute toxicity in the young adult and the newborn mouse. *Toxic App Pharmacol, 7:*486, 1965.

33. Judah, J.D.: Studies on the metabolism and mode of action of DDT. *Brit J Pharmacol, 4:*120, 1949.

34. Kanagaratnam, K.; Boon, W.H., and Hoh, T.H.: Parathion poisoning from contaminated barley. *Lancet, 1:*538, 1960.

35. Keplinger, ML.L.; Lanier, G.E., and Deichmann, W.B.: Effects of environmental temperature on the acute toxicity of a number of compounds in rats. *Toxic Appl Pharmacol, 1:*156, 1959.

36. Laug, E.P., and Prickett, C.S.: Occurrence of DDT in human fat and milk. *Arch Ind Hyg Occup Med, 3:*245, 1951.

37. Laug, E.P., and Kunze, F.M.: Effect of carbon tetrachloride on the toxicity and storage of methoxychlor in the rat. *Fed Proc, 10:*318, 1951.

38. Lee, M.; Harris, K., and Trowbridge, H.: Effects of the level of dietary protein on the toxicity of dieldrin for the laboratory rat. *J Nutr, 84:*136, 1964.

39. Lu, F.C.; Jessup, D.C., and Lavallee, A.: Toxicity of pesticides in young rats versus adult rats. *Food Cosmet Toxic, 3:*591, 1965.

40. Maier-Bode, H.: DDT im Korperfett des Menschen. *Med Exp (Basel), 1:*146, 1960.

41. Milby, T.H.; Ottoboni, F., and Mitchell, H.W.: Parthion residue poisoning among orchard workers. *JAMA, 189:*351, 1964.

42. Murphy, S.D., and DuBois, K.P.: The influence of various factors on the enzymatic conversion of organic thiophosphates to anticholinesterase agents. *J Pharmacol Exp Ther, 124:*194, 1958.

43. Murphy, S.D.; Anderson, R.L., and Du Bois, K.P.: Potentiation of toxicity of malathion by triorthotolylphosphate. *Proc Soc Exp Biol Med, 100:*483, 1959.

44. Ortega, P.; Durham, W.F., and Mattson, A.: DDT in the diet of the rat. *Public Health Monogr, 43:* (484) 1956.

45. Quinby, G.E., and Lemmon, A.B.: Parathion residues as a cause of poisoning in crop workers. *JAMA, 166:*740, 1958.

46. Read, S.I., and McKinley, W.P.: DDT and and DDE content of human fat. *Arch Environ Health, 3:*209, 1961.

47. Robinson, J.; Richardson, A.; Hunter, C.G., and Crabtree, A.N.: Organocholrine insecticide content of human adipose tissue in southeastern England. *Brit J Ind Med, 22:*220, 1965.

48. Sauberlich, H.E., and Baumann, C.A.: Effect of dietary variations upon the toxicity of DDT to rats or mice. *Proc Soc Exp Biol Med, 66:* 642, 1947.

49. Street, J.C.: DDT antagonism to dieldrin storage in adipose tissue of rats. *Science, 146:*1580, 1964.

50. Street, J.C.; Wang, M., and Blau, A.D.: Drug effects on dieldrin storage in rat tissue. *Bull Environ Contam Toxic, 1:*6, 1966.

51. TINSLEY, I.J.: Nutritional Interactions in Dieldrin Toxicity. Abstr. Meeting Amer. Chem. Soc., Phoenix, Ariz., 1966.
52. WASSERMANN, M.; GON, M.; WASSERMANN, D., and ZELLERMAYER, L.: DDT and DDE in the body fat of people in Israel. *Arch Environ Health 11*:375, 1965.
53. WASSERMANN, M.; ZELLERMAYER, L., and GON, M.: L'etude de la toxicologie des pesticides en climat subtropical. I. L'intensite de l'exposition toxique pendant l'application des pesticides. Excerpta Med., Internat. Congress Ser. No. *62*, Proc. 14th Internat. Congress on Occupational Health, Madrid, 1963.
54. WEINBERG, M.S.; MORGAREIDGE, K., and OSER, B.L.: The Influence of Dispersing Agents on the Acute Oral Toxicity of Pesticides. Abstr. Meeting Amer. Chem. Soc., Phoenix, Ariz., 1966.
55. YEARY, R.A., and BENISH, R.A.: A comparison of the acute toxicities of drugs in newborn and adult rats. *Toxic Appl Pharmacol, 7*:504, 1965.
56. ZAVON, M.R.; HINE, C.H., and PARKER, K.D.: Chlorinated hydrocarbon insecticides in human body fat in the United States. *JAMA, 193*: 837, 1965.

DISCUSSION

BERG: You mentioned the potentiation of malathion by another insecticide (EPN). Could you explain the mechanism?

DURHAM: It is thought that the mechanism is inhibition by one of the organophosphates of the enzyme which degrades the other organophosphate. There is a phosphatase present in the body which breaks down these chemicals and is apparently a target for one of them.

BERG: Would this be analogous to the way in which Dr. Welch used carbon tetrachloride in one of his experiments?

WELCH: There are esterases in the liver and other organs which are not microsomal. They can hydrolyze and detoxify organophosphates like malathion and EPN. Other organophosphates such as triorthocresyl phosphate can inactivate these esterases and cause and increase in toxicity of malathion and EPN, as Dr. Durham mentioned. Triorthocresyl phosphate can inhibit the breakdown of malathion and increase its toxicity some hundredfold [Murphy, S. D., et al.: *Proc Soc Exp Biol Med, 100:*483, 1959]. These esterases are more generally distributed in the body than are the oxidative enzymes I was discussing, which are found in the microsomes of the liver.

FASSETT: With regard to the influence of light on the physiological reactions, dermatologists are reporting more and more cases of phototoxicity and photosensitization to various chemicals. There is the famous example of the outbreak in Turkey caused by eating hexachlorobenzene-treated seeds. There were intense light-induced skin reactions and excretion of porphyrins in the urine. Dermatologists are quite concerned about this and have reported occasional phototoxic responses to a variety of material, for example, to certain halogenated anti-biotics in soap.

WEST: When phenothiazine tranquilizers are contraindicated in the treatment of phosphate ester poisoning, is it generally known what mechanism is responsible? Are the phosphates themselves potentiated, or the symptoms of phosphate poisoning exacerbated indirectly?

DURHAM: I do not think we really know this. There are some indications of direct potentiation.

WILLS: I think it is wrong to say that phenothiazines are contraindicated. Rather, the administration of an organophosphate to a person who has been taking a tranquilizer regularly is contradicted. If the person has not been taking a tranquilizer, chloropromazine actually serves as a fairly good antichlorinergic drug; if he has been taking chloropromazine for a period of time before he is exposed to an organophosphate, then he is more sensitive to the organophosphate than a person who has not been taking tranquilizers. What the mechanism of the combined action is, I am not sure; it seems to be related to the central nervous system, since the animals that have been subjected to chronic exposure to chloropromazine before administration of an organophosphate show more prominent central nervous effects than unexposed animals.

Chapter 24

DETECTION AND CONTROL OF PESTICIDE POISONING: A PUBLIC HEALTH PROBLEM

I. WEST

ABSTRACT

From a public health standpoint, trying to control pesticide poisoning consists too often of reacting to untoward events rather than being able to change the basic circumstances which lead to poisoning. This age of technology is providing a great number of useful and valuable chemicals. Economic incentives produce much greater pressure for their wide distribution and use than pressures for education and training of persons who will be using them, or for devising practical protective gear, equipment, and procedures so that hazardous chemicals can be handled safely.

Further, this wide and rapid distribution of chemicals finds the practicing physician in the difficult position of recognizing and treating the chemical casualties. The same economic pressures leading to the wide use of modern chemicals do not carry equal pressures for setting up medical toxicological information and consultation services commensurate with today's need. There are an estimated four hundred to six hundred different chemicals used for pest control alone. They are formulated into almost sixty thousand mixtures and trade-named preparations. There is no way for the practicing physician to keep up with this mass of rapidly changing data which he may suddenly need to have access to when faced with a possible case of poisoning.

In spite of the tremendous amount of research which has been carried out with respect to pesticides, it has not necessarily been directed toward preventing or controlling many of the human health problems which do arise. The methodology for pursuing the more complex human health questions does not seem to be available.

REPORT

Trying to detect and control pesticide poisoning is a frustrating but fascinating activity. One is suddenly confronted with such questions as:

1. What happens to parathion if it is baked in a sweet roll at 350°F for 30 minutes?

2. What should be decided about a city's drinking water after its only reservoir has been generously sprayed by air with DNOC (dinitro ortho cresol) ?

3. What help can you offer the highway patrol trying to untangle a traffic jam caused by a swath of very irritating dust laid across a freeway by a long-gone agricultural aircraft?

4. What explanation can you offer the physician caring for two farm workers who had been applying a fertilizer all day? Before going home they stopped at a bar. After half a beer, they collapsed and were taken to the hospital by ambulance.

5. Should a farmer taking phenothiazine tranquilizers work with parathion?

6. What advice and help do you give a trucking firm which has been delivering consumer goods to a number of retail establishments only to find a drum of Systox leaking in the middle of the van?

7. What action should be taken where a warehouse containing agricultural chemicals is burning and it's near a populated area? What should be done with the contaminated ground and partly burned chemicals?

8. What are the health hazards to swimmers and others from adding organic mercury algicides to swimming pools?

9. Why is not a harmless dye added to poisonous pesticides, to warn that foods, consumer goods and workers have been contaminated?

10. When a victim of tetraethylpyrophosphate (TEPP) poisoning dies within 20 minutes, why are red cell and plasma cholinesterase values considerably higher than expected?

11. Should we follow Japan's example and bar parathion and TEPP?

12. Why are certain toxic ingredients in pesticide formulations which are very important in treatment of poisoning not identified on the label and often not included in the data given out by Poison Information Centers?

13. Must veterinarians dispense highly toxic pesticides (example, dog dip) in medicine bottles?

14. Why do peach pickers become sick from parathion poisoning

when the parathion on the peaches is within legal tolerance? Has parathion remaining on foliage oxidized to more stable, more toxic paraoxon? If so, is the residue going undetected on the fruit?

15. What are the health problems of agricultural aircraft personnel and the public posed by ultra low volume pesticide application?
16. How much and what kind of alkaline material should be used to decontaminate 60 gallons of parathion and methyl parathion spilled on the highway?
17. What action is indicated when a farmer backsiphons 50 gallons of parathion into his well?
18. Why is not a truck carrying 999 pounds of TEPP considered hazardous enough by the Federal Transportation Agency for a "Poison" placard?

From these questions it can be seen that the great quantities of research being carried out on agricultural chemicals are not necessarily addressed to the real life human health problems which may arise. Health agencies spend much of their time reacting to various small and large pesticide emergencies as well as they can. With certain limited exceptions, health departments do not participate in the legal machinery which makes basic decisions about how a pesticide may be used and about the technical data which is required before registration. In addition, health agencies have limited responsibility over the circumstances leading to poisoning, whether in the home, at work, or in the transportation, storage, sale, and disposal of toxic materials.

The only area which has received considerable research and regulatory attention concerns the pesticide residues on crops to which pesticides have been applied and on dairy and meat products.

California State Health Department Activities

The State Health Department has legal responsibility for protecting domestic water supplies, shellfish beds, recreation water, foodstuffs, drugs, and comestics from contamination, whether from pesticides or from other harmful materials. Holding or transporting foods, drugs, or cosmetics, in proximity to poisons is not permitted. In bakeries and restaurants, foodstuffs must be stored separately from pesticides and the latter must be plainly labeled.

Otherwise, poison prevention activities are carried out in a limited manner in the areas of surveillance, investigation, consultation, education and research. Most of these activities are carried out by the Bureau of Occupational Health, which must cover many occupational health problems in addition to pesticides. However, more attention is given to pesticides since they cause the state's most serious occupational disease hazard in the state's largest industry, agriculture.

Surveillance

Surveillance activities are limited to analyzing death certificates and Doctor's First Reports of Work Injury. The latter are required by law when an employed person covered by Workmen's Compensation Laws is treated for an occupational injury or illness. There is no surveillance with respect to nonfatal pesticide poisoning in any other segment of the population.

With respect to mortality there were twenty-six deaths from pesticides in California in 1966. Eighteen were suicide, including one worker who apparently drank the pesticide which he was applying. Eight were accidents. Of the accidental poisoning deaths, five were children and three were workers. Pesticide deaths, particularly from phosphate esters, can easily be missed. There could be additional cases which went undetected.

With respect to morbidity among workers from pesticide poisoning, there were 233 cases in 1966. Most of these workers were employed in agriculture (1). 173 of these cases were attributed to phosphate ester pesticides, of which parathion and Phosdrin accounted for a majority. About one thousand additional workers were adversely affected by pesticides in 1966 because of dermatitis, conjunctivitis, and other miscellaneous ailments.

This occupational disease surveillance system usually receives only reports of obvious effects from easily identified exposures. If there are chronic, delayed, or unusual effects on health, this system would not necessarily detect them, although certain leads may present themselves which are worthy of follow-up study. One example was the reporting of a number of dermatitis cases from malathion which was not supposed to cause skin irritation. A project investigating this point demonstrated that malathion is a skin sensitizer capable of producing allergic responses usually after prolonged contact.

From the findings in a project undertaken in 1960, it is possible to estimate that about three thousand California children received medical attention during the year because they had swallowed a pesticide. All were not necessarily sick.

Investigation

Investigations are frequent. One example of an investigation was the subject of a report published in the *JAMA* (4). It concerned parathion poisoning in ninety peach pickers and significant cholinesterase reduction in the blood of half of the sample of the other eight thousand pickers who were not sick. The source of the exposure was skin contact with parathion remaining on the heavy foliage. It had been subject to unusually heavy spraying in the preceding months. As a result, spray schedules and the pre-harvest interval have since been revised (time interval between last spraying and picking).

More recently, workshops on human safety in the transportation and storage of pesticides have been held with representatives of the ten state and federal agencies with responsibilities in this area. The purpose was to investigate all of the pertinent regulations, policies and procedures now in operation and to suggest practical improvements in the protection of human health.

Consultation

Consultation on specific problems related to pesticides is the most frequent service requested of the Bureau. The subject matter of some of these requests may be of interest. Requests for technical help from lawyers who are handling personal injury cases alleged to be due to pesticides have tripled to quadrupled in the last year.

Requests for help from physicians who have puzzling cases apparently resulting from pesticide exposure have increased considerably. Insurance companies have inquired about the risks of working with pesticides, apparently in connection with issuing life insurance policies. The most notable increase in the character of inquiries concerning pesticides has been in the area of accidental spills during transportation and storage.

Education

Education and information services have focused primarily on two categories of persons — the users of pesticides and the physicians called upon to diagnose and treat pesticide posioning, including pathologists who must be on the alert or they may misdiagnose deaths from this cause.

It is obvious from our investigation and surveillance data that the actual user of hazardous pesticides, particularly in agriculture, too often has insufficient knowledge and equipment for his job. Pressures to "get the job done" often prevent use of the necessary precautions which are frequently awkward and time consuming. There is a great need to develop for the pesticide applicator more practical protective gear and safety equipment to be used in the air and on the ground.

This age of technology is providing a great number of useful and economically valuable chemicals. Economic incentives produce much greater pressures for their wide distribution than (a) pressures for education and training of the persons who will be using them or (b) the devising of methods to place hazardous chemicals only in the hands of those who can use them safely. It will be very difficult to revise the haphazard practices of 20 years' standing; however, if substantial improvement in pesticide safety is to be expected, there is obviously no other choice. To assure responsible handling of hazardous materials, it has been pointed out by many experienced observers that there is a need for an organized system of licensing the actual users of toxic pesticides on the basis of evidence that such persons have the knowledge and the equipment for safe operations, and with the provision that they will retain their license only on the basis of safe performance. When legal standards for safety performance are required of all users, there is a much greater chance of combating the economic incentives which tend to militate against safe operations.

One example of an educational effort followed a tragic 15-month period in California when six Spanish-American farm workers died because they mistook a pesticide for drinking water or another beverage. In one case, two identical unlabeled bottles were on a spray rig; one contained drinking water and the other the pesticide, paraquat. The spray operator took a drink out of the wrong bottle. Warnings in Spanish and English were dispersed through all mass media, and into Spanish publications and organizations serving these workers in the hope of reaching all farm workers.

Special educational efforts have been directed toward agricultural aircraft personnel who apply about half the pesticides used in agriculture. Their ground personnel, usually teenagers, have an exceptionally high incidence of pesticide poisoning. Our recommendations to raise the age limit from 16 to 20 years for workers handling hazardous agricultural chemicals have fallen on deaf ears.

Bulletins and published articles on pesticide poisoning have been regularly directed to California's 31,000 physicians over a 17 year period (2, 3, 4, 5, 6, 7). Still it is obvious the present mechanisms for bringing information and education to physicians are not geared to meet today's rapid introduction of hundreds of new chemicals of potentially dangerous effect. It is entirely unrealistic to expect every physician in general practice to keep up with what is known and unknown about the toxicologic properties of modern chemicals. Since technology produced this urgent problem, it is only fair to expect it should be used to devise imaginative new procedures to bring to the physician, when he needs it, effective help and up-to-date information to assist in the diagnosis and treatment of poisoning and other conditions resulting from exposure to pesticides. A national computerized center where medical and toxicological consultation can be provided and case data recorded has been suggested. Also urgently needed is the development of more clinical chemical laboratory tests to help the physician confirm a diagnosis. For many chemicals, such tests either do not exist are or not available locally. Circumstantial and clinical evidence alone are often quite inadequate to arrive at a sound diagnosis. As a result there are probably more unsound diagnoses and more missed diagnoses in chemical poisoning than in most other areas of medicine.

Nevertheless, California physicians have demonstrated remarkable acumen in recognizing pesticide poisoning. Of all of the incidents of pesticide poisoning occurring throughout the world because foodstuffs or other cargo became contaminated in a spill of a pesticide during transportation or storage, the only two instances in which the pesticide poisoning was immediately suspected because of the appearance of patients on admission to the hospital, occurred in California.

Research

Research activities are limited to a series of projects now under way by the California Community Studies on Pesticides, operating on federal funds and under the direction of the Bureau of Occupational Health. One of these studies is to assist the Federal Bureau of Aviation Safety in the investigation of fatal agricultural aircraft accidents with respect to any contribution by agricultural chemicals. Of three fatal crashes investigated during 1967, agricultural chemicals caused the death of one of the pilots by poisoning, contributed to the fire causing the pilot's death in the second case, and may have contributed to the pilot error which apparently caused the third crash. This pilot had a lowered blood cholinesterase probably due to phosphate ester pesticides he had been applying on the day preceding the accident.

One of the most discussed subjects for pesticide research is determining whether or not there are significant effects upon human health from long-term low-level exposure. Desirable as it may be to obtain this information, satisfactory methodology for this task does not appear to be available. For one thing, there are no controls since everyone is receiving these exposures. In addition, even if it could be determined just which of the many variables to study and how to study them, the observations could take a lifetime. Much more sophisticated laboratory research and methodology needs to be developed to come closer to answering these questions. At the same time there are many lost opportunities for making observations on human subjects who are exposed to sizeable quantities of pesticides in their work. Developing better procedures, both legal and medical, for utilizing human volunteers for certain kinds of pesticide studies is another avenue for research deserving more attention.

Summary

Investigations of pesticide poisoning in California suggest areas where poison prevention could be more effective. The most important concerns the competence of the person in charge of or otherwise using a toxic pesticide. His lack of knowledge and practical safety equipment, and the economic pressures "to get the job done" too often prevent the use of the very precautions and procedures which could assure safe use of the material. However, the blame does not rest so much with the user as with the system which places these materials in his hands without first making certain he is equipped and motivated to handle them safely.

Medical and public health considerations should be given greater attention in setting up and implementing laws and regulations covering registration and use of pesticides. More practical research needs to be directed toward helping the user perform safely. The technical competences involved in decisions about effectiveness of pesticides are quite different from those involving human safety. Too often the latter are in the hands of the former and are overlooked or receive inexpert attention.

REFERENCES

1. CALIFORNIA DEPARTMENT OF PUBLIC HEALTH, BUREAU OF OCCUPATIONAL HEALTH: Occupational Disease in California Attributed to Pesticides and Other Agricultural Chemicals (annually 1951 to 1965). 2151 Berkeley Way, Berkeley, Calif. 94704.
2. CALIFORNIA DEPARTMENT OF PUBLIC HEALTH, BUREAU OF OCCUPATIONAL HEALTH: Diagnosis and Treatment of Phosphate Ester Poisoning. Technical Bulletin for Physicians, 1967.

3. LEACH, P.: Organic phosphorous poisoning in general practice. *Calif Med, 78:*491, 1953.
4. MILBY, T.H.; OTTOBONI, F., and MITCHELL, H.: Parathion residue poisoning among orchard workers. *JAMA, 189:*351, 1964.
5. WEST, I.: Control of insecticide exposure in employment. *Calif Med, 86* (5) :325, 1957.
6. WEST, I.: Pesticide induced illness. *Calif Med, 105* (4) :257, 1966.
7. WEST, I.: Occupational disease of farm workers. *Arch Environ Health, 9:*92, 1964.

DISCUSSION

RUDD: In California there is an increasing use of sodium fluoroacetate which is distributed by air and is a very toxic, water-soluble substance. Is there any example of a public health hazard from this use?

WEST: It poses a serious potential hazard. No health agency personnel were involved in the decision to use this material.

SPENCER: There have been considerable changes in labeling procedure since the interdepartmental review went into effect three years ago. All labeling is subject to review, including all brochures that accompany the shipment or that reach the user. Even if the printed material is shipped via a different route, it is subject to review and the manufacturer of the product is responsible for all statements. All antidotes, medical instructions, etc., occurring on the label are reviewed by the United States Public Health Service, who advise the Department of Agriculture as to acceptable statements (see Appendix).

BERG: There are some rules concerning the shipment of dynamite. Why aren't they extended to such things as toxicants?

SPENCER: There are new regulations published by the Department of Transportation, in the *Federal Register,* December 28, 1967, Vol. 32, page 251, which became effective January 10, 1968. This has now been followed by a manual entitled *Safety Guide for Warehousing Parathions* published by the National Agricultural Chemicals Association.

WEST: The Department of Transportation's new ruling is that food, when shipped with Class B poisons, must be in non-permeable containers. Their very latest ruling is that food, drugs, and cosmetics may not be stored or shipped in proximity to any poisonous materials.

WELCH: Under what circumstances would anyone spray TEPP and if it is necessary, how does one do it?

WEST: By aircraft and by ground spraying; it's done both ways in California.

HUNT: They can spray it on crops within 24 to 36 hours of harvest and have no residues present. Other materials may leave a residue; therefore, TEPP is used when a pest hits a crop near harvest. Phosdrin is in the same category.

WEST: And they mix TEPP with Phosdrin.

WELCH: Is there any harm from it?

WEST: With workers, yes. The pilot of the second to last agricultural aircraft that crashed in California was found to be fatally poisoned with his own spray, TEPP and Phosdrin.

HUNT: The newest practice in aerial spraying is the use of low volume spraying devices, which allow the use of very concentrated formulas of organophosphates. Pilots find this a serious problem, for a number of reasons. Competition and pressure by a salesman force them to carry increasingly toxic materials, not just malathion. An aircraft full of concentrated organophosphate can be deadly to the pilot and is a great hazard to others if it falls. When valleys are sprayed, temperature inversions make it impossible to aim the spray, which may actually drift upwind from the plane.

WILLS: If the pilots want to stand on legal grounds, they can because only malathion is approved for that type of use.

HUNT: I think five chemicals are in use in California now as low volume applications. I should like to make one additional comment relating to the use of Azodrin. Most of the cotton fields in the Imperial Valley are in communities which draw their water supply from the irrigation ditches. Azodrin is a water-soluble phosphate and does not disappear like others. What is the answer to this water pollution problem? We will have to find an answer in a few months.

JOHNELS: I heard it said that the country's best chemist

should be in charge of the garbage dump. What is done with used containers of pesticides?

WEST: That is a real problem in California and there are certain very specific safety recommendations that people follow, but I am not sure they do very often. There are some containers that are returnable. I know of one mosquito abatement district that has a very good system: they have a locked cage for all of their used containers which they decontaminate before either returning or crushing or burying in a dump which receives only hazardous material. They also have a special place to burn the empty pesticide sacks. Do you know of any problems, Dr. Hunt, that have arisen with fish and game because of improper disposal?

HUNT: Yes. In the last two or three months we've had fish losses in lakes. We found in several instances that boat floats at private docks have been made from parathion-containing drums, with material still in the drums leaking slowly out into the water. As Dr. West pointed out, there are some channels through which one can take care of these containers, but they are not always used.

BERLIN: I have proposed to the Swedish Board of Poison and Pest Control that one should require at registration an answer from the applicant to the question of how the user will dispose of the package and the remainder of the unused pesticide.

BERGLUND: Did you come across any public health problems from pesticide poisoning *in utero?* I am referring to the sensitivity of the fetus to methylmercury when compared to adults. In Minamata twenty-two out of 121 cases of intrauterine intoxication were recorded, and in Niigata there were thirty-five cases of poisoning and not one case of intrauterine intoxication. The detection of neurological poisoning in newborn or young babies is a very real and difficult problem of diagnosis. Perhaps we don't even have the means for detecting this type of neurological damage.

SUZUKI: People in the Minamata area were accustomed to make more use of hospitals; especially the pregnant women who went to the clinic. There was more opportunity for diagnosis of poisoning in babies. Today the population is quite conscious of the dangers of poisoning, and the obstetrician and pediatrician may have more of a job in separating real cases from imagined ones than in finding unsuspected cases.

WEST: There is an annual publication by our department entitled, "Occupational Disease in California Attributed to Pesticides and Other Agricultural Chemicals." This covers about two-thirds of the work force but the self-employed are excluded. Also for the agricultural workers in California who have a high incidence of occupational disease, particularly from pesticides, we have another special bulletin, "Occupational Health of Agricultural Workers in California." If anyone would like copies they should write to the following address:

Bureau of Occupational Health
California Department of Public Health
2151 Berkeley Way
Berkeley, California 94704

Another bulletin for physicians is entitled "The Diagnosis and Treatment of Phosphate Ester Poisoning." Because of the tremendous need for every physician in California to know how to recognize and treat this poisoning we have widely advertised this bulletin with the help of the California Medical Association. If anyone is interested, he can write to the above address for a copy.

WURSTER: The first publications on wildlife disasters from DDT began appearing in 1944, in fact, as soon as it went into usage. In 1946 Cottam and Higgins published a paper [*J Econ Entom,* pp. 39-44, 1946] in which they outlined rather well what could be expected from DDT. They turned out to be quite the prophets although they still underestimated how serious the situation would eventually become.

I must admit that I wish I could share Dr. Spencer's confidence in the various agencies and bureaus to which he suggests we address our complaints. I know from first-hand experience what happens when you complain to the United States Department of Agriculture about wildlife damage; nothing happens. We've got twenty-five years of history of what happens within the environment and to wildlife from the use of DDT, and we have not profited by it. Those agencies have not served as Dr. Spencer said they serve.

WEST: As a bureaucrat I certainly understand why they sometimes don't, but I can certainly agree with Dr. Wurster. I have complaind many times about these labels, I think my first complaint was registered in 1954, and I have all this in writing. I might publish a book one of these days.

Chapter 25

EFFECTS OF CHLORINATED HYDROCARBONS ON SMALLER ANIMALS AS GUIDES IN THE DESIGN OF EXPERIMENTS WITH HUMAN VOLUNTEERS

J. H. WILLS

ABSTRACT

After a general discussion of criteria for the initiation of a study of toxicity in human subjects, the effects of methoxychlor on experimental animals are used as guides in the design of a human study. Completion of the planned study revealed that daily doses of up to 2 mg/kg of methoxychlor had none of the effects in men and women that had been found in animal experiments to be the most sensitive indicators of toxic activity.

REPORT

THE CRITERIA APPLICABLE to the institution of a human experiment fall into three more or less logically successive stages, each consisting of one or more distinct steps:

Stage I — Definition of Items for Study

Step 1. Principal effects of procedure are identified by animal experiments; the appropriate methods for detecting and measuring these effects in human subjects developed.

Step 2. Potential risks to human subjects are defined by animal experiments.

Step 3. Reversibility of effects to be studied in human subjects are demonstrated in animal experiments.

Stage II — Decision to Use Human Subjects

Step 1. Information to be obtained from experiments on human subjects is important; it is *not* trivial.

Step 2. Information to be obtained from experiments on human subjects can not be obtained readily in other ways.

461

Stage III — Obtaining Human Subjects

Step 1. Inform the possible subjects of both the possible benefits to mankind and the possible dangers to himself that may arise from his taking part in the proposed experiment.

Step 2. Subjects must be entirely voluntary. Tangible rewards, if any, for taking part in the experiment should not be so large as to outweigh the reward arising from feelings of fulfilled social responsibility and should not be sufficient to influence the subject to neglect his own welfare during the experiment; each subject must feel free, and must have this feeling reinforced during the experiment, to withdraw from the experiment at any time at which the continuation of the experiment does not bring some satisfaction to him.

I shall try to illustrate not only these general principles but also some practical procedures in some work with methoxychlor, carried out largely by our Institute of Experimental Pathology and Toxicology at Albany Medical College. We start, of course, with animal experiments; here we make use not only of our own work but also of results derived from research by others and published by them.

Oral administration of methoxychlor to laboratory animals seems to be an ineffective method of administration (2, 5). Even by such other routes of administration as application to the clipped skin of the rabbit and inhalation of a dust, methoxychlor is less toxic than its analog DDT (4). Only when injected intravenously does methoxychlor become equal to DDT in toxic action (3). Excretion of substances derived from methoxychlor in the bile is rapid (6); the importance of the liver in detoxifying methoxychlor and removing it from the body is underscored by Barnes' finding (1) that methoxychlor fed to rats with livers damaged by previous administration of CCl_4 produces tremors similar to those produced by DDT.

In animal studies carried out in the Institute of Experimental Pathology and Toxicology of Albany Medical College the LD50 dose of methoxychlor for the mouse by a single administration via the stomach tube was 1.5 ± 0.2 g/kg body weight. The signs of toxicity observed in these animals were respiratory depression, ptosis of the eyelids, tremor, hyperreflexia, clonic convulsions, and death. Daily doses for 28 days led to lowering of the concentration of triglyceride in the livers of the animals.

In the rat, daily methoxychlor doses of up to 2.5 g/kg body weight administered through a stomach tube (5x/wk for 14 wk) produced slight salivation immediately after each dose, some depression of growth and some loss of hair. In immature male rats (50-70 g), daily doses of 2.5 g/kg of methoxychlor on 5 successive days produced a significant decrease in the mean weight of the testes and delayed maturation of the spermatids. The mitochondria of epithelial lining cells of the duodenum, the jejunum, and the liver became markedly swollen, and these same cells contained numerous distended vesicles of endoplasmic reticulum. There was a significant increase in the demethylase activity of the liver (by nearly 70%) .

In the monkey, administration by stomach tube of daily doses of up to 2.5 g/kg of methoxychlor seemed to have no detectable effect on the general health of the animals. There was some decrease in the concentration of triglycerides in the liver accompanied by increased incorporation of linoleate-1-^{14}C into the triglyceride fraction. Methoxychlor seemed to have no enzyme-inducing effect in this species. The mitochondria of hepatic cells underwent changes somewhat similar to those seen in the rat, with an increase in size and loss of transverse cristae.

These experiments in laboratory animals demonstrated the principal things to be looked for in human experiments: possible regressive action on the testis of the male, possible alteration of menstrual cycle in the female (both these possible effects arising from a postulated estrogen-like action of methyoxychlor) , possible changes in the structures of mitochondria in intestinal mucosa and liver, possible decrease in the concentration of triglycerides in the liver and, perhaps, in other fat-rich loci. The methods to be used in carrying out the indicated studies were all well known. The studies in animals showed also that the effects induced by methoxychlor disappeared when administration of the compound was discontinued. Thus, the experiments in animals apparently had satisfied all 3 Steps of Stage I.

The information to be derived from experimental studies with methoxychlor in human subjects was considered to be important because (a) methoxychlor is being advocated fairly strongly as an especially desirable substitute for DDT and (b) widespread use of methoxychlor in food-growing agriculture could result conceivably in exposure of many millions of people to some intake of this compound or its degradation products. Information, as distinct from opinion, about the safety of ingestion of methoxychlor by man was held to be obtainable only from planned experiments with human subjects. Only by such experiments can one say reliably that a particular dose of a chemical can be taken by man without any deleterious effect, whereas some larger dose produces a deleterious, although perhaps not actually harmful, effect.

Because of the suggestion from experiments in laboratory animals of an estrogen-like activity by methoxychlor, the desirability of using both male and female subjects was evident. Because methoxychlor had effects in laboratory animals on both intestinal mucosa and liver and might affect other lipid-rich sites, collection of biopsy ˜pecimens of intestinal mucosa, liver, fat, bone marrow and testis seemed desirable.

Groups of potential volunteers from both male and female inmates of the Alabama State Penitentiary were given by the prison physician, Dr. E. Stough, a resumé of the existing knowledge about the actions of methoxychlor and were invited to ask for clarification of any statements they had not understood. When all questions had been answered, the groups were invited to volunteer for the experiment. Each volunteer signed a release form and then had a careful physical examination in which the liver was palpated especially carefully and the medical histories were examined for any indications of prior endocrine or liver disorders. Sixteen subjects of each sex were selected, being told that he or she could withdraw from the experiment at any time without prejudice. Subjects were given $0.50 per day for so long as they continued to take part in the experiment. Submission to special procedures, such as biopsies, was compensated by single payments of considerably larger size graded according to rough estimates of the hazards to the subjects inherent in the various procedures.

The subjects were separated into four groups, each of which contained four members of each sex. One group took an inactive placebo, a second took 0.5 mg/kg/day of methoxychlor, a third took twice that dose of methoxychlor and the last group took 2.0 mg/kg/day. All doses were taken in hard gelatin capsules. Three men from the control group and three from the group ingesting the largest daily dose of methoxychlor agreed to permit the collection of biopsy samples of intestinal mucosa, of subcutaneous fat, of bone marrow, of testis and of liver.

Biopsies of mucosa from the small intestine were obtained with either Shiner or Crosby instruments, passed under direct fluoroscopy, from four subjects. Fat was aspirated from a buttock with a syringe and needle from six subjects, rich sternal marrow was obtained from sternal punctures of five subjects, fragments of testicular tissue were removed surgically from five subjects, and liver samples were aspirated through a Meneghine needle from six subjects. Appropriate histological and histochemical procedures were applied to these various biopsy specimens.

Neither the routine blood chemistries performed on all subjects (glucose, urea nitrogen, uric acid, total cholesterol, Na, K, total protein, glutamic-oxalacetic transaminase, glutamic-pyruvic transaminase, lactic dehydrogenase, alkaline phosphatase, acid phosphatase) nor the special studies carried out on the biopsy specimens from six subjects (regular and electron microscopic study of small intestine, bone marrow, liver and testis and determination of the distribution of fatty acids (C14-C20:5) in subcutaneous fat) disclosed any significant differences between the subjects taking the placebo and those taking daily doses of methoxychlor. No changes in the menstrual cycles of the female subjects were noted.

Accordingly, we have concluded that daily doses up to 2 mg/-kg/day of methoxychlor are safe for ingestion by both males and females of the genus *Homo sapiens*. This dose had none of the actions in human subjects that had been found in the animal experiments to be the most sensitive indicators of toxic activity by the same chemical.

REFERENCES

1. BARNES, J.M.: Quoted by Winteringham, F.P.W., and Barnes, J.M.: *Physiol Rev, 35:*701, 1955.
2. BROWNING, H.C.; FRASER, F.C.; SHAPIRO, S.K.; GLICKMAN, I., and DUBRULE, M.: *Canad J Res, 26:*282, 1948.
3. DOMENJOZ, R.: *Helvet Chim Acta, 29:*1317, 1946.
4. HAAG, H.B.; FINNEGAN, T.K.; LARSON, P.S.; RIESE, W., and DREYFUSS, M.L.: *Arch Int Pharmacodyn Ther, 83:*491, 1950.
5. LAUG, E.P., and KUNZE, F.M.: *Fed Proc, 10:*381, 1951.
6. WEIKEL, J.H., JR.: *Arch Int Pharmacodyn Ther, 110:*423, 1957.

DISCUSSION

DURHAM: Did you take any blood levels of methoxychlor on these subjects?

WILLS: I do not believe so; at any rate, we did not as a routine.

BERGLUND: You mentioned that these subjects were given doses for a period of 4 to 6 weeks. How near was this dose to a nontoxic dose of highest no-effect dose in experimental animals, was it 0.1 or 0.01? I question the general design of experiments on people. The experiments are necessary as a matter of principle, since they duplicate accepted human exposures and allow laboratory tests. For example, DDUP is used to fumigate passenger airplanes for mosquitoes on international routes, so that crews are chronically exposed, and one has to know whether this has any effect on them.

WILLS: The dose was very small compared with the no-effect dose in animals. The no-effect dose in animals is about 0.9 g/kg. Our largest dose was 2 mg/kg.

BERGLUND: This would be less than 0.01 of the no-effect dose in animals. This is actually a check that the safety factor is acceptable.

WILLS: Yes, that is correct.

WELCH: The estrogenic effects of methoxychlor are interesting. Is this information from your work?

WILLS: From our observations on festicular growth in rats.

WELCH: Berlington and Lindeman too have shown that DDT decreases testicular size in the cockerel [*Proc Soc Exp Biol,* 74:48, 1958]. Did you test for liver enzyme induction in humans after feeding methoxychlor? Did you check fat samples for residue levels of methoxychlor or metabolites?

WILLS: No, we did not examine enzyme induction in humans. Fat samples were analyzed in a GLC for methoxychlor, using an electron capture detector.

WELCH: I understand that the methoxychlor administered orally to women did not affect the menstrual cycle.

WILLS: That is correct.

WELCH: How much of the methoxychlor administered was excreted as metabolites?

WILLS: It was excreted in the bile, about half in the form of methoxychlor and half converted to other forms.

WELCH: From the microscopic examination of the liver, did you observe any effect on the endoplasmic reticulum of the human?

WILLS: No. The human liver and intestine looked completely normal. The only change we observed in the testes was a shortening of the neck of the spermatozoon.

BERG: Could the shortening of the neck of the spermatozoa be an effect on mitochondria?

WILLS: Yes, it could be.

Chapter 26

GENERAL DISCUSSION: SESSION V

ORGANOCHLORINES

WILLS: Mr. Herman would like to introduce the comments that Dr. Risebrough will make concerning Dr. Robinson's paper.

HERMAN: I'd like to begin by thanking the Conference Chairman for the opportunity to make these remarks. I'd also like to apologize to the conferees for the fact that our criticism of Dr. Robinson's paper has been the source of some confusion. Dr. Risebrough and I had a preprint of Dr. Robinson's paper before this Conference convened. We had hoped that everyone at the Conference would have had an opportunity to read this paper prior to the presentation of our criticism. Accordingly, we wrote our criticism, only to find that indeed only a few conferees have had access to the paper. We did, however, present a written copy of the criticism to Dr. Robinson before he gave his verbal presentation. His verbal presentation was much shorter than his written paper (as has been the case with all of us) and it excluded much of the portion we had criticized. Furthermore, he didn't answer our criticism, even in private conversation. These, then, are some of the reasons for the confusion. What we wish to make clear now, however, are our motives in criticizing Dr. Robinson's paper.

There is now, and there has been for some 20 years, a fairly well defined controversy concerning the effects of pesticides on wildlife. The controversy has been between the pesticide industry on one side, and conservationists, including many ecologists, on the other side. We feel that a criticism of Dr. Robinson's paper is very important because his approach typifies and is almost symbolic of the industry position.

It is clear now that global contamination by pesticides is not only a scientific problem. It is also a social problem, and it is our contention that those of us concerned with the scientific study of

this contamination would do well to concern ourselves to a greater extent with this social responsibility.

Now we—and by "we" I mean conservationists and ecologists
—have not always been able to get a fair and reasonable hearing.
I think many of the people here would agree with that statement,
but I'd also like to give an example, for the benefit of those present who have not been aware of the nature of the controversy. I
choose this example for two reasons. First, because it concerns
the peregrine falcon, a species dealt with yesterday by Dr. Wurster and also the main subject of our disagreement with Dr. Robinson. Secondly, because there are at least two other persons in
this room who have concerned themselves previously with the
evidence indicating that peregrines are in fact suffering a population decline as a result of contamination with biocides. Both
Dr. Spencer and Dr. Stickel were present, as I was, in Madison,
Wisconsin, in 1965 when Dr. Joseph Hickey convened a conference to consider the widespread decline of the peregrine. I would
like to ask each of these gentlemen to correct me, as I go along,
if I make any mistake in telling this story.

The Peregrine Conference drew participants from all parts of
the United States and much of the world. We considered all kinds
of evidence—disease, parasites, weather changes, prey availability,
human persecution of the species, and so on. One of the sessions
was to be devoted to the consideration of the possibility that pesticides might be involved in the decline. As the conference proceed-
ed toward this particular session, it became quite obvious that a
majority of the conference was interested, very interested, in
hearing an open and extensive discussion of the pesticide issue as
it pertains to peregrines.

The Chairman of this pesticide session set the tone of his
reign in his opening remarks, when he directed the participants to
use the term "occurrence" rather than the term "contamination"
when commenting on the presence of biocides in these birds and
their eggs.

However, Dr. Ratcliffe, the distinguished British student of
the peregrine, was there, and he presented very convincing evidence showing a close correlation, both spatial and temporal, be-

tween contamination of peregrines and their reproductive success. (This is the same gentleman who 2 years later discovered the same correlation as manifested in eggshell thinning, about which Dr. Wurster spoke last night.) The outcome was ambivalent. Following Dr. Ratcliffe's presentation, and the equally impressive presentation of Dr. Presst, his colleague, there was a general discussion of the problem. I remember Dr. Spencer standing and expressing what appeared to be a great deal of interest in the peregrine falcon, but then finishing his comments with a statement to the effect that what had been presented was not proof by any means, but that he would like to "help," if only there were some facts available. Now Dr. Spencer, for whom were you working at that time?

SPENCER: I had spent 34 years as a research biologist with the United States Fish and Wildlife Service; I was at the time of the Peregrine Falcon Conference, Chief Staff officer—Animal Biology in the Pesticide Regulation Division (USDA), with the responsibility of reviewing registration for the use of pesticides with respect to their side effects on wildlife and the environment.

HERMAN: And was this before or after your involvement with The National Agricultural Chemicals Association?

SPENCER: I was a career scientist in Government Service, nothing else. Does it make any difference as to where a man works?

HERMAN: Yes, and that's a good part of our point. But, to reiterate, you said at that time you would like to help if there could be provided any hard evidence, any proof, that there was a relationship between pesticides and the peregrine decline. Is that right?

SPENCER: Yes.

HERMAN: Now, our question is this: What constitutes proof? It's not a new question, and it's not going to be resolved by this conference or many more conferences like it, but, that, in fact, is our question. Thank you.

RISEBROUGH: Steve Herman is a co-author of these remarks. Our basic criticism of Dr. Robinson's paper is that it is irrelevant. We can perhaps show that cigarette smoking does not cause cancer in nonsmokers. Dr. Robinson has shown that the collared dove is doing well in Britain. The collared dove is a nonsmoker, a species which because of its position in the ecosystem does not accumulate high residues of chlorinated hydrocarbons.

Dr. Robinson has developed an equation, the significant conclusion of which is that any contaminant in the world's ecosystem will eventually reach a finite level and therefore won't accumulate indefinitely. We think this is also irrelevant since the important problems will arise much before that point is reached.

Dr. Robinson mentioned the potentially worldwide distribution of very small residues. Petrels and shearwaters now have higher residue concentrations than almost all land birds.

We wish to confine our comments almost entirely to one part of Dr. Robinson's text, where he discusses the status of the peregrine falcon in Great Britain. In the first mention of the peregrine Dr. Robinson discusses Treleaven's study of a decline in Cornwall in the period 1955 to 1960 and repeats the latter's suggestion that an "excess of females might be an important factor" and that "disturbance by man and increased predation of the eggs and young were also possible contributory causes." The reported excess number of female birds is an abnormal situation. Treleaven wrote: "After the 1939 to 1945 war, during which period its numbers were much reduced and breeding prevented by the official campaign of destruction the peregrine (*Falco peregrinus*) quickly recovered in Cornwall until by 1955 there were almost as many pairs as in the 1930's." Dr. Robinson also fails to mention that throughout the previous century the species had fared well in Cornwall despite intense persecution by game keepers and egg collectors and the use of the species in falconry.

Then, after admitting the decline of both the sparrow hawk and the peregrine in an area of southern England after 1952, and the possibility that this may be correlated with the use of insecticides, Dr. Robinson concludes that "no firm conclusions can be drawn." At least three, firm, irrefutable conclusions can, however, be drawn: (a) the populations of both species were in good shape before the onset of the chlorinated hydrocarbons; (b) the decline of both species and the history of use of the insecticides were synchronous; (c) no other factor has adequately explained the decline.

Later Dr. Robinson lists the conclusions reached by Dr. Ratcliffe after several years of detailed study of the peregrine. "The evidence given by Ratcliffe does not seem sufficient to warrant the firm conclusions that he draws." Dr. Robinson then claims that Ratcliffe has misquoted other sources and that he has made sweeping conclusions on the basis of little evidence. The details follow: Dr. Robinson quotes from Ratcliffe "that several southern districts depleted by wartime persecutions showed only a partial recovery to the prewar level of population." Robinson continues "Ferguson-Lees gives only one example of partial recovery, namely that eight to twelve pairs were occupying 16 miles of cliffs in Sussex prewar but only five to six pairs were in occupation in 1954." The text from Ferguson-Lees reads as follows: "So that, although the species was considerably reduced under the wartime control, when the official destruction ceased there was in most cases a remarkable increase in the number of breeding pairs (36% over the end-war low level). On this basis, one might have expected the population to have regained its prewar strength by now, but that does not seem to be the case, and a fair estimate of the present population would be eighty-five, just possible ninety, pairs. To give one example of a county which is still below par, Sussex before the war had eight to twelve pairs in 16 miles of cliff, with sometimes three aeries in the space of a mile [Walpole-Bond, J.: *A History of Sussex Birds,* p 237]—figures which must indicate almost the maximum possible density—but in 1954 [*Sussex Bird Report,* p. 10, 1954], only five to six pairs were recorded."

Dr. Ratcliffe's interpretation is quite correct. In the narrow sense Dr. Robinson is correct in stating that only one example was given but the information he has imparted is wrong; Ferguson-Lees also supports the conclusion with an estimation of the then-current total population of the area. Dr. Robinson's statements are here completely irrelevant to any point at issue and can be construed only as an attempt to discredit Dr. Ratcliffe's integrity.

Dr. Robinson's next sentence reads: "Ratcliffe then asserts that by 1961 'there were clear signs of decrease in nesting populations in southern England' and cites only Treleaven (82), whose observations were summarized above, in support of this assertion." This statement is false. Ratcliffe noted that he had spent the entire nesting season of 1961 in the field; the paper to which Dr. Robinson refers consists largely of results from Ratcliffe's personal observations and inquiries and this is abundantly clear in the text.

Dr. Robinson continues: "Again, Ratcliffe states that 'egg-eating, which I believe to be a first symptom of decline, was first noticed in 1951 and has been a common occurrence ever since.' " Dr. Robinson then quotes several instances of egg-breaking which had occurred before 1951, apparently again to disqualify Ratcliffe as a competent observer. We must first note a certain distinction between egg-eating and egg-breaking. Even *Archaeopteryx* undoubtedly suffered on occasion from broken eggs. Those of us who have climbed to Peregrine aeries, frequently against our better judgment, can appreciate how a peregrine egg might be broken. Dr. Robinson must, however, show that egg breakage is a phenomenon which extends back before the introduction of the chlorinated hydrocarbon into the global ecosystem. This he does by quoting from a study of the peregrines which nested between 1936 and 1952 on the Sun Life Building in Montreal and quotes "an apparent loss of eleven out of fifty eggs from 1940 to 1952 inclusive." It is not convenient to mention that almost all of the eggs disappeared in 1949 or later, when chlorinated hydrocarbons were already present in the environment. The more detailed history follows: Of the fifty eggs, twenty-six hatched; in 1942 two eggs disappeared from the nest, in 1948 the author was out of town, one egg was seen but apparently no brood was raised; in 1949 one egg was eaten for the first time, four others were laid and disappeared; in 1950, four were laid, one disappeared, three hatched; in 1951 four eggs were laid, three disappeared one by one, the chick in the fourth died in the shell when ready to hatch; in 1952 four eggs were laid, one disappeared, the other three hatched. These Canadian observations paralleled those of Ratcliffe in England and supported the conclusion that egg-eating and a large proportion of broken eggs were an abnormal, postwar phenomenon, yet Robinson uses these data to support the opposite conclusion.

Peregrines no longer nest in Montreal.

In discussing the dramatic changes in eggshell weights which have occurred after 1945, Robinson states, "If organochlorine insecticides are the casual agents involved, then the date of onset of the decrease in eggshell weight indicates that only DDT or BHC could be involved." Dieldrin was used in Great Britain only after 1955. Dr. Robinson, however, has ignored a whole area of research dealing with the effects of chlorinated hydrocarbons on the molecular level, the level where one must look in proving whether or not the relationship is casual. Dieldrin, like DDT, the DDT derivatives, and probably most chlorinated hydrocarbons induce steroid hydroxlase enzymes in both mammals and birds. Another whole area of research which has been equally ignored is the tie-up between chlorinated hydrocarbons and calcium metabolism. Why should bird-watchers be considered more unreliable than biochemists?

In interpreting the residue data in peregrine eggs and their relevance to nesting success, Dr. Robinson states, "There are no significant differences between the residues in the eggs from successful and unsuccessful aeries in the area (northern England and southern Scotland) in which both types were found . . . As there were no significant differences between the concentrations of organochlorine insecticides in the eggs from successful and unsuccessful aeries within a geographical area it does not seem plausible to attribute the differences in success to these compounds." Dr. Robinson would, we are sure, admit that a sufficiently high level of organochlorine residues would kill the adults or effectively prevent them from breeding. What might we predict if a sublethal level of chlorinated hydrocarbons did affect reproductive success? Would a given level of contamination necessarily cause the same degree of reproductive success in each pair of peregrines? Not at all: some would succeed, others would fail, depending upon the variability inherent in biological systems at this level of organization. What Dr. Robinson has done is to compare those which failed with those which succeeded and concluded that because no statistical difference exist in the eggs, the residue levels cannot be associated with reproductive success. We should not expect necessarily a statistical difference, particularly when the range is relatively limited. Dr. Ratcliffe did find higher residues in eggs from unsuccessful nests, but the differences were not significant. Dr. Robinson also did not mention that the sample size of eggs available for analysis was exceptionally small.

Dr. Robinson mentions that the eggs from the Central Scottish Highlands are significantly lower in both total DDE and total chlorinated hydrocarbon content. He does not, however, mention that the nesting success in this region is much higher than in the rest of Britain and in both 1965 and 1966 accounted for 52 per cent of the successful nestings in all of Great Britain. Nor does he mention that eggs from this region are more "normal" in terms of eggshell thickness.

ROBINSON: Dr. Risebrough criticizes my paper on the grounds that (a) much of the evidence I have discussed is irrelevant, (b) I have quoted from various authors, particularly Ratcliffe, in a partial manner, and (c) my discussions of Ratcliffe's conclusions can be considered only as an attempt to discredit Dr. Ratcliffe's integrity. I am indebted to Dr. Risebrough for his comments; and he is obviously extremely concerned at the possibility that my paper will cause harm. I will try to be brief in commenting on his criticisms.

First I will discuss the criticism of irrelevance. I consider that my reference to the exponential increase in the number of collared doves in Britain, in the particular context in which it was quoted, was extremely relevant. I quoted the population figures for this species in Britain merely as an example of the exponential phase of the increase of a bird population, and I did not make any inferences regarding the relevance or otherwise of this population increase in relation to organochlorine insecticides. It is pertinent, however, to consider Dr. Risebrough's comments on this species in some detail since they indicate either that he has expressed his ideas very badly, or that he has a number of false ideas on the relationship between the use of organochlorine insecticides and the occurrence of residues in birds. It is stated that the collared dove is "a species which because of its position in the ecosystem does not accumulate high residues of chlorinated hydrocarbons." The diet of the collared dove consists largely of grain and seeds, and it also eats small fruits, i.e. it is in the second trophic level. Consequently I agree with Dr. Risebrough that it does not accumulate residues as a result of their transfer through a succession of organisms. However, it would be false to conclude that those species, whose position in the ecosystem is such that they accumulate residues by this mode, are the only ones in which high residues occur. I think Dr. Risebrough will agree that any organism, whatever the trophic level in which it is found, may accumulate residues by direct contact either with insecticidal sprays, deposits, etc. Consequently high residues may occur in birds in the second trophic level, and an example of this is provided, of course, by the occurrence of high residues in some grain-eating birds, such as woodpigeons, in Britain. Can this type of exposure be excluded in the case of the collared dove? The habitat and sources of food of the collared dove and the woodpigeon show considerable differences, and it would therefore be wrong to argue by analogy that as both are grain-eating birds, and as high residues are found in some woodpigeons, therefore it is probable that high residues will be found in some collared doves. On the other hand it would also be wrong to conclude that high residues will never be found in collared doves merely on the grounds of the position of this species in an ecosystem. So far as I am aware, the results of the analysis of only one collared dove have been published. The concentration of HEOD (dieldrin) in the liver of this bird was 18 ppm (a high level which, incidentally, is inconsistent with Dr. Risebrough's hypothesis). The paucity of information on residues of organochlorine insecticides on this species was the very reason why I did not make any inferences concerning the relationship between the exponential increase of this species and the use of organochlorine insecticides, in spite of the fact that the initial increases were occurring during the period (1956-1961) in which poisoning of some grain-eating birds by cyclodiene insecticides were occurring in Britain. I assume that Dr. Risebrough does not regard these incidents including grain-eating birds as irrelevant to the general topic of my paper, and I assume that it was an oversight on his part to create the impression that high residues can only occur as a result of an accumulation process involving food chain transfers.

The second example of an allegedly irrelevant argument cited by Dr. Risebrough is concerned with the equation I developed on the basis of a proposed model for the dynamics of organochlorine insecticides in ecosystems. This equation predicts that there are finite upper limits to the accumulation of compounds in a particular phase of an ecosystem, an upper limit that is related to the rates of entry and removal of the compound from that phase. I did not infer that the upper limit of concentration is necessarily less than biologically effective concentrations, but if the model is valid then it is apparent that there is a rate of usage of a compound which will not have a particular effect. It should also be noted that I did not argue on *a priori* grounds that the steady state has been achieved in all phases, but I quoted experimental evidence which indicates that the upper limits of accumulation have been achieved in some phases. In the context of the discussion of the model which I have proposed, I consider that, again, my comments on the implications of the equation under discussion were quite relevant, and I cannot accept that the various topics discussed in my paper are irrelevant.

I will now consider the charge that my quotations, from Ratcliffe and others, are either partial or misleading or both, particularly in relation to the status of the peregrine falcon in Great Britain. The first alleged case of this relates to Treleaven's paper on the peregrine in Cornwall. I am at a loss to understand the alleged deficiencies in my summary of this paper and my mystification is intensified by Dr. Risebrough's statement that I fail "to mention that throughout the previous century the species had fared well in Cornwall—." I can find no statement to this effect in Treleaven's paper.

The next alleged example of incomplete and/or misleading quotations is concerned with Ratcliffe's papers. I said that "The evidence given by Ratcliffe does not seem sufficient to warrant the firm conclusions he draws," and "that there are weaknesses in the arguments used by Ratcliffe to support at least some of his conclusions." Neither of my statements is couched in terms of a categorical rejection of Ratcliffe's conclusions; they do not, I submit, support Dr. Risebrough on his contention that I claim Ratcliffe has "made sweeping conclusions on the basis of little evidence." Further, I *did not* claim, as Dr. Risebrough states, that Ratcliffe has misquoted other sources. Perhaps Dr. Risebrough means that my quotations are such as to imply that I considered (without saying so explicitly) Ratcliffe had misquoted his sources. Let us consider the first quotation from Ratcliffe in full: "Yet, by the mid-1950's it was evident (e.g., Ferguson-Lees, 1957) that several southern districts depleted by wartime persecution showed only a partial recovery to the prewar level of population. By the time the present enquiry was launched in 1961, there were clear signs of decrease in nesting populations in southern England [e.g., Treleaven, 1961], and very few young were being reared by the remaining pairs." Dr. Risebrough agrees that, as regards the partial recovery of the peregrine population in southern districts, only one example was given by Ratcliffe, but goes on to state that "Ferguson-Lees also supports the conclusions with an estimation of the then-current total population of the area." Ferguson-Lees *does not* give estimates of the then current population of the area (namely, the southern districts). He does say that in *England* the population, which on the basis of the recovery between 1944 and 1949, "one might have expected the population to have regained its prewar strength by now" (i.e. the mid-1950's), "but that does not seem to be the case." He then gives the numbers of birds on the Sussex coast, in a passage which I summarized in my paper and which Dr. Risebrough quotes in full. I submit that my summary does not alter the meaning of the passage, and that Ferguson-Lees does not give information providing further support to Ratcliffe's generalization. In my paper I stated that Ratcliffe quoted only Treleaven in support of his conclusion "there were clear signs of decrease in nesting populations in southern England." That this is correct is evident from the full quotation given above, but Dr. Risebrough says: "This statement is false, Ratcliffe noted that he had spent the entire nesting season of 1961 in the field;—and this is abundantly clear from the text." Any unprejudiced reader of the full quotation given above will appreciate that this is irrelevant to the text in question since Ratcliffe says "By the time the present enquiry was launched in 1961 there were clear signs of decrease—."

I consider that the detailed discussion of these specific examples are sufficient to demonstrate that Dr. Risebrough has failed to substantiate his charges that I have given partial and/or misleading quotations. It would be tedious to continue the detailed examination of my alleged deficiencies in this regard, and, in fact, an extended refutation of the charge would only give the impression that I was only interested in a polemical demonstration of Dr. Risebrough's misquotations or misunderstandings of my paper.

I think it is more important t   ...t the claim that I have attempted to discredit Dr. R   ...e's integrity (and disqualify him as a competent observer). In some ways I regard this as the most serious of the charges made by Dr. Risebrough. I can only say that my criticisms of Dr. Ratcliffe's arguments were not made with any such intention; any such impression formed by Dr. Risebrough is the result, I consider, of overreaction on his part. My meager knowledge of the history of scientific ideas suggests to me that controversy in science has been brisk and unsparing during the last hundred years and that lack of controversy can to some extent be equated with stagnation of knowledge. Polanyi has shown that the advance of scientific knowledge is not a smooth, wholly objectively determined activity. (Polanyi, M.: *Personal Knowledge.* London, Poutledge & Kegan, Paul, 1958) Disagreements between scientists are not necessarily indicative of malice on one side or the other. It is perhaps appropriate to quote one of Samuel Johnson's dicta, "I have found you an argument: but I am not obliged to find you an understanding," and to point out that *argumentum ad hominem* is one of the classical fallacies in logic. Further point is given to the dangers of attributing differences of opinion to malice by a number of studies by psychologists in recent years. The paper by Sherwood and Nataupsky can be cited as one example of these studies. (Sherwood, J. J., and Nataupsky, M.: *J Personality Soc Psychol, 8:*53, 1968.)

I will close my discussion of Dr. Risebrough's comments by some brief comments on "three, firm, irrefutable conclusions" that he gives. "The populations of both species (i.e. sparrow hawks and peregrines) were in good shape before the onset of the chlorinated hydrocarbons." This "irrefutable" conclusion requires considerable qualification. For example, before the onset of the chlorinated hydrocarbons (by which, I assume, Dr. Risebrough means "before the introduction of organochlorine insecticides") it is difficult to accept that the population of peregrines in Britain was in good shape since wholesale destruction of the peregrines was organized between 1939 and 1945. Widespread usage of DDT and BHC did not start until 1946 at the earliest. "The decline of both species and the history of the use of the insecticides were synchronous." I discussed this very point in detail in my paper and demonstrated, I consider, that this is also not an "irrefutable" conclusion. None of the criticisms made by Dr. Risebrough, in my opinion, weaken my discussion of this topic. The third conclusion, "no other factor has adequately explained the decline," I consider to be particularly weak. Ratcliffe himself states that "several bird species have undergone marked and often unexplained changes in distribution and numbers in Britain within the last few decades."

I think I am justified in pointing out that my discussion of the relationship between the use of organochlorine insecticides and the status of birds of prey, particularly the peregrine falcon, was based not only upon the changes in the numbers of these birds, but also upon the residues found in their eggs and body tissues. I am not certain whether Dr. Risebrough agrees with my general treatment of this data, but he does discuss in detail one set of residue results, namely, the residues in the peregrine falcon. I found Dr. Risebrough's comments on the significance of the residues most interesting and his suggestions obviously have considerable merit. Taken together with Dr. Stickel's comments on the residues in the eggs of peregrines in Alaska and Canada, populations which appear to be reproducing normally, it is apparent that we have a complex situation and it is difficult to draw firm conclusions about the significance of the residues in the eggs of British peregrines.

WURSTER The paper by Robinson has reached many subjective conclusions, either unsupported by evidence or supported by certain evidence while other data have been ignored. Much of the discussion is correct, but irrelevant to the subject of the paper.

Specific commentary about specific points is as follows:

1. Robinson has referred to sixty kinds of birds endangered by extinction and has pointed out that chlorinated hydrocarbon insecticides have been mentioned in connection with the rarity of only two of them, one of which was the Bermuda petrel. This rare and endangered list is purely arbitrary. Presumably the peregrine is not on the list, nor is the British sparrow hawk, yet both have become either rare or extinct over large portions of their ranges. It is more relevant to consider the prevention of these species becoming so reduced in numbers as to be placed on this list. This means removal of the factors that cause the population decline, and there is strong evidence that the chlorinated hydrocarbon insecticides represent such a factor.

2. It is incorrect to conclude that farmland species are at the greatest risk from the use of chlorinated hydrocarbons in agriculture. Risk is determined in large part by the characteristics of the food chain that supports the bird, and many farmland species are herbivores that receive relatively low exposure. An exception to this, of course, lies with the insecticide treatment of seeds which may be subsequently eaten by herbivores. It is my impression that this practice has been largely discontinued.

3. It is not safe to conclude that since the usage of chlorinated hydrocarbons varies geographically, the effects on bird populations will vary similarly. Such differences between regions are often minimized by: (a) the mobility of pesticides (b) the mobility of the birds, and (c) the mobility of other organisms in the birds' food chains. Each of these factors must be considered before concluding that effects will be restricted to the area of application.

4. Robinson has indicated that chlorinated hydrocarbons have

not affected the blackbird or chaffinch. Has anyone suggested that these birds have been affected? The American robin, in the same genus as the British blackbird, has often been a victim of insecticides, but its population increases for other reasons and in spite of the application of insecticides. The chaffinch is primarily an herbivore and not a target species for chlorinated hydrocarbons.

5. Jefferies and Presst (*Brit Birds, 59:*49, 1966) did not conclude that p,p′-DDE was "not toxicologically significant" as Robinson has quoted them. They said it was "of little importance in birds when compared to dieldrin," that its effect was in addition to the other compounds present and that 70 ppm DDE is equivalent to 1 to 2 ppm dieldrin. Probably each of these authors minimizes the effects of DDE, which is one-third to one-half as toxic as DDT to birds of several species, is present in large amounts in many species, and is a powerful enzyme inducer.

You also concluded that the deaths of the five British peregrines "were not caused by poisoning by HEOD or dieldrin." This conclusion was based on a discussion of residues in liver with brief mention of the brain, and little attention given to DDE. In the original reference, Jefferies and Presst [*Brit Birds, 59:*49, 1966] gave brain analyses for two of these peregrines. They contained 45 and 44.1 ppm DDE. Assuming that DDE is one-half as toxic as DDT to birds, and this has been shown in several species not including the peregrine (*US Fish Wildl Serv, Resource Pub, 1:*43, 1967), these two birds contained about 75 per cent of a lethal dose, based on the lethal concentration of DDT and DDD in the brain being 30 ppm (Stickel, L. F., *et al.: Science, 151:*1549 1966). Brains of these birds also contained 3.5 and 7.8 ppm dieldrin, however. Based on your data (*Life Sci, 6:*1207, 1967) this represents 29 per cent and 65 per cent of a lethal dose of dieldrin. Combining these values indicates that these two peregrines contained 104 per cent and 140 per cent of the combined lethal doses of DDT and dieldrin, and probably died of poisoning by these two compounds.

One cannot automatically assume that the toxicity of two different compounds is additive. That the mechanisms of these compounds are similarly related supports the assumption, however.

Based on the best knowledge we presently have, it seems reasonable to conclude that these two peregrines died of the combined action of dieldrin and DDE.

ROBINSON: Dr. Wurster states on the one hand that I have reached many conclusions, either unsupported by evidence or supported by certain evidence while other data have been ignored; on the other hand, he says that much of the discussion is correct but irrelevant to the subject of the paper. There appears to be some inconsistencies in Dr. Wurster's reaction to my paper. The reference to the sixty kinds of birds endangered by extinction is said to be irrelevant. In the context in which I made the statement, namely, a general discussion of the population dynamics of birds and some factors affecting them, it did not seem inappropriate to quote them. I am in full agreement with Dr. Wurster that we should investigate the factors involved in preventing species becoming so reduced as to be placed on this list, and my paper was an attempt to assess the evidence available in relation to the effects of the use of the organochlorine insecticides in Britain.

Dr. Wurster states that it is incorrect to conclude that farmland species are at greatest risk from use of chlorinated hydrocarbons in agriculture. I have already discussed certain aspects of this topic in my comments on Dr. Risebrough's criticisms. I think it is fair to point out that I did not say that farmland birds were at greatest risk: I said that one would expect them to be amongst those at maximum risk. If it is Dr. Wurster's opinion that fears concerning the effects of organochlorine insecticides upon farmland birds have been exaggerated then I would agree with him. However, such fears have existed, and still exist in certain quarters. The Common Bird Census was set up specifically (at the request of the Nature Conservancy) to investigate the possible harmful effects of toxic seed-dressings, etc. and various environmental changes (Williamson, K.: *Handbook of the Society for the Promotion of Nature Reserves,* 1966, 1). It is pertinent to my inclusion of the results of this Census in my paper that the latest Report of the Joint Committee on Toxic Chemicals also gives a summary of the results of this Census (Cramp, C., and Olney, P.J. S.: *6th Rpt Joint Committee Brit Trust Ornithology & Royal*

Society Protection Birds Tox Chem, 1968). Dr. Wurster asks if anyone has suggested that blackbirds or chaffinches have been affected by chlorinated hydrocarbons. If by affected he means allegedly killed the answer is yes; this is apparent from the various reports of the Joint Committee on Toxic Chemicals. Once again I would agree with Dr. Wurster if he is suggesting that these fears were misplaced and I am grateful for his support.

Dr. Wurster gives another example of alleged misquotation on my part; he says I quoted Jefferies and Presst as concluding pp′DDE was "not toxicologically significant." There is no such quotation in my paper. Presumably he is referring to the passage in which I stated "Jefferies and Presst suggest that the concentrations of pp′-DDE in these birds were not toxicologically significant." This was an attempt to summarize their views. They state "The high pp′-DDE contents of the livers of peregrines "144" and "172" have been ignored in the above analysis—," they continue "According to published data—70 ppm DDE can only be equivalent to 1 to 2 ppm dieldrin. This is supported by the finding that 59.7 ppm pp′-DDE plus 69.2 ppm pp′-DDT in the liver of a Bengalese finch not only did not kill the bird but did not prevent it from breeding successfully." My summary of their findings does not appear to misrepresent their views. The various charges of misquotation become rather wearisome, and I regret that my critics appear to be unable to avoid claiming that I do so.

I find it difficult to assess the validity of Dr. Wurster's statement that DDE is one-third to one-half as toxic as DDT. I assume he means that the residues of pp′-DDE in the brain at death, following poisoning of birds by this compound, are two to three times as high as the residues of pp′-DDT found in birds killed by dosing with pp′-DDT. I have seen no paper giving the results of such a comparison, but perhaps I have missed the relevant paper. With regard to the residues of HEOD I can only repeat that I find the arguments of Jefferies and Presst unconvincing. Furthermore, Dr. Wurster's attempt to vindicate their arguments seems to involve some inconsistency. Dr. Wurster argues that two of the peregrines almost surely died of poisoning by DDT + dieldrin. Jefferies and Presst emphasized, however, that one of these peregrines, "144" had not ingested pp'DDT, adducing the absence of pp'-DDD in support of this conclusion. Further, Jefferies and Presst conclude that large falcons are probably some four times more sensitive to organochlorine insecticides than pigeons, whereas, Dr. Wurster says that deaths can be attributed to poisoning by DDT + dieldrin without making the assumption that peregrines are more sensitive.

RISEBROUGH: In an area of northern England and southern Scotland peregrines were not reproducing very well. Ratcliffe obtained some eggs from nests which later produced young: for example, he took one egg and left two. He also sampled eggs from nests which were not successful, where eggs did not hatch. He then compared the residues in the two sets of eggs; the residue levels from the unsuccessful nests were higher than those from the successful nests, but the differences were not significant.

Robinson concludes from this that because these differences are not significant there is no relationship. There are two questions. First, would a given level of contamination necessarily cause the same failure of reproduction in each pair of peregrines? We know that there is a range of doses where a toxic compound damages some individuals, and not others, depending on biological variability. Second, can we base a conclusion on a lack of statistical significance? In this case the range of doses, expressed in ppm in the eggs, was so small that a significant correlation with effects could hardly be expected, whether the doses were toxic or not.

Dr. Robinson states that the residue levels in eggs from the central Scottish Highlands where land is relatively uncontaminated, are significantly lower than those in eggs from the south of England, in both DDE and total chlorinated hydrocarbon content. But he did not at all mention that the nesting success in this region is much higher than in the rest of Britain and in both 1965 and 1966 accounted for more than half of the successful nestings in all of Great Britain. Nor did he mention that the eggs from this region are also more normal in terms of eggshell thickness and calcium content.

STICKEL: Dr. Welch and Dr. Wurster deserve a great deal of credit for their presentations and for the amount of light they are shedding on this problem. As they have shown, the current work on enzymes, hormones, and calcium metabolism stems in part from Derek Ratcliffe's discovery of eggshell-thinning (*Nature, 215*:208, 1967). Shell-thinning need not be subtle; it can be so manifest that the difference in springiness is detected when halves of shells are gently squeezed between the fingers. Differences in shell thickness are readily measured with a micrometer. We are, therefore, talking about some very concrete effects that are rooted in physiology.

It is clear that we must take the new physiological findings and theories with great seriousness. Perhaps Dr. Risebrough is correct in saying that "ppm toxicology" based on high doses of pesticide may soon be obsolete. But I do not believe that we have all pieces of the puzzle and I would like to mention some of the evidence in this direction.

Much of the pesticide controversy centers on the peregrine falcon. In 1966, three parties went to the Arctic to take samples of peregrines, peregrine eggs, and food materials. One party went to Alaska and two went to different areas of Canada (Enderson, J. H., and Berger, D.D.: *Condor, 70*:149, 1968) (Cade, T.J.; White, C.M., and Haugh, J.R.: *Condor, 70*:170, 1968). They found very substantial residues, often hundreds of parts per million, in peregrine fat. These residues were large enough to account for any amount of enzyme induction. But in each of the three populations, reproduction was normal and the population was still up to the original levels shown by previous studies. So here are three populations of peregrines with normal reproduction and normal population levels, but with highly substantial residues. The implication, of course, is that some type of homeostasis or compensation in the animals' physiology allowed them to surmount any trouble caused by residues, at least in large part. It seems clear that the defense mechanisms of these peregrines were not swamped by the pesticide levels that were present. Efforts to attribute dire effects to even smaller residues seem hardly wise.

The graph of eggshell size-weight indices presented by Ratcliffe (ibid) and the similar data compiled by Hickey and Anderson (unpublished) indicate that shell thicknesses were substantially the same throughout history until suddenly, in the late 1940's, they decreased. Thereafter they continued on a new plateau, with no really significant fluctuation from year to year. This may indicate that we are dealing with a threshold effect beyond which an increased effect is not easily produced by a greater dosage.

The thinning of eggshells in Britain occurred in the late 1940's, years before dieldrin was used. Hence DDT or its metabolites were taken to be the probable cause of shell thinning and there is experimental evidence that this can happen. However, the marked decline and reproductive failure of the peregrine and certain other raptorial birds in Britain began in the mid-1950's at a time when there was much mortality of seedeating birds and their predators because of dieldrin used on grain seed. Thus the shell-thinning and the population crash may have had different causes. The shell-thinning may have contributed to the crash but it was not the sole factor and it may not have been the major factor. Dieldrin, although not present when shell-thinning began, remains suspect as the major factor in the crash.

Another bit of evidence points in the same direction. In 1967, the British peregrine population began to rise and to have slightly better breeding success, according to Ratcliffe *(Bird Study, 14:238, 1967)*. This improvement was accompanied by a drop in residues of dieldrin and heptachlor epoxide in eggs rather than by any great drop in DDE. Thus it looks as if the main decrease in peregrine numbers came at the time of maximum trouble with dieldrin, and the population rose when dieldrin levels dropped.

Shell-thinning probably is one important factor in the problem of declining birds; quite possibly it is associated with enzyme-induction effects on directly estrogenic effects of pesticides. It may not be the whole story, however; there may be mechanisms that are still unknown and there may be factors other than pesticides involved. As the recent discovery of polychlorinated biphenyls in animals has emphasized we still do not know all pollutants we should be studying.

I have been asked to tell about the kestrel experiment at Patuxent. The kestrel or American sparrow hawk is congeneric with the peregrine; they are *Falco sparverius* and *Falco peregrinus*. We began studying the effects of pesticides on reproduction of experimental kestrels 3 years ago to help solve the peregrine problem. One group received 1 ppm dieldrin and 5 ppm DDT on a dry-weight basis in the diet. Another group received 3 ppm dieldrin and 15 ppm DDT. We used both chemicals because birds commonly get both in the wild and because it would take too many years to test them separately and together. The "light" dosage represents a degree of contamination that birds could get at various places in nature. The "heavy" dosage represents an unusually heavy, but still possible, contamination in nature. A higher level of dosage would have killed the birds; the heavy dosage did kill a few.

The experiment gave a good test of reproduction, including reproductive behavior, for the birds had to go through the whole chain of reactions involved in mating, incubation, and care of young. The results obtained in 1967 indicate that reproductive troubles in these kestrels occurred at the same points at which they occur in peregrines and certain other declining species in the wild. What was found was not dying young, which are characteristic of most experiments in which heavily contaminated gallinaceous birds have poor reproduction. Instead there was failure during the period of the egg. There was no problem about egg production and there was no significant loss of young as compared with controls. The trouble was failure to hatch and disappearance of eggs. Control birds had a hatching success of 84 per cent equivalent to that in nature. The "light" dosage group had hatching success of 61 per cent and the "heavy" dosage group had success of 59 per cent (Porter, R.D., and Wiemeyer, S.N., *in press*). The drop in hatching success appeared to be accompanied by shell-thinning, but the 1967 data on this are not extensive.

It was of interest that results were so nearly alike in "light" and "heavy" treatment groups. Further work should reveal whether this was due to chance or to a threshold effect. It was also of interest that results were not more sweeping in view of the severity of the dosages. The chemical challenge surely was strong enough and long enough to permit maximal enzyme induction or directly estrogenic effects. Apparently the homeostatic mechanisms of many individuals enabled them to compensate for relatively severe intoxication.

As one looks at a large body of field and experimental data from pesticide studies, it is difficult to avoid the impression that homeostatic, compensatory, or threshold effects are common when residues are below lethal levels. In many instances, as with the Arctic peregrines mentioned earlier, they appear to be highly successful.

YOUNG: Has anyone looked at bird shells from before the 1940's and the newer ones with regard to the presence of strontium 90? Dr. Clarkson last evening suggested that it was about at this time that the United States began atomic testing and one of the chief products of fallout from this testing was strontium 90. Strontium has a similar chemistry to calcium and I wonder if this might have something to do with the bird shells.

CLARKSON: I don't know that there is any evidence as to how strontium 90 would do this, except of course that it is accumulated in calcifying tissue.

WELCH: Can anyone tell me anything about the relationship between estrogen in the bird and calcium? I understand that calcium is mobilized from the bone to the egg.

WURSTER: Estrogen increases calcium absorption from the diet and decreases its excretion; therefore the organism exhibits a positive calcium balance. The difference between uptake and excretion, the net uptake, is stored in the marrow of the bones, especially the femur and several other limb bones. At the time of calcification of the eggshell, which is fairly rapid and at a late stage of egg formation, there is a transfer of calcium from the marrow of the bones as well as from the diet. If there is an inadequate amount of calcium in the diet, thin shells are formed (Romanoff, A. L., and Romanoff, A. J.: *The Avian Egg.* New York, Wiley, 1949).

RISEBROUGH: PCB's were already in the environment in the late '30s, but came in to widespread use only after the war. Since Dr. Peakall has just showed that they all have enzyme-inducing capacities, their chemical structure does have similarities to that of DDT. I think it is necessary to begin to consider all kinds of environmental pollution, rather than just DDT, as causing some of these effects on birds.

CLARKSON: I understand, that as far as implicating DDT is concerned, the critical point in time is from 1946 and 1947. Would it have been distributed efficiently enough by that time throughout the world to have caused this effect?

STICKEL: We know now that the pesticides have essentially worldwide atmospheric distribution; there could have been submicrogram quantities getting into almost everything, with effects showing up only in the organisms that gave the most sensitive endocrine responses.

WELCH: A small fraction of the total dose of DDT could be converted to an estrogen-like by-product and could produce this effect. Only a few gamma of metabolite would be necessary to provide this type of pharmacologic action. Furthermore, this small amount of metabolite, or whatever it might be, could be missed in the usual detection methods. One is usually correlating DDE levels with decrease in reproductive capacity. This might be a mistake, it might be something else.

CLARKSON: If we know the mechanism of these actions, we can then perhaps say with some confidence what is the cause; DDT in this case might be the cause of the shell-thinning, it can explain its action in terms of estrogen. Otherwise we're always against this difficulty that we ask what happened in 1946? There may be many other things released into the atmosphere.

WURSTER: There is a good reason to expect DDT to have done it based on what is known about enzyme induction and the normal function of estrogen in a bird.

DDT could have been very widely dispersed within weeks of its first usage. There is an interesting study (Cole, H., *et al.*: *Bull Environ Contam Toxicol*, 2:127, 1967) of a rather large and remote wilderness tract in Pennsylvania that had never been treated with chlorinated hydrocarbons; it was to be treated with DDT. Prior to spraying, however, the soil was found to contain DDT,

DDE, and dieldrin. None of these was found in the water, but the trout also contained these three compounds and furthermore they contained about as much as those from areas that had previously been treated. When spraying of one watershed began, DDT appeared in the water of a different unsprayed watershed. There is no question of the mobility of DDT; it is very rapidly dispersed.

DeFREITAS: The estrogenic action of DDT is only hypothetical with respect to birds, is it not? I believe that Dr. Welch had done his work only with rats.

WURSTER: The first slide I showed last night was Dr. Peakall's work on pigeons.

DeFREITAS: Not on the estrogenic effect?

WELCH: We have experiments on birds which I cannot release, but which show this action. We have the thinning out of the eggshell in response to relatively low doses of some organochlorines.

WURSTER: You do not need the estrogenic effect to explain the change in eggshell. Enzyme induction will reduce the level of estrogen. Estrogen-like action would only enhance the effect.

WELCH: There is no evidence even in the rat that enzyme induction will cause a lowering in the endogenous circulating estrogen. It is true that in our studies we administered doses of estrogen which approximate those found physiologically by other endocrinologists. And treatment with organochlorines hastened the removal of this material considerably. But control study is still absolutely necessary in birds in order to establish a change in endogenous hormones. The only evidence now available in birds, is the report of Burlington and Lindemann that the testicular size of the cockerel is decreased. This is consistent with a possible estrogenic effect of DDT.

WURSTER: Another indication: The delay in ovulation of Bengalese finches (Jefferies, D. J.: *Ibid., 109*:266, 1967). The effect seems related to that of contraceptive pills.

BUTLER: I would like to point out that the American Defense Department had spread large quantities of DDT across North Africa and Southern Europe by 1942 and 1943.

WILLS: Also in India and China.

ORGANOMERCURIALS

BERLIN: Dr. Berglund and I do not want our figure for the acceptable daily intake (ADI) of methylmercury to be used as a precedent for others. The methylmercury problem is not limited to Sweden. This problem will occur in many places. I don't want people to accept our ADI value as a final one. It is a practical value and it should be looked upon as a practical estimate, forced by a critical situation.

There are a number of factors which we have not been able to take into consideration due to the lack of knowledge. A very important factor behind the estimated ADI is the amount of fish consumed in Sweden. It would not be possible to use this calculation for other countries where the fish consumption is different. I want to mention a few other points which illustrate the weakness in our calculation. For example, the rate of elimination of mercury in relation to the total body burden is a very critical factor in our calculation. If this is changed in any direction it would lead to quite different values. We can, for example, consider the possibility of a decrease in the rate of elimination with age; this would completely change the ADI calculation. The possibility of anti-mitotic action should also be investigated more thoroughly. There are other obvious missing factors.

In Sweden we now have a special committee which is dealing with this problem, trying to promote more research and gathering all relevant facts, with the hope that we will be able to design a more valid tolerance limit in the future.

EXPERIMENTS ON HUMANS

WILLS: In discussing the relationship between respiratory and dermal toxicities, you pointed out, Dr. Durham, that you studied only parathion. Is it possible that water-soluble or predominantly water-soluble materials might behave differently from parathion?

DURHAM: The only basis yet for making a comparison is that one might expect the chemicals which have shown high dermal toxicity in experimental animals to behave the same way in man. Parathion penetrates animal skin rapidly. We studied parathion since it is very widely used; furthermore, there's a good method for measuring the metabolite of parathion in urine. We could also do DDT in this way because we could measure DDA in urine. With other compounds we have no good way of knowing what the total absorption has been.

WILLS: Yes, except that if TEPP, for example, were studied, the blood level of chlorinesterase could be used as an approximate measure of absorption of the toxic material. Such work has not been done yet, so far as I know, with TEPP but it has been done with some other materials, for example, (Grob, D., *et al.*: *Bull Hopkins Hosp, 81*:217, 1947; *81*:245, 1947; *81*:257, 1947) (Moeller, H. C., and Rider, J. A.: *Toxic Appl Pharmacol, 4*:123, 1962) (Hartwell, W. V., and Hayes, G. R., Jr.: *Arch Environ Health, 11*: 564, 1965). This is a rather complicated procedure because the rate at which the material enters the blood determines the amount of lowering of cholinesterase activity.

BUTLER: Dr. Wills, may I direct a question to you? I recognize the difficulty in working with human volunteers, but would you care to comment on the validity of interpretation of data obtained from such small samples recognizing the individual variation that we run into?

WILLS: Well, of course one is never as sure in using eight people in a group as you would be in using, say, sixteen or twenty-four. And one can figure the relationship between them with·the eight — if you use sixteen people instead of eight people you would have about four times the reliability.

FASSETT: Dr. Wills, may I comment on the use of human subjects? I think that it is true that there are problems and that one can't do the same sort of study as with animals. But in my view many of our mistakes and difficulties in the past have come about because we have not, relatively early in the development of a toxicologic study, found out what the human does with these substances; in other words, the metabolic pathways and the rates of excretion remain to be solved.

This is why I feel that our Swedish friends have shown us a very good example of the sort of study that should have been done long before these substances were used. I think we are beginning to realize that we cannot plan proper animal experiments without knowing the rate and pattern of metabolism in the human compared to whatever species we are going to be dealing with. Of course, one cannot always find a species that exactly mimics the human. Nevertheless, in my opinion, it is most important that some knowledge of the human metabolic pattern be developed early in the study of the safety of any new substance.

There are other reasons for this. For example, we have to know something about metabolism in order to investigate properly any accidental poisonings, in which we need to get blood or tissue levels of the substance or its metabolite. We also need this information in order to evaluate occupational exposures or effects. We cannot learn anything about dosage and effects in humans without this and it should be done early in an investigation, as soon as it is felt safe to do so based on animal tests.

BUTLER: Are you suggesting that we should first work on a few human beings and then do the animal experiments?

FASSETT: No, I think we have to do some animal work first to assure safety. But bear in mind that with our analytical techniques today it is increasingly possible to go down to very low dosages so that with appropriate animal data this can be done safely.

BERGLUND: One should, as early as possible, when testing a new drug, obtain data from man. But those data are not toxicological data, they are metabolism data which can often be obtained with tracer doses. One then selects the animal species in which the metabolism of the potential drug is most similar to that in man. This philosophy is reflected in suggestions on the use of safety factors in WHO publications (*WHO Tech Rep Ser, 348*:8, 1967). In early publications, the safety factor was supposed to be 100 to 1 between the most sensitive test species and man, which of course is quite impossible sometimes. Lately, they have recommended that we apply these safety factors on the "most appropriate species," meaning the species where the metabolism of that particular compound is similar to that in man.

A preliminary testing program does not remove every hazard.

I doubt that there is any drug where you really can find out every effect on people before it has been in use for a considerable length of time.

FASSETT: I think the example of the question of drugs is very pertinent, but I think it is necessary to keep in mind a fundamental distinction between deciding safety compared to toxicity and hazard. I would again like to refer to the stress-strain. With a drug, we are really studying the right-hand part of the curve. In other words, we want to find out what dose is necessary to produce a given amount of homeostasis and altered physiology, whereas in our evaluation of the safety of environmental materials, we're really talking about the left hand part of the curve. In other words, how far are we from an actual toxic effect and how much readjustment of homeostatic mechanisms is possible? It seems likely that there may be fundamental differences between the materials that persist in the body and the short-lived substances. With the persistent materials, many of which aren't immediately producing any readily detected enzyme or toxic effects, it is probable that there may be a rather complex but minor readjusting of homeostatic mechanisms. And, therefore, with this type of substance it is especially important to know the blood and tissue levels etc.

BERGLUND: Who is going to judge whether or not the requirements for safety and health are fulfilled? In Sweden we have a few "ethical committees" at the universities which should be consulted in such cases and give their opinions. The experimenter still has the whole responsibility, but the "ethical committees" might help him.

WILLS: We use a similar sort of system here; every medical school doing human experimentation under the aegis of NIH is required to have a general advisory committee. In the New York State Health Department we also have such a group, including a judge, clergymen, and various other people. One group of which I know includes a lawyer.

HODGE: This business of administering a candidate pesticide to a human has a number of fairly fundamental philosophical questions attached to it, that do not attach to a candidate drug for human use. First, there are unavoidable exposures of people to a material not intended for people. The chemist who synthesizes the material — it is almost impossible for him to do so without having some exposure. It is very difficult to find out how much he gets or where it goes at this stage. So this is a very difficult one to handle. If a candidate pesticide gets far enough so that it is going to be tested technologically in pilots, the man that applies it is going to be very hard to measure, but he is the person who is exposed in the process of developing the material. The same for the man who runs the pilot plant, etc. There are people who are unavoidably exposed, and these people are subjects, it seems to me, who should not be overlooked to analyze their blood, urine, etc.

I know that this raises questions because as soon as a toxicologist goes to a plant and asks for urine samples, blood samples, a liver biopsy, etc. the question posed to the toxicologist by the employee would be, "Is there something unusually dangerous about this?"

So there may be problems connected with trying to get data from unavoidable exposures. But I would like to submit that this is a very profitable place to look and one place where we can get early information.

FASSETT: One cannot really do this properly in these unavoidably exposed subjects without knowing a little bit of basic information about the pattern and rate of metabolism. You have to have a method of some sort in order to obtain this extremely valuable information from accidents, suicides, occupationally-exposed people, etc.

HODGE: I'm submitting that it's difficult, but possible. Secondly, we have no legal framework to support the experimenter who wishes to treat the human volunteer, and very little basis in legal precedent.

But, on the "drug" side we have centuries of responsibility demonstrated on the part of the medical profession, accepted and defined over and over by court action; there are also well defined codes of ethics, which have been accepted in the Nuremberg code, etc. There is a whole structure of protection for the man who starts with phase one on an investigational new drug to find out what the human pharmacology is. The investigator, the medical man, has a time-tested and developed structure of responsibility for the best interest of his patient. There is no such structure at all for the candidate pesticide.

FASSETT: This is due largely to the fact that there have not been many legal tests but the need is just as great as in the case of drug investigation. The initial human metabolism studies require a high degree of skill and experience, and it is desirable that they are done in university or governmental medical centers.

HODGE: I am convinced that we have to know what the pattern of metabolism is in the human to choose an animal species wisely; there is no argument about that position at all. But suppose you take one of the pesticides that has been discarded within the past five years because extensive studies have shown it to be a carcinogen. What do we say to the family of a man who took a voluntary injection of that material before the long-term work was done, and the man developed a cancer that has no relation to the cancer that developed in the dog or cat in the long term tests and may be purely coincidental? Who is responsible? The Department of Agriculture? The State Department of Health? The physician who administered it? The school who employed the physician? The company that supplied the candidate pesticide? There is a very complicated and important philosophical and maybe a practical question here.

CONCLUDING STATEMENT

RUDD: I'm an ecologist concerned with the external environment, and not necessarily with the medical environment. You spoke of the advantages of time-tested procedures in medical practice.

We have had 7,000 years of agriculture in which, through empirical methods and decisions, we have acquired an essentially stable and useful producing agro-ecosystem. What has happened in the last two or three decades? We discarded much of the tested procedure and are now exposing the whole system to new procedures and chemicals. No one accepts the ethical responsibility for this changeover. The process is divided from the start, so that increases of production are demanded by people who do not accept responsibility for the damaging effects produced. These side effects are not necessarily the kinds of things that we've been discussing here more or less parochially, such as damage to peregrine falcons. There is, for example, the very important phenomenon of pest resistance which commits us to further development of an agro-ecosystem in a direction that looks catastrophic. In other words we are headed for the kind of damage that we will not be able to control and we are, I think, close to that point in California right now. There are other examples. Who is responsible for DDT in Antarctica? And the discussion of ethical committees: I believe farmers are ethical men but who umpires the ethics of the great agricultural experiment? It seems to me that our presentations of the subjects in this conference were entirely too limited. We are really talking about phases of the same thing, but what we have done was to compartmentalize it more or less for our own convenience and understanding. The advantage of a conference like this is that we can somehow communicate the similarity of problems. I don't think the problem was broadly enough defined. Nor do I think we have the corrective system appropriately understood.

The lack of a corrective system was illustrated in Dr. West's paper. What is accomplished by the advisory committees, the regulatory officials, the legislative representation? There is no system that will install correctives for the very fast-moving environmental changes taking place. All we get instead, and I have participated in this work, is a process of challenging change by imputing damages, fixing guilt, and generally resting our case upon essential legal grounds. It seems to me that the ethical example is the one we're losing. We should not be thinking in terms of payoff and hazard. We could make far better judgments if we had a wider kind of identification with the fate of the human community and other living communities, and if we accepted the responsibility for their stability and health.

None of us is a proper prophet, but I think that an extrapolation of the payoffs and hazards of the use of, in this case, chlorinated hydrocarbon insecticides, portends rather ill. One can look at the example of the peregrine falcon and say, "Well, that's interesting!" Some people are interested in peregrine falcons. That's one way of interpreting it. They are beautiful birds. One can look at peregrine declines as a biological phenomenon and say we are really examining a process of extinction; this is what we're talking about. Or one can take the reverse side of the coin, all of those parts are related, and say that what we are doing is making a prediction about the likelihood of food production in the future. We're talking about simplification of entire biological systems, and with the worldwide simplification that seems to be before us, we are in serious danger in my view, of inhibiting our basic welfare.

We scientists by and large seem to serve other scientists. Scientists, in my view, are in fact an instrument of society and there should be communication both ways between scientists and their social community. In other words there should be an announcement to a decision-making, ethical group, if you want to call it that, and a feedback to direct it. We're not really identifying the problems appropriately to this social decision-making group, however constituted. We use a system of information and education which simply is archaic in a social environment that is indeed moving very rapidly.

If we're going to be talking about systems, all we have to do is pick up any newspaper or remember the tragic events of recent days and months and look at what is happening socially. It seems to me that two elements are involved: one is credibility, the other responsibility. I see no reason why we cannot apply terms like "credibility gap" to our subject, to the way in which we dispense information and discharge our responsibility to society. We ought to reach farther to the future, right now.

I don't think we have to be metaphysicians or cranks to predict from good information with a likely kind of base, what can happen. It's easy on a graph paper to project that in 1978 there will be extinction of the peregrine falcon, or something of this sort. But if we do not widen our view to the whole problem very soon we won't even know what has already happened.

Notice the key words in the title of the Presidential Report on

"Restoring the Quality of our Environment"; "restoring" and "quality." These mean that we have to put something back into the environment. Now, this can be looked upon as an atavistic notion that we wish things were as simple as they used to be. We've lost a great deal that we value, even in this age of technology, and we wish to put it back. But what we really wish to put back is a *working control system*. This is what we want.

Appendix

THE PROCEDURE FOR PROCESSING INFORMATION ON PESTICIDES

D. A. SPENCER

THIS CONFERENCE HAS expressed marked interest in the unintentional contamination of the environment with pesticide chemicals. Of particular interest are those pesticides whose degradation under natural conditions is slow enough to permit the accumulation of residues following certain specific practices.

The registration procedures which govern the sale and use-practices of pesticide chemicals in the United States make every effort to foresee these difficulties and prevent their occurrence. The environment, however, is such a complex medium and the number of species of plants and living animals so large that it is beyond the limits of practicability to study each situation before permitting a new chemical on the market. During the period of subsequent use, additional information on performance and side effects becomes available. Adequate, even elaborate, provisions have been made for the evaluation of data on pesticide chemicals and the registration can be rather quickly modified or withdrawn if new information justifies a change.

It will be informative to briefly review the requirements for registration of a new chemical pesticide — particularly those items having a bearing on residue in wildlife and the environment. The manufacturer making application for registration of a product used on crops must provide the following:

1. Physical and chemical properties of the pesticide.
2. Analytical method for determining residues in biological materials.
3. Pharmacological data.
   a. Acute mammalian studies.
      1) Oral — usually rats.
      2) Dermal — rabbits.

501

    3) Inhalation.

    4) Eye and skin irritation.

  b. Subacute mammalian studies.

    1) Oral — 90-day intake in diet.

    2) Dermal — 21-day exposure on skin.

    3) Inhalation — 14-day exposure airborne.

  c. Other studies.

    1) Effects on reproduction.

    2) Synergism or potentiation with other compounds.

    3) Metabolic products of decomposition.

    4) Toxicity to birds and fish.

4. Effects on area treated.

  a. Phytotoxicity.

  b. Translocation within the plant or animal treated.

  c. Persitence in soil, water, plants or animals.

  d. Compatability with other chemicals.

If the data in any one of the above categories is particularly damaging then the requirement in terms of tests conducted and species studied is considerably expanded.

If the pesticide chemical is to be applied to a growing crop (or a food-animal) then a tolerance for residues occurring in the food must be obtained. The petition to the Food & Drug Administration for the tolerance must be supported by the following data:

1. Acute toxicity.

  a. LD_{50} in at least two species of animals.

  b. A description of the signs of toxicity.

2. Short-term toxicity (subacute toxicity).

  a. At least two species, one a nonrodent.

  b. Duration, 90 days.

  c. At least three dosage levels plus a control group; one dosage level should be toxic.

  d. When 90-day studies are designed to continue for long-term toxicity studies, sufficient animals must be started to supply the required number for the long-term studies.

  e. Observations: growth, food consumption, general appearance and behavior, mortality, organ weights, clinical-laboratory tests (blood and urine, organ function, enzymatic and metabolic) , and gross and microscopic examinations. All animals dying before termination should be examined grossly and microscopically. All tumors should be recorded and examined microscopically.

f. In the case of organophosphorus and carbamate pesticides, cholinesterase inhibition and demyelination studies must be made.

3. Long-term toxicity (chronic toxicity) .

a. At least two species, one a nonrodent.

b. Duration, usually 2 years.

c. At least three dosage levels plus a control group; one dosage level should be toxic.

d. The groups at the start of the experiment should be sufficiently large to assure an adequate number of survivors at termination. For rodents, this may vary from strain to strain so that numbers to be used are contingent on the judgment of the investigator.

e. Observations: growth, food consumption, general appearance and behavior, mortality, organ weights, clinical-laboratory tests, gross examination of animals throughout experimental period and tissues at autopsy, and comprehensive microscopic examination of tissues. Animals dying before termination should be examined grossly and microscopically. All tumors should be noted and examined microscopically.

4. Biochemical data.

a. Data from at least two species of animals.

b. Observations: absorption, distribution, metabolic transformation, elimination, possible accumulation, and the effect of enzymes which should be examined because of the nature of the chemical under study.

c. Information on metabolism of pesticides and their other conversion products in treated plants. At least acute and short-term toxicity data on the significant plant metabolites or other conversion products. If the conversion products are different from metabolites of the test animals, long-term toxicity data on the pesticide metabolites or other conversion products may be required.

5. Reproduction studies.
   a. At least one species, preferably two.
   b. At least two dosage levels plus a control group.
   c. One dosage level should be definitely toxic; one should have no effect.
   d. Usually three successive generations in the rat; two generations are satisfactory if the results are conclusive.
   e. Preferably two litters per generation.
   f. Both the males and females should be treated for 60 days prior to breeding. The second and third generations are treated from weaning throughout the breeding period.
   g. Observations: fertility, length of gestation, live births, still births, survival at 4 days and at weaning, sex of newborn and of weanlings, body weights, gross abnormalities, and microscopic and skeletal examination of young in last generation.

6. Data on man.
   a. From industrial exposure.
   b. From accidental poisonings and suicides.
   c. From controlled experiments in special cases.
   d. Biochemical data of the type indicated under 4 b. above are particularly useful.

The interlocking review of the above data by competent scientists is depicted in the accompanying figure. The application is received by the Pesticides Regulation Division and funneled to the appropriate Product Evaluation Staff. Here the claims for the pesticide and the manufacturer's data on effectiveness and safety are reviewed by a professional staff representing four separate disciplines:

1. Either an entomologist, plant pathologist, plant biologist, animal biologist or microbiologist.
2. Organic chemist.
3. Pharmacologist.
4. A wildlife biologist.

They determine the acceptability of the claims for possible registration. The manufacturer's data and product claims are then submitted concurrently to three other Federal Government Pesticide Review Staffs, each studying the information from a viewpoint that permits somewhat different interpretation.

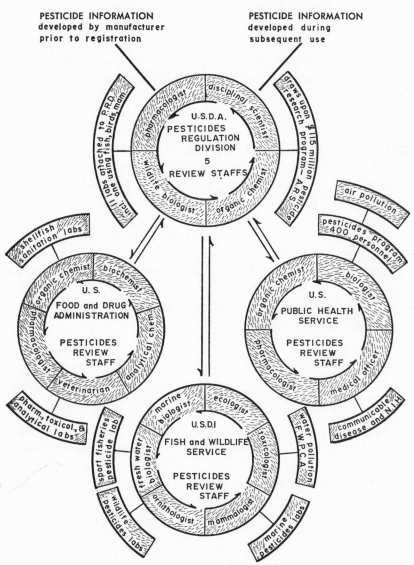

FIGURE Appendix-1.

At the Department of the Interior, biologists specializing in six different fields study the application for registration as to possible side effects on fish, wildlife, and the environment. This is a balanced team directed by an ecologist, Dr. Frederick H. Dale, and includes a toxicologist, a mammalogist, an ornithologist, a freshwater-fisheries biologist, and a marine biologist.

At the Public Health Service, there is a professional staff from four different disciplines. Dr. Thomas Harris, organic chemist, directs a staff made up of medical officers (M.D.'s), a pharmacologist, and a biologist, in the evaluation of the reliability of the data from the viewpoint of public health. Environmental contamination, of course, has an important bearing on public health.

Both of these review staffs make recommendations to the Department of Agriculture, accepting, modifying or rejecting the proposed registration. Despite the fact that the Department of Agriculture alone has final responsibility for registration of pesticides, seldom, if ever, have they disregarded the advice of the members of the interdepartmental review.

The Food and Drug Administration enters the program when it is necessary to establish (a) an exemption, (b) a negligible residue or (c) a finite tolerance for residues found in and on raw agricultural products (including food-animals). Their review staff consists of scientists from at least five disciplines:

1. An organic chemist.
2. An analytical chemist.
3. A biochemist.
4. A pharmacologist.
5. A veterinarian.

The Food and Drug Administration has final say on the tolerance established and without it registration of the pesticide for use on food crops cannot be granted by the Department of Agriculture.

Each of the four Federal Government organizations participating in the acceptance of pesticide chemicals for registration, apart from the regulatory staffs, is engaged in research on some phase of pesticide chemical development or evaluation. These programs are indicated on the accompanying chart by the outer segments tied to each staff circle. Except in the case of the eleven laboratories operated by the Pesticides Regulation Division of the Department of Agriculture, these activities are not under the control of the Pesticides Review Staff but simply serve as a source of in-house information on pesticides that can materially contribute to the caliber of the regulatory activities. Please note that some of these research programs are very large. For example, the annual budget for the pesticide investigations by the Department of Agriculture is in the neighborhood of $115 million.

It is reasonable to assume that most of the investigations undertaken by Federal Departments, other than the United States Fish and Wildlife Service, have only an indirect bearing on fish and wildlife. Surprisingly enough, a few studies involving pesticides and wildlife are funded by almost every Federal organization in the research grant field. Thus we find that in 1967 of the 117 grants for research sponsored by the National Communicable Disease Control Center, United States Public Health Service, thirty-eight projects (32 per cent of the total) dealt with pesticides and included such titles as the following:

1. Kinetics and Distribution of Pesticides in the Ecosystem.
2. Wind-borne Pesticides in California Mountains.
3. Biological Concentration of Pesticides by Algae.
4. Influences of Chemical Pesticides on Forest communities.
5. Effect of Selected Pesticides on Intertidal Biota.
6. Effects of Pesticides on Estuarine Organisms.

Conclusion

There exists an extensive provision for the protection of wildlife and the environment from adverse effects of pesticides. The system of cross checks and balances assures capable and thorough evaluation of any data provided. It only remains that when new information develops from the registered use of pesticides that data be promptly submitted to the Pesticides Regulation Division, United States Department of Agriculture, for insertion into the evaluation system which has authority to translate sound information into corrective action.

AUTHOR INDEX

(Italicized page numbers indicate discussion.)

SUBJECT INDEX

(Bold type indicates Figure or Table)